Springer Series in Reliability Engineering

Series Editor

Hoang Pham, Department of Industrial and Systems Engineering, Rutgers University, Piscataway, NJ, USA

More information about this series at http://www.springer.com/series/6917

Cao Wang

Structural Reliability and Time-Dependent Reliability

 Springer

Cao Wang ⓘ
School of Civil, Mining
and Environmental Engineering
University of Wollongong
Wollongong, NSW, Australia

ISSN 1614-7839   ISSN 2196-999X   (electronic)
Springer Series in Reliability Engineering
ISBN 978-3-030-62507-8   ISBN 978-3-030-62505-4   (eBook)
https://doi.org/10.1007/978-3-030-62505-4

This Springer imprint is published by the registered company Springer Nature Switzerland AG
The registered company address is: Gewerbestrasse 11, 6330 Cham, Switzerland

To my wife Wenwen Tang for her love and support.

Preface

The failure of important engineering structures and infrastructure facilities, caused by either environmental or anthropogenic extreme events, may lead to substantial economic losses to the facility owner or occupant and further a ripple effect in the surrounding community. Recently, public awareness has been raised significantly regarding structural performance in the context of community resilience, for which the research community and engineers are seeking advanced implementations for building and construction practice. With this regard, structural reliability analysis is an essential tool in terms of evaluating and managing structural safety and serviceability, aimed at providing quantitative information on whether a structure can withstand extreme events with an acceptable level of reliability during its future service life. The design specifications in currently enforced standards and codes, which are for the most part reliability-based, also demonstrate the up-to-date applications of reliability theory in structural design and safety evaluation.

The various sources of uncertainties arising from the structural performance and the external loads, as well as the computational models, are at the root of the structural safety problem of the structures. In an attempt to measure the safety of a structure, it is necessary to quantify and model these uncertainties with a probabilistic approach so as to further determine the structural reliability. In fact, the identification of the probability distribution of random variables is a crucial ingredient in structural reliability assessment.

Engineered structures are often exposed to severe operating or environmental conditions during their service life, which are responsible for the deterioration of structural strength and stiffness with time. There are many classes of structures for which degradation in service is either known or believed to have an impact on structural safety. Due to socio-economic constraints, many degraded structures are still in use. Motivated by this, it has been an important topic in the research community over the past decades to conduct time-dependent reliability and damage assessment of aging structures.

The aim of this book is to provide an accessible introduction to structural reliability assessment and a solid foundation for problem-solving. Totally, five Chapters are included in this book. Chapter 1 introduces the background and presents an overview of the book. Chapter 2 deals with the probability models for random variables, providing the relevant fundamental knowledge for the subsequent Chapters. Readers who are familiar with the probability theory may skip this Chapter. Chapter 3 introduces the simulation techniques for single random variables, random vectors consisting of different (correlated) variables, and stochastic processes. Chapter 4 addresses the analytical approaches for structural reliability assessment, including the reliability models for a single structure and those for multiple structures (series system, parallel system, and k-out-of-n system). Chapter 5 discusses the approaches for structural time-dependent reliability assessment in the presence of discrete and continuous load processes.

This book has been prepared as a textbook for undergraduate and postgraduate courses of structural reliability and design. This book is also suitable to serve as an individual reading for a rather diverse and varied audience who is interested in the structural reliability theory. Keeping that in mind, more than 170 worked-out Examples (with detailed solutions) and more than 150 Problems have been included in this book to better illustrate the relevant theory.

This book has collected many materials from the author's recent research outputs. The author wishes that this book will provide not only the fundamental knowledge in structural reliability, but also a snapshot of the latest development in the field of reliability engineering.

Wollongong, NSW, Australia   Cao Wang

Contents

Chapter 1
Introduction

Abstract This chapter introduces the background and presents an overview of this book. Basically, engineering structures are required to satisfy predefined performance targets such as successfully resisting external load actions. Due to the randomness nature of both the structural property and the external attacks, however, the "absolute safety" of a structure cannot be achieved in engineering practice. Rather, the practical strategy is to control the probability of violating the performance requirements (e.g., structural safety) under an acceptable level. To that end, some probability-based approaches are essentially needed to quantify the occurrence possibility of such undesired consequences. Under this context, the mathematical formulation of analytical tools for structural reliability assessment is the topic of this book.

1.1 Performance Requirement of Engineering Structures

Engineering structures are expected to satisfactorily meet the requirements associated with their functionalities. The requirements may include the structural safety against collapse, limited deflection and others [8, 16], defining a *limit state* for the structure. For instance, consider a portal frame that is subjected to a horizontal load F (e.g., wind load or earthquake excitation), as illustrated in Fig. 1.1a. Given the geometry information as well as the material properties, the relationship between the horizontal displacement at roof, Δ, and F can be derived with some numerical or software-based techniques, as illustrated in Fig. 1.1b (see the solid line, e.g., [28]). With this, if the target for the frame is not to collapse under the action of F, then the performance requirement is that the ultimate yielding strength F_{u1} exceeds F (i.e., $F_{u1} > F$). However, if the target for the frame is that the horizontal displacement does not exceed a predefined limit $\Delta_{\lim}$, accounting for the *structural serviceability*, then the performance requirement is that $F_{s1} > F$, where F_{s1} is the load with which the horizontal displacement is $\Delta_{\lim}$. Clearly, the structural requirement is dependent on the specific performance target.

Existing structures may be subjected to performance (e.g., resistance, stiffness) deterioration due to the aggressive environmental or operational factors during their

C. Wang, *Structural Reliability and Time-Dependent Reliability*,
Springer Series in Reliability Engineering,
https://doi.org/10.1007/978-3-030-62505-4_1

Fig. 1.1 Force-displacement
relationship of a portal frame

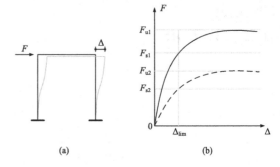

(a)   (b)

normal service [1, 5, 13, 18, 21–24]. Subsequently, the deterioration may impair
the safety level of the structure. Consider the example in Fig. 1.1 again, where the
F-Δ relationship becomes that in dashed line due to the structural performance
deterioration. In such a case, in the presence of the limit state of structural collapse,
if $F_{u2} < F < F_{u1}$, then the structure is deemed to survival when not considering
the performance deterioration, which clearly yields an overestimate of the structural
behaviour because the frame does have collapsed. Moreover, if the horizontal load is
also time-variant (e.g., it increases with time), then the overestimate of the structural
reliability may be even enhanced if not considering the impact of the load time-
variation. As such, it is essentially important to take into account the potential time-
dependence of the structural performance and/or external loads when estimating the
safety and serviceability of a structure.

In the built environment of a community, more than one structure/facility would
be involved with interactions and dependencies with each other, jointly providing
physical supports to the functionalities of the community. Illustratively, consider
the linkage between departure point D and arrival point A, as shown in Fig. 1.2.
Let, in each subfigure, the box represent a bridge, and the solid line be a road. The
performance of the bridges within the roads (linkages) is key to the connectivity
between D and A. Suppose that the post-hazard failure probability of the bridges is
equally 0.5 in Fig. 1.2, with the probability of connectivity between D and A is 0.5,
0.25, 0.75 and 0.375 respectively for Fig. 1.2a–d (The background information for
the calculations can be found in Chap. 2). One can use a simple, series, parallel or
mixed model to describe the behaviour of Fig. 1.2. These models serve as a basis for

Fig. 1.2 Illustration of the
performance requirements of
a structure in different
circumstances

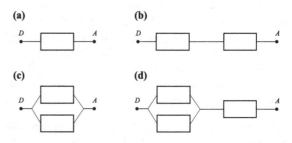

more sophisticated cases of a complex bridge network. Note that the performance of individual structures may be mutually correlated in an uncertain environment. Under this context, the reliability problem should not only consider the behaviour of a single structure, but also the similarity (correlation) between different components.

1.2 Measure of Uncertainty

The *failure* of a structure, i.e., the violation of the limit state as stated before, is usually rare for well-designed structures. However, the occurrence of structural failure may usually lead to a catastrophic consequence. As such, it is necessary to measure the possibility of the failure occurrence quantitatively, taking into account the uncertainty associated with the relevant variables (e.g., structural geometry, material properties, and others).

Uncertainties arise in structural resistance and the external load effects due to different sources such as the inherent randomness of the variables, the model error, or the statistical uncertainty [7, 10, 20]. Illustratively, Fig. 1.3 shows the histogram for the resistance (moment-bearing capacity) of a bridge girder [14], which indicates that the structural resistance cannot be simply described in a deterministic manner due to its uncertainty. The statistics in Fig. 1.3 can be used to determine which distribution type can "best" fit the yielding strength (lognormal distribution type for this example, c.f. Sect. 2.2.5), which will be further used in structural reliability assessment.

It is typical to distinguish the uncertainties as either *aleatory* or *epistemic* [7, 20]. The former is associated with the intrinsic randomness of the variable/observation while the latter is due to the lack or knowledge. One simple example is the modelling of the occurrence of future hazardous events (say, a tropical cyclone). The occurrence time for each event cannot be exactly predicted due to the inherent randomness (aleatory uncertainty). On the other hand, the choice of a probability model (e.g., a Poisson process, c.f. Sect. 2.2.3) to describe the occurrence process is associated with the epistemic uncertainty. Both uncertainties should be considered in a probability-based framework for structural reliability assessment. The statistical uncertainty is

Fig. 1.3 Histogram and distribution of bridge girder resistance, R. Here, "PDF" means probability density function, see Sect. 2.1.3

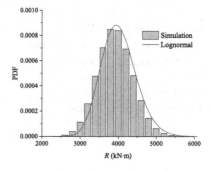

an important source of the epistemic uncertainty, representing the difference between the probability model of a random variable inferred from limited sampled data and the "true" one [6, 7, 9, 19, 27].

The identification of the distribution type of a random variable is the key in measuring its uncertainty, which is, unfortunately, difficult or even impossible to realize in some cases due to information or observations. With this regard, one may use a family of candidate probability distributions rather than a single known distribution function to capture the probabilistic behaviour of the variable [2, 4, 11, 15, 17, 25, 26]. Furthermore, the reliability assessment problem should be modified accordingly in the presence of the imprecise variables.

1.3   Structural Reliability Assessment

For an engineered structure, keeping in mind the performance target of structural safety, it is impossible to achieve a "definitely safe" design, taking into account the uncertainties associated with both the structural performance and the external attacks (c.f. Sect. 1.2). On the other hand, a naïve ambition of pursuing an "as-safe-as-possible" structure will often result in an overconservative design. In engineering practice, it is more reasonable to examine the safety of a structure from a view of probability (that is, the chance of the occurrence of an event). Table 1.1 presents some selected indicators of failure probability for building structures in relation to a limit state of collapse, as adopted from [16]. Ensuring the probability of structural failure below an acceptable level [3, 8, 12] would make the design and normal-use of a structure be guided in a more reliable manner.

The assessment of structural reliability, i.e., the probability of structural performance not violating the predefined limit state, should be performed, ideally, based on many repeated observations of the outcomes of structural state (i.e., survival or failure). However, this is often not the case in practice; alternatively, the reliability assessment is preferably conducted based on the knowledge (degree of belief) of the

Table 1.1 Selected Typical failure rates for building structures [16]

Structure type	Data cover	Average life (years)	Estimated lifetime failure probability
Apartment floors	Denmark	30	3×10^{-7}
Mixed housing	The Netherlands	/	5×10^{-4}
Controlled domestic housing	Australia (NSW)	/	10^{-5}
Mixed housing	Canada	50	10^{-3}
Engineered structures	Canada	/	10^{-4}

problem itself, where the probabilistic models are used to describe the randomness of the undeterminable quantities.

As has been mentioned before, there may be different types of limit state, depending on the specific performance targets. That being the case, the structural reliability assessment in the presence of different limit states could be solved in a universal framework, given that they can be modelled in a similar form.

The term structural reliability can be categorized into two types, time-independent (classical) reliability and time-dependent reliability. The former does not consider the time-variant characteristics of the uncertain quantities (or simply represent the time-variation with a time-invariant variable), while the latter focuses on the structural performance over a reference period (for example, the service life) of interest. Due to the distinction between the two types of reliability problems, different analytical approaches would be used. Intuitively, the latter would require more skills and computational costs due to the additional consideration of the time scale.

1.4 Overview of This Book

We have briefly introduced some background information, in Sects. 1.1, 1.2 and 1.3, for structural reliability assessment. More details for each topic will be given in the subsequent chapters. Naturally, the following questions will arise,

1. How to model the uncertain (random) quantities arising from structural engineering?
2. How to describe the joint behaviour of different random variables with similarity?
3. How to quantitatively measure the structural reliability (or probability of failure) in the presence of a specific limit state?
4. Given a desirable safety level, how to determine a reasonable design criterion for a structure?
5. How to capture the time-variant characteristics of both structural performance (e.g., stiffness, strength) and the external loads on the temporal scale?
6. How to estimate the structural reliability over a service period of interest?

As the scope of this book is on the mathematical tools for structural reliability assessment, the above questions will be addressed in the following chapters from a view of probability-based modelling.

Chapter 2 will address questions 1 and 2 by introducing probability models for random variables. The probabilistic behaviour of a random variable can be described by some fundamental tools such as the moment information, distribution function, moment generating function and characteristic function. The joint behaviour of different variables is also discussed, which is the basis for reliability assessment incorporating correlated variables.

Chapters 3 and 4 deal with questions 3 and 4, where the approaches for structural reliability assessment, both analytical and simulation-based, are discussed. Specially,

Chap. 3 addresses the simulation techniques for a single random variable, a random vector consisting of different (correlated) variables, and a stochastic process (a collection of random variables at different times). Chapter 4 presents a systemic introduction to structural reliability assessment, including the reliability models for a single structure and those for multiple structures (series system, parallel system, and k-out-of-n system). Also, the reliability-based limit state design approaches are also included, which consist of two aspects: (1) Given the structural configuration and load information, perform reliability assessment to determine the structural safety level; (2) Given a predefined safety level, establish a criterion for structural design. Furthermore, the evaluation of structural reliability in the presence of imprecisely-informed random variables is considered.

For questions 5 and 6, Chap. 5 discusses the approaches for structural time-dependent reliability assessment. The modelling techniques of the resistance deterioration and the external load processes are included. Subsequently, the time-dependent reliability assessment approaches associated with both a discrete load process and a continuous one are addressed and compared.

References

1. Akiyama M, Frangopol DM, Yoshida I (2010) Time-dependent reliability analysis of existing RC structures in a marine environment using hazard associated with airborne chlorides. Eng Struct 32(11):3768–3779. https://doi.org/10.1016/j.engstruct.2010.08.021
2. Alvarez DA, Uribe F, Hurtado JE (2018) Estimation of the lower and upper bounds on the probability of failure using subset simulation and random set theory. Mech Syst Signal Process 100:782–801. https://doi.org/10.1016/j.ymssp.2017.07.040
3. Ang AS, De Leon D (1997) Determination of optimal target reliabilities for design and upgrading of structures. Struct Saf 19(1):91–103. https://doi.org/10.1016/S0167-4730(96)00029-X
4. Baudrit C, Dubois D, Perrot N (2008) Representing parametric probabilistic models tainted with imprecision. Fuzzy Sets Syst 159(15):1913–1928. https://doi.org/10.1016/j.fss.2008.02.013
5. Bhattacharya B, Li D, Chajes M (2008) Bridge rating using in-service data in the presence of strength deterioration and correlation in load processes. Struct Infrastruct Eng 4(3):237–249. https://doi.org/10.1080/15732470600753584
6. Der Kiureghian A (2008) Analysis of structural reliability under parameter uncertainties. Prob Eng Mech 23(4):351–358. https://doi.org/10.1016/j.probengmech.2007.10.011
7. Der Kiureghian A, Ditlevsen O (2009) Aleatory or epistemic? Does it matter? Struct Saf 31(2):105–112. https://doi.org/10.1016/j.strusafe.2008.06.020
8. Ellingwood BR (2005) Risk-informed condition assessment of civil infrastructure: state of practice and research issues. Struct Infrastruct Eng 1(1):7–18. https://doi.org/10.1080/15732470412331289341
9. Ellingwood BR, Kinali K (2009) Quantifying and communicating uncertainty in seismic risk assessment. Struct Saf 31(2):179–187. https://doi.org/10.1016/j.strusafe.2008.06.001
10. Faber MH (2005) On the treatment of uncertainties and probabilities in engineering decision analysis. J Offshore Mech Arct Eng 127(3):243–248. https://doi.org/10.1115/1.1951776
11. Ferson S, Kreinovich V, Ginzburg L, Myers DS, Sentz K (2003) Constructing probability boxes and dempster-shafer structures. Technical Report, SAND2002–4015, Sandia National Laboratories

12. Frangopol DM, Lin KY, Estes AC (1997) Life-cycle cost design of deteriorating structures. J Struct Eng 123(10):1390–1401. https://doi.org/10.1061/(ASCE)0733-9445(1997)123:10(1390)
13. Kumar R, Cline DB, Gardoni P (2015) A stochastic framework to model deterioration in engineering systems. Struct Saf 53:36–43. https://doi.org/10.1016/j.strusafe.2014.12.001
14. Li Q, Wang C (2015) Updating the assessment of resistance and reliability of existing aging bridges with prior service loads. J Struct Eng 141(12):04015072. https://doi.org/10.1061/(ASCE)ST.1943-541X.0001331
15. Limbourg P, De Rocquigny E (2010) Uncertainty analysis using evidence theory-confronting level-1 and level-2 approaches with data availability and computational constraints. Reliab Eng Syst Saf 95(5):550–564. https://doi.org/10.1016/j.ress.2010.01.005
16. Melchers RE, Beck AT (2018) Structural reliability analysis and prediction, 3rd edn. Wiley. https://doi.org/10.1002/9781119266105
17. Möller B, Beer M (2013) Fuzzy randomness: uncertainty in civil engineering and computational mechanics. Springer Science & Business Media, Berlin
18. van Noortwijk JM, van der Weide JA, Kallen MJ, Pandey MD (2007) Gamma processes and peaks-over-threshold distributions for time-dependent reliability. Reliab Eng Syst Saf 92(12):1651–1658. https://doi.org/10.1016/j.ress.2006.11.003
19. Oberkampf WL, Helton JC, Joslyn CA, Wojtkiewicz SF, Ferson S (2004) Challenge problems: uncertainty in system response given uncertain parameters. Reliab Eng Syst Saf 85(1–3):11–19. https://doi.org/10.1016/j.ress.2004.03.002
20. Paté-Cornell ME (1996) Uncertainties in risk analysis: six levels of treatment. Reliab Eng Syst Saf 54(2–3):95–111. https://doi.org/10.1016/S0951-8320(96)00067-1
21. Sanchez-Silva M, Klutke GA, Rosowsky DV (2011) Life-cycle performance of structures subject to multiple deterioration mechanisms. Struct Saf 33(3):206–217. https://doi.org/10.1016/j.strusafe.2011.03.003
22. Stewart MG, Suo Q (2009) Extent of spatially variable corrosion damage as an indicator of strength and time-dependent reliability of RC beams. Eng Struct 31(1):198–207. https://doi.org/10.1016/j.engstruct.2008.08.011
23. Wang C, Li Q, Zou A, Zhang L (2015) A realistic resistance deterioration model for time-dependent reliability analysis of aging bridges. J Zhejiang Univ Sci A 16(7):513–524. https://doi.org/10.1631/jzus.A1500018
24. Wang C, Zhang H, Li Q (2017) Reliability assessment of aging structures subjected to gradual and shock deteriorations. Reliab Eng Syst Saf 161:78–86. https://doi.org/10.1016/j.ress.2017.01.014
25. Wang C, Zhang H, Beer M (2018) Computing tight bounds of structural reliability under imprecise probabilistic information. Comput Struct 208:92–104. https://doi.org/10.1016/j.compstruc.2018.07.003
26. Zhang H (2012) Interval importance sampling method for finite element-based structural reliability assessment under parameter uncertainties. Struct Saf 38:1–10. https://doi.org/10.1016/j.strusafe.2012.01.003
27. Zhang H, Mullen RL, Muhanna RL (2010) Interval monte carlo methods for structural reliability. Struct Saf 32(3):183–190. https://doi.org/10.1016/j.strusafe.2010.01.001
28. Zhang H, Liu H, Ellingwood BR, Rasmussen KJ (2018) System reliabilities of planar gravity steel frames designed by the inelastic method in AISC 360-10. J Struct Eng 144(3):04018011. https://doi.org/10.1061/(ASCE)ST.1943-541X.0001991

Chapter 2
Probability Models

Abstract This chapter provides a brief introduction to some important probability models. It starts from the basic concept of probability space and random variables, cumulative distribution function, probability density function/probability mass function, moment generating function and characteristic function, followed by selected frequently-used distribution types with illustration of their applications in practical engineering. The joint probabilistic behaviour of different random variables is also discussed, including the use of copula function to construct the joint distribution functions of dependent random variables.

2.1 Probability Theory and Random Variables

2.1.1 Probability Space

The *probability space* is an important mathematical concept to describe events with uncertainties (i.e., the outcome of the event cannot be exactly predicted in advance). There are three fundamental elements in a probability space, namely sample space (denoted by Ω), event space ($\mathcal{E}$) and probability function ($\mathbb{P}$). The sample space consists of all the possible outcomes of an experiment, and the event space, which is a subset of Ω, defines events in such a way that an event is deemed to occur if the outcome of the experiment lies in $\mathcal{E}$.

For two events E_1 and E_2 with event spaces of $\mathcal{E}_1, \mathcal{E}_2 \subset \Omega$ respectively, their *union*, $E_1 \cup E_2$, is such an event that occurs if either E_1 or E_2 occurs (i.e., the outcome of the union lies in $(\mathcal{E}_1 \cup \mathcal{E}_2) \subset \Omega$). Similarly, their *intersection*, $E_1 \cap E_2$ or simply $E_1 E_2$, occurs when E_1 and E_2 occur simultaneously (that is, the outcome lies in $(\mathcal{E}_1 \cap \mathcal{E}_2) \subset \Omega$). Specifically, If $E_1 \cap E_2 = \varnothing$, then E_1 and E_2 are deemed as *mutually exclusive*.

Example 2.1

(1) If an experiment is to roll a die, then the sample space $\Omega = \{o_1, o_2, o_3, o_4, o_5, o_6\}$, where the outcome o_i denotes that i appears on the die for $i = 1, 2, \ldots, 6$. If the event space is $\mathcal{E} = \{o_1, o_3\} \subset \Omega$, the corresponding event E means that either "1" or "3" appears on the roll. Furthermore, if event F has a space of $\{o_1, o_4, o_5\}$, then $E \cap F = \{o_1\}$, and $E \cup F = \{o_1, o_3, o_4, o_5\}$.

(2) If an experiment is to measure the lifetime of a structure (in years), then the sample space Ω consists of all positive numbers, i.e., $\Omega = (0, +\infty)$. For an event space of $\mathcal{E} = (30, +\infty) \subset \Omega$, the corresponding event E is that the service life of the structure is greater than 30 years. If event F has a space of $\mathcal{E}' = (20, +\infty)$, then $E \cap F = E$ and $E \cup F = F$ since $\mathcal{E} \subset \mathcal{E}' \subset \Omega$.

The probability function $\mathbb{P}$ provides a mapping from $\mathcal{E}$ to the interval $[0, 1]$, with which the probability of the corresponding event E varies between 0 and 1 (i.e., $0 \leq \mathbb{P}(E) \leq 1$). Specifically, if E has a space of $\varnothing$, $\mathbb{P}(E) = 0$; on the contrary, if the event space for E is Ω, then $\mathbb{P}(E) = 1$. Furthermore, for n mutually exclusive events $E_1, E_2, \ldots, E_n$, the probability function satisfies [28]

$$\mathbb{P}\left(\bigcup_{i=1}^{n} E_i\right) = \sum_{i=1}^{n} \mathbb{P}(E_i) \tag{2.1}$$

Equation (2.1) indicates that the probability function $\mathbb{P}$ is countably additive.

Example 2.2

In the die tossing experiment as in Example 2.1, assuming that all the six numbers are equally likely to appear, then $\mathbb{P}(E) = \frac{2}{6} = \frac{1}{3}, \mathbb{P}(F) = \frac{3}{6} = \frac{1}{2}, \mathbb{P}(E \cap F) = \frac{1}{6}$, and $\mathbb{P}(E \cup F) = \frac{4}{6} = \frac{2}{3}$.

Based on the aforementioned properties of the probability function, we have the following corollaries.

(1) If the complement of E is $\overline{E}$ with an event space of $\Omega \setminus \mathcal{E}$,

$$\mathbb{P}(\overline{E}) = 1 - \mathbb{P}(E) \tag{2.2}$$

Proof By noting that E and $\overline{E}$ are mutually exclusive, with Eq. (2.1), $\mathbb{P}(E) + \mathbb{P}(\overline{E}) = \mathbb{P}(E \bigcup \overline{E}) = 1$ since $E \bigcup \overline{E}$ has an event space of Ω. Thus $\mathbb{P}(\overline{E}) = 1 - \mathbb{P}(E)$. $\square$

(2) For events E_1 and E_2,

$$\mathbb{P}(E_1 \cup E_2) = \mathbb{P}(E_1) + \mathbb{P}(E_2) - \mathbb{P}(E_1 \cap E_2) \tag{2.3}$$

Proof Since $E_2 = E_1 E_2 \cup \overline{E}_1 E_2$, according to Eq. (2.1),

$$\mathbb{P}(E_2) = \mathbb{P}(E_1 E_2 \cup \overline{E}_1 E_2) = \mathbb{P}(E_1 E_2) + \mathbb{P}(\overline{E}_1 E_2) \tag{2.4}$$

which further gives

$$\mathbb{P}(\overline{E}_1 E_2) = \mathbb{P}(E_2) - \mathbb{P}(E_1 E_2) \tag{2.5}$$

Thus, by noting that $E_1 \cup E_2 = E_1 \cup \overline{E}_1 E_2$,

$$\begin{aligned}
\mathbb{P}(E_1 \cup E_2) = \mathbb{P}(E_1 \cup \overline{E}_1 E_2) &= \mathbb{P}(E_1) + \mathbb{P}(\overline{E}_1 E_2) \\
&= \mathbb{P}(E_1) + \mathbb{P}(E_2) - \mathbb{P}(E_1 \cap E_2)
\end{aligned} \tag{2.6}$$

which completes the proof. $\qquad\square$

(3) Based on (2), a generalized statement is that for n events E_1, E_2 through E_n,

$$\begin{aligned}
\mathbb{P}\left(\bigcup_{i=1}^{n} E_i\right) = \sum_{i=1}^{n} \mathbb{P}(E_i) &- \sum_{1 \leq i < j \leq n} \mathbb{P}\left(E_i \cap E_j\right) + \sum_{1 \leq i < j < k \leq n} \mathbb{P}\left(E_i \cap E_j \cap E_k\right) - \cdots \\
&+ (-1)^{n+1} \mathbb{P}\left(E_1 \cap E_2 \cap \cdots \cap E_n\right)
\end{aligned} \tag{2.7}$$

Equation (2.7) is known as the *inclusion-exclusion identity*.

Proof We will prove Eq. (2.7) inductively. First, for case of $n = 2$, the formula holds, as has been shown above. Next, we assume that Eq. (2.7) holds for $n = k$ and consider the case of $n = k + 1$. With Eq. (2.3),

$$\begin{aligned}
\mathbb{P}\left(\bigcup_{i=1}^{k+1} E_i\right) &= \mathbb{P}\left[\left(\bigcup_{i=1}^{k} E_i\right) \bigcup E_{k+1}\right] = \mathbb{P}\left(\bigcup_{i=1}^{k} E_i\right) + \mathbb{P}(E_{k+1}) - \mathbb{P}\left[\left(\bigcup_{i=1}^{k} E_i\right) \bigcup E_{k+1}\right] \\
&= \sum_{i=1}^{k} \mathbb{P}(E_i) - \sum_{1 \leq i < j \leq k} \mathbb{P}\left(E_i \cap E_j\right) + \sum_{1 \leq i < j < l \leq k} \mathbb{P}\left(E_i \cap E_j \cap E_l\right) - \cdots \\
&\quad + (-1)^{k+1} \mathbb{P}\left(E_1 \cap E_2 \cap \cdots \cap E_k\right) + \mathbb{P}(E_{k+1}) \\
&\quad - \sum_{i=1}^{k} \mathbb{P}(E_i \cap E_{k+1}) + \sum_{1 \leq i < j \leq k} \mathbb{P}\left(E_i \cap E_j \cap E_{k+1}\right) \\
&\quad - \sum_{1 \leq i < j < l \leq k} \mathbb{P}\left(E_i \cap E_j \cap E_l \cap E_{k+1}\right) - \cdots \\
&\quad + (-1)^{k+2} \mathbb{P}\left(E_1 \cap E_2 \cap \cdots \cap E_k \cap E_{k+1}\right)
\end{aligned} \tag{2.8}$$

Note that

$$\sum_{1\leq i<j\leq k} \mathbb{P}\left(E_i\cap E_j\right) + \sum_{i=1}^{k}\mathbb{P}(E_i\cap E_{k+1}) = \sum_{1\leq i<j\leq k+1} \mathbb{P}\left(E_i\cap E_j\right)$$

$$\sum_{1\leq i<j<l\leq k} \mathbb{P}\left(E_i\cap E_j\cap E_l\right) + \sum_{1\leq i<j<l\leq k} \mathbb{P}\left(E_i\cap E_j\cap E_l\cap E_{k+1}\right)$$

$$= \sum_{1\leq i<j<l\leq k+1} \mathbb{P}\left(E_i\cap E_j\cap E_l\right)$$

$$\vdots$$

$$(-1)^{k+2}\mathbb{P}\left(E_1\cap E_2\cap\cdots\cap E_k\cap E_{k+1}\right) = (-1)^{(k+1)+1}\mathbb{P}\left(E_1\cap E_2\cap\cdots\cap E_k\cap E_{k+1}\right) \tag{2.9}$$

Thus, Eq. (2.8) becomes

$$\mathbb{P}\left(\bigcup_{i=1}^{k+1} E_i\right) = \sum_{i=1}^{k+1}\mathbb{P}(E_i) - \sum_{1\leq i<j\leq k+1} \mathbb{P}\left(E_i\cap E_j\right) + \sum_{1\leq i<j<k\leq k+1} \mathbb{P}\left(E_i\cap E_j\cap E_k\right) - \cdots$$

$$+ (-1)^{(k+1)+1}\mathbb{P}\left(E_1\cap E_2\cap\cdots\cap E_{k+1}\right) \tag{2.10}$$

which completes the proof.

(4) For two events E and F, if $E = \cup_{i=1}^{n} E_i$ and $E_i\cap E_j = \varnothing$ for $i\neq j$, then

$$\mathbb{P}(F) = \sum_{i=1}^{n}\mathbb{P}(F\cap E_i) \tag{2.11}$$

Proof Note that $\bigcup_{i=1}^{n}(E_i\cap F) = F$, and $E_iF\cap E_jF = \varnothing$ for $i\neq j$. With this, according to Eq. (2.1),

$$\mathbb{P}(F) = \mathbb{P}\left(\bigcup_{i=1}^{n}(E_i\cap F)\right) = \sum_{i=1}^{n}\mathbb{P}(F\cap E_i) \tag{2.12}$$

which completes the proof. $\square$

Example 2.3

Using the inclusion-exclusion identity as in Eq. (2.3), show that for n events $E_1, E_2, \ldots, E_n$,

(1) (Bonferroni's inequality) $\mathbb{P}(E_1\cap E_2) \geq \mathbb{P}(E_1) + \mathbb{P}(E_2) - 1$.
(2) (Boole's inequality) $\mathbb{P}\left(\bigcup_{i=1}^{n} E_i\right) \leq \sum_{i=1}^{n}\mathbb{P}(E_i)$.

Solution

(1) According to Eq. (2.3),

$$\mathbb{P}(E_1) + \mathbb{P}(E_2) = \mathbb{P}(E_1\cup E_2) + \mathbb{P}(E_1\cap E_2) \leq 1 + \mathbb{P}(E_1\cap E_2) \tag{2.13}$$

holds since $\mathbb{P}(E_1 \cup E_2) \leq 1$. Thus, $\mathbb{P}(E_1 \cap E_2) \geq \mathbb{P}(E_1) + \mathbb{P}(E_2) - 1$.

(2) We use the method of induction. For the starting case of $n = 1$, $\mathbb{P}(E_1) \leq \mathbb{P}(E_1)$ holds. Next, we consider the case of $n = k + 1$ provided that $\mathbb{P}\left(\bigcup_{i=1}^{k} E_i\right) \leq \sum_{i=1}^{k} \mathbb{P}(E_i)$. According to Eq. (2.3),

$$\mathbb{P}\left(\bigcup_{i=1}^{k+1} E_i\right) = \mathbb{P}\left[\left(\bigcup_{i=1}^{k} E_i\right) \bigcup E_{k+1}\right] = \mathbb{P}\left(\bigcup_{i=1}^{k} E_i\right) + \mathbb{P}(E_{k+1}) - \mathbb{P}\left[\left(\bigcup_{i=1}^{k} E_i\right) \bigcap E_{k+1}\right]$$

$$\leq \mathbb{P}\left(\bigcup_{i=1}^{k} E_i\right) + \mathbb{P}(E_{k+1}) \leq \sum_{i=1}^{k} \mathbb{P}(E_i) + \mathbb{P}(E_{k+1}) = \sum_{i=1}^{k+1} \mathbb{P}(E_i)$$

$$(2.14)$$

which completes the proof.

2.1.2 Conditional Probability

Let E and F be two events. In some cases, the occurrence of one event (say, E) will affect the probability of the other event (F). For instance, consider the quality check of a batch of steel bars. Examining the yielding strength of seven (7) samples gives 233, 210, 223, 214, 243, 226, 241 MPa respectively. For an untested steel bar, its yielding strength is roughly assumed to be one of the seven testing results with an equal probability. Let events E and F denote that an untest bar has a strength greater than 220 and 230 MPa respectively, with which $\mathbb{P}(E) = \frac{5}{7}$ and $\mathbb{P}(F) = \frac{3}{7}$. Now, given that event E occurs, the conditional probability of F, $\mathbb{P}(F|E)$, becomes $\frac{3}{5}$ (instead of $\frac{3}{7}$), because the conditional sample space of F becomes $\{233, 223, 243, 226, 241\}$ MPa.

Mathematically, the conditional probability of event F on E is defined as follows,

$$\mathbb{P}(F|E) = \frac{\mathbb{P}(E \cap F)}{\mathbb{P}(E)} \tag{2.15}$$

with which

$$\mathbb{P}(E \cap F) = \mathbb{P}(E) \cdot \mathbb{P}(F|E) \tag{2.16}$$

Note that Eq. (2.15) holds only when $\mathbb{P}(E) \neq 0$. In the aforementioned case of steel bar quality check, $E \cap F = F$ and thus $\mathbb{P}(E \cap F) = \mathbb{P}(F) = \frac{3}{7}$. With Eq. (2.15), $\mathbb{P}(F|E) = \frac{\mathbb{P}(E \cap F)}{\mathbb{P}(E)} = \frac{\frac{3}{7}}{\frac{5}{7}} = \frac{3}{5}$, consistent with the previous result.

Specifically, if

$$\mathbb{P}(E \cap F) = \mathbb{P}(E) \cdot \mathbb{P}(F) \tag{2.17}$$

then E and F are said to be statistically independent of each other, and vice versa. With Eq. (2.15), one can also claim that E and F are independent if

$$\mathbb{P}(E|F) = \mathbb{P}(E) \tag{2.18}$$

Equation (2.18) further indicates that $\mathbb{P}(F|E) = \frac{\mathbb{P}(EF)}{\mathbb{P}(E)} = \frac{\mathbb{P}(E) \cdot \mathbb{P}(F)}{\mathbb{P}(E)} = \mathbb{P}(F)$.

Example 2.4

For events $E_1, E_2, \ldots, E_n$, show that

$$\mathbb{P}\left(\bigcap_{i=1}^{n} E_i\right) = \mathbb{P}(E_1)\mathbb{P}(E_2|E_1)\mathbb{P}(E_3|E_1 E_2)\ldots\mathbb{P}(E_n|E_1 E_2 \ldots E_{n-1}) \tag{2.19}$$

Solution
According to Eq. (2.16),

$$\mathbb{P}(E_1)\mathbb{P}(E_2|E_1)\mathbb{P}(E_3|E_1 E_2)\ldots\mathbb{P}(E_n|E_1 E_2 \ldots E_{n-1})$$
$$= [\mathbb{P}(E_1)\mathbb{P}(E_2|E_1)]\mathbb{P}(E_3|E_1 E_2)\ldots\mathbb{P}(E_n|E_1 E_2 \ldots E_{n-1})$$
$$= [\mathbb{P}(E_1 E_2)\mathbb{P}(E_3|E_1 E_2)]\ldots\mathbb{P}(E_n|E_1 E_2 \ldots E_{n-1}) \tag{2.20}$$
$$= \cdots = \mathbb{P}\left(\bigcap_{i=1}^{n} E_i\right)$$

which completes the proof.

Remark

It is noticed that in Example 2.4, generally speaking, $\mathbb{P}(E_1)\mathbb{P}(E_2|E_1)\mathbb{P}(E_3|E_2)\ldots\mathbb{P}(E_n|E_{n-1})$ does not necessarily equal $\mathbb{P}\left(\bigcap_{i=1}^{n} E_i\right)$, unless $\mathbb{P}(E_k|E_{k-1}) = \mathbb{P}(E_k|E_1 E_2 \cdots E_{k-1})$ holds for $\forall k = 2, 3, \ldots, n$. Specially, for a sequence of events $E_1, E_2, \ldots, E_n$ that satisfy $E_n \subset E_{n-1} \subset E_{n-2} \subset \cdots E_2 \subset E_1$, one can show that $\mathbb{P}(E_1)\mathbb{P}(E_2|E_1)\mathbb{P}(E_3|E_2)\ldots\mathbb{P}(E_n|E_{n-1}) = \mathbb{P}\left(\bigcap_{i=1}^{n} E_i\right) = \mathbb{P}(E_n)$. An application of this observation can be found in Sect. 3.5.4.

Example 2.5

In a traffic network, there are 20 steel bridges and 25 reinforced concrete (RC) bridges that were constructed before year 2000. For the post-2000s ones, the numbers of steel and RC bridges are x and y respectively. Show that when $x/y = 4/5(= 20/25)$, the bridge's type and service history are independent when randomly selecting a bridge from the network.

Solution

We first summarize the bridge portfolio as follows,

	steel	RC
Before 2000	20	25
After 2000	x	y

Let B = the event that a bridge was constructed before year 2000, A = the event that a bridge was constructed after 2000, S = the event of selecting a steel bridge, and R = the event of selecting an RC bridge. With this, one has

$$
\mathbb{P}(B) = \frac{45}{45 + x + y}, \quad \mathbb{P}(A) = \frac{x + y}{45 + x + y}
$$

$$
\mathbb{P}(S) = \frac{x + 20}{45 + x + y}, \quad \mathbb{P}(R) = \frac{25 + y}{45 + x + y}
$$

$$
\mathbb{P}(BS) = \frac{20}{45 + x + y}, \quad \mathbb{P}(AS) = \frac{x}{45 + x + y} \tag{2.21}
$$

$$
\mathbb{P}(BR) = \frac{25}{45 + x + y}, \quad \mathbb{P}(AR) = \frac{y}{45 + x + y}
$$

The independence between a bridge's type and service history is equivalent to

$$
\begin{cases}
\mathbb{P}(BS) = \mathbb{P}(B) \cdot \mathbb{P}(S) \\
\mathbb{P}(BR) = \mathbb{P}(B) \cdot \mathbb{P}(R) \\
\mathbb{P}(AS) = \mathbb{P}(A) \cdot \mathbb{P}(S) \\
\mathbb{P}(AR) = \mathbb{P}(A) \cdot \mathbb{P}(R)
\end{cases}
\Leftrightarrow
\begin{cases}
\dfrac{20}{45 + x + y} = \dfrac{45(x + 20)}{(45 + x + y)^2} \\[2mm]
\dfrac{25}{45 + x + y} = \dfrac{45(25 + y)}{(45 + x + y)^2} \\[2mm]
\dfrac{x}{45 + x + y} = \dfrac{(x + y)(x + 20)}{(45 + x + y)^2} \\[2mm]
\dfrac{y}{45 + x + y} = \dfrac{(x + y)(25 + y)}{(45 + x + y)^2}
\end{cases}
\Leftrightarrow \frac{x}{y} = \frac{20}{25} = \frac{4}{5}
\tag{2.22}
$$

Thus, if $x/y = 4/5$, the bridge's type and service history are independent when selecting a bridge from the network randomly.

Based on the conditional probability (c.f. Eq. (2.15)), the *law of total probability* states that for two events E and F, if $E = \bigcup_{i=1}^{n} E_i$ and $E_i \bigcap E_j = \varnothing$ for $i \neq j$,

$$
\mathbb{P}(F) = \sum_{i=1}^{n} \mathbb{P}(F \cap E_i) = \sum_{i=1}^{n} \mathbb{P}(F|E_i) \cdot \mathbb{P}(E_i) \tag{2.23}
$$

Proof Note that $F = \bigcup_{i=1}^{n} E_i F$. According to Eqs. (2.1) and (2.16),

$$\mathbb{P}(F) = \mathbb{P}\left(\bigcup_{i=1}^{n} E_i F\right) = \sum_{i=1}^{n} \mathbb{P}\left(E_i F\right) = \sum_{i=1}^{n} \mathbb{P}(F|E_i) \cdot \mathbb{P}(E_i) \qquad (2.24)$$

which yields Eq. (2.23).   □

With Eqs. (2.16) and (2.23), if $E = \bigcup_{i=1}^{n} E_i$ and $E_i \cap E_j = \varnothing$ holds for $i \neq j$, it further follows

$$\mathbb{P}(E_i|F) = \frac{\mathbb{P}(E_i F)}{\mathbb{P}(F)} = \frac{\mathbb{P}(F|E_i) \cdot \mathbb{P}(E_i)}{\sum_{i=1}^{n} \mathbb{P}(F|E_i) \cdot \mathbb{P}(E_i)} \qquad (2.25)$$

Equation (2.25) is known as the *Bayes' theorem*.

Example 2.6

In the quality control of a factory's produced steel bars, procedure A results in a qualifiedness probability of p and with procedure B, the probability of qualifiedness is q. If the ratio of the productions associated with procedures A and B is $r/(1-r)$, where $0 < p, q, r < 1$, find the probability that a qualified steel bar is produced with procedure A.

Solution

Let Q = the event that a steel bar is qualified, P_A = the event that a steel bar is produced with procedure A, and P_B = the event that a steel bar is produced with procedure B. With the Bayes' theorem as in Eq. (2.25),

$$\begin{aligned} \mathbb{P}(P_A|Q) &= \frac{\mathbb{P}(P_A Q)}{\mathbb{P}(Q)} \\ &= \frac{\mathbb{P}(Q|P_A) \cdot \mathbb{P}(P_A)}{\mathbb{P}(Q|P_A) \cdot \mathbb{P}(P_A) + \mathbb{P}(Q|P_B) \cdot \mathbb{P}(P_B)} \\ &= \frac{pr}{pr + q(1-r)} \end{aligned} \qquad (2.26)$$

Example 2.7

A portal frame is subjected to a vertical load V (e.g., gravity load) and a horizontal load H (e.g., wind load), as shown in Fig. 2.1a. During a reference period of interest, the occurrence probabilities of V and H are q_v and q_h respectively. Assume that V and H occur independently. Under the action of V only, the frame may fail in a "beam" mode with a probability of p_{v1} (c.f. Fig. 2.1b). Conditional on the occurrence of H only, the failure mode is called "sway", as illustrated in Fig. 2.1c, with a probability of p_{h2}. In the presence of the joint actions of V and H, however, the frame may fail in one of the following three modes: (1) "beam", (2) "sway" and (3) the combination of "beam" and "sway" (c.f. Fig. 2.1d), and the failure probabilities associated with the three modes are p_{vh1}, p_{vh2} and p_{vh3} respectively.

(1) Calculate the failure probability associated with mode 2 (i.e., "sway").

Fig. 2.1 Failure modes of a portal frame

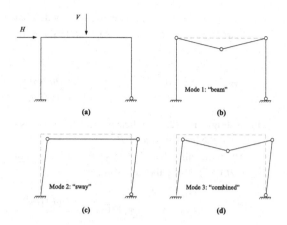

(a)

Mode 1: "beam"

(b)

Mode 2: "sway"

(c)

Mode 3: "combined"

(d)

(2) Determine the probability that H occurs only conditional on the failure mode of "sway".

(3) Determine the probability that H occurs only conditional on the failure of the frame.

(4) Show that the probability obtained in (2) is always greater than that in (3).

Solution

For simplicity, we let in the following $V =$ the event that the vertical load occurs, $H =$ the event that the horizontal load occurs, $F =$ the event of frame failure, F_1, F_2 and $F_3 =$ the events of frame failure associated with mode 1 ("beam"), mode 2 ("sway") and mode 3 ("combined"), respectively. Since V and F are independent, according to Eq. (2.17),

$$\mathbb{P}(VH) = \mathbb{P}(V) \cdot \mathbb{P}(H) = q_v q_h \tag{2.27}$$

Further, we summarize the load occurrence and frame failure probabilities as follows.

Load case	Probability of load case	Conditional $\mathbb{P}(F_1)$	Conditional $\mathbb{P}(F_2)$	Conditional $\mathbb{P}(F_3)$
V only ($V\overline{H}$)	$q_v(1-q_h)$	p_{v1}	0	0
H only ($\overline{V}H$)	$q_h(1-q_v)$	0	p_{h2}	0
Both V and H (VH)	$q_v q_h$	p_{vh1}	p_{vh2}	p_{vh3}
No load ($\overline{V} \cap \overline{H}$)	$(1-q_h)(1-q_v)$	0	0	0

A quick check shows that the sum of the probabilities of all the four load cases equals $q_v(1-q_h) + q_h(1-q_v) + q_v q_h + (1-q_h)(1-q_v) = 1$.

(1) The probability of failure due to mode 2 ("sway", c.f. Fig. 2.1c) is obtained with the total probability theorem (c.f. Eq. (2.23)) as follows,

$$\mathbb{P}(F_2) = \mathbb{P}(F_2|V\overline{H}) \cdot \mathbb{P}(V\overline{H}) + \mathbb{P}(F_2|\overline{V}H) \cdot \mathbb{P}(\overline{V}H) + \mathbb{P}(F_2|VH) \cdot \mathbb{P}(VH)$$
$$+ \mathbb{P}(F_2|\overline{V} \cap \overline{H}) \cdot \mathbb{P}(\overline{V} \cap \overline{H})$$
$$= 0 \cdot q_v(1-q_h) + p_{h2} \cdot q_h(1-q_v) + p_{vh2} \cdot q_v q_h + 0 \cdot (1-q_h)(1-q_v)$$
$$= p_{h2} \cdot q_h(1-q_v) + p_{vh2} \cdot q_v q_h$$

$$\tag{2.28}$$

(2) The probability that H occurs only conditional on the failure mode of "sway" is estimated by $\mathbb{P}(H\overline{V}|F_2)$ (rather than $\mathbb{P}(H|F_2)$). With the Bayes' theorem (c.f. Eq. (2.25)), it follows,

$$\mathbb{P}(H\overline{V}|F_2) = \frac{\mathbb{P}(F_2|H\overline{V}) \cdot \mathbb{P}(H\overline{V})}{\mathbb{P}(F_2)} = \frac{p_{h2}q_h(1-q_v)}{p_{h2}q_h(1-q_v) + p_{vh2}q_vq_h} = \frac{p_{h2}(1-q_v)}{p_{h2}(1-q_v) + p_{vh2}q_v}$$
(2.29)

Interestingly, Eq. (2.29) indicates that $\mathbb{P}(H\overline{V}|F_2)$ is independent of the occurrence probability of the horizontal load H.

(3) The probability that H occurs only conditional on the failure of the frame is assessed by $\mathbb{P}(H\overline{V}|F)$. Using the Bayes' theorem,

$$\mathbb{P}(H\overline{V}|F) = \frac{\mathbb{P}(F|H\overline{V}) \cdot \mathbb{P}(H\overline{V})}{\mathbb{P}(F)}$$
(2.30)

where

$$\begin{aligned}
\mathbb{P}(F) &= \mathbb{P}(F|V\overline{H}) \cdot \mathbb{P}(V\overline{H}) + \mathbb{P}(F|\overline{V}H) \cdot \mathbb{P}(\overline{V}H) + \mathbb{P}(F|VH) \cdot \mathbb{P}(VH) \\
&\quad + \mathbb{P}(F|\overline{V} \cap \overline{H}) \cdot \mathbb{P}(\overline{V} \cap \overline{H}) \\
&= p_{v1} \cdot q_v(1-q_h) + p_{h2} \cdot q_h(1-q_v) + (p_{vh1} + p_{vh2} + p_{vh3}) \cdot q_vq_h \quad (2.31) \\
&\quad + 0 \cdot (1-q_h)(1-q_v) \\
&= p_{v1} \cdot q_v(1-q_h) + p_{h2} \cdot q_h(1-q_v) + (p_{vh1} + p_{vh2} + p_{vh3}) \cdot q_vq_h
\end{aligned}$$

Thus,

$$\begin{aligned}
\mathbb{P}(H\overline{V}|F) &= \frac{p_{h2} \cdot q_h(1-q_v)}{p_{v1} \cdot q_v(1-q_h) + p_{h2} \cdot q_h(1-q_v) + (p_{vh1} + p_{vh2} + p_{vh3}) \cdot q_vq_h} \\
&= \frac{p_{h2} \cdot (1-q_v)}{p_{v1} \cdot q_v/q_h \cdot (1-q_h) + p_{h2} \cdot (1-q_v) + (p_{vh1} + p_{vh2} + p_{vh3}) \cdot q_v}
\end{aligned}$$
(2.32)

(4) With Eqs. (2.29) and (2.32), since

$$p_{v1} \cdot q_v/q_h(1-q_h) + p_{h2} \cdot (1-q_v) + (p_{vh1} + p_{vh2} + p_{vh3}) \cdot q_v > p_{h2}(1-q_v) + p_{vh2}q_v$$
(2.33)

It is easy to see that

$$\mathbb{P}(H\overline{V}|F_2) > \mathbb{P}(H\overline{V}|F)$$
(2.34)

---

*Example 2.8

[*Readers may refer to Sect. 2.1.3 for relevant background information.*]

The use of proof loading information to update the current structural resistance is illustrated in this example [11, 16]. Suppose that the resistance of a bridge girder, R, has a probability density function of $f_R(r)$ and a cumulative distribution function of $F_R(r)$. Given the bridge's post-loading survival subjected to

(1) a deterministic proof load of s;
(2) a proof load S with a cumulative distribution function of $F_S(s)$,

determine the "updated" distribution of bridge resistance for the two cases respectively.

Solution

(1) Suppose that the post-loading probability density function of R is $f_{R'}(r)$. With $\Delta r \to 0$, we have $\mathbb{P}(r < R \le r + \Delta r | R > s) = f_{R'}(r)\Delta r$ (c.f. Eq. (2.53)). According to the Bayes' theorem (c.f. Eq. (2.25)),

$$\mathbb{P}(r < R \le r + \Delta r | R > s) = \frac{\mathbb{P}(r < R \le r + \Delta r \cap R > s)}{\mathbb{P}(R > s)} \tag{2.35}$$

Since $\mathbb{P}(r < R \le r + \Delta r) = f_R(r)\Delta r$, and $\mathbb{P}(R > s) = 1 - F_R(s)$, Eq. (2.35) becomes

$$\mathbb{P}(r < R \le r + \Delta r | R > s) = \frac{\mathbb{P}(r < R \le r + \Delta r \cap R > s)}{\mathbb{P}(R > s)} = \begin{cases} \dfrac{f_R(r)\Delta r}{1 - F_R(s)}, & r \ge s \\ 0, & r < s \end{cases} \tag{2.36}$$

Thus,

$$f_{R'}(r) = \begin{cases} \dfrac{f_R(r)}{1 - F_R(s)}, & r \ge s \\ 0, & r < s \end{cases} \tag{2.37}$$

(2) The updated distribution of R, $f_{R'}(r)$, is obtained with a similar approach as in (1). As before,

$$f_{R'}(r)\Delta r = \frac{\mathbb{P}(r < R \le r + \Delta r \cap R > s)}{\mathbb{P}(R > s)} \tag{2.38}$$

where $\mathbb{P}(r < R \le r + \Delta r \cap R > s) = F_S(r)f_R(r)\Delta r$, and $\mathbb{P}(R > S) = \int_{-\infty}^{\infty} F_S(\tau) f_R(\tau)d\tau$ (using the law of total probability, c.f. Eq. (2.23)). Thus,

$$f_{R'}(r) = \frac{F_S(r)f_R(r)}{\int_{-\infty}^{\infty} F_S(\tau) f_R(\tau)d\tau} \tag{2.39}$$

2.1.3 Random Variable

In the presence of the events defined in the sample space Ω, a random variable provides a mapping from Ω to real numbers, which, in many case, offers a convenient approach to describe the outcomes of an experiment. For example, recall

Example 2.1. For case (1), since $\Omega = \{o_1, o_2, o_3, o_4, o_5, o_6\}$, we introduce a random variable X, which has a value of i if "i" appears on the die, $i = 1, 2, \ldots, 6$. With this, one has

$$\mathbb{P}(o_i) = \mathbb{P}(X = i) = \frac{1}{6}, \quad i = 1, 2, \ldots, 6$$

$$\mathbb{P}(E) = \mathbb{P}(X = 1) + \mathbb{P}(X = 3) = \frac{1}{6} + \frac{1}{6} = \frac{1}{3}$$

$$\mathbb{P}(F) = \mathbb{P}(X = 1) + \mathbb{P}(X = 4) + \mathbb{P}(X = 5) = \frac{1}{6} + \frac{1}{6} + \frac{1}{6} = \frac{1}{2}$$

$$\mathbb{P}(E \cap F) = \mathbb{P}(X = 1) = \frac{1}{6}$$

$$\mathbb{P}(E \cup F) = \mathbb{P}(X = 1) + \mathbb{P}(X = 3) + \mathbb{P}(X = 4) + \mathbb{P}(X = 5) = \frac{1}{6} + \frac{1}{6} + \frac{1}{6} + \frac{1}{6} = \frac{2}{3}$$
$$\tag{2.40}$$

Recall that an event with an event space of Ω has a probability of 1. This can be used to verify the correctness of Eq. (2.40), with which $\sum_{i=1}^{6} \mathbb{P}(X = i) = \frac{1}{6} \times 6 = 1$.

For case (2) of Example 2.1, let random variable Y denote the structural lifetime (in years), with which it follows,

$$\mathbb{P}(E) = \mathbb{P}(X > 30), \quad \mathbb{P}(F) = \mathbb{P}(X > 20)$$
$$\mathbb{P}(E \cap F) = \mathbb{P}(E) = \mathbb{P}(X > 30), \quad \mathbb{P}(E \cup F) = \mathbb{P}(F) = \mathbb{P}(X > 20) \tag{2.41}$$

A random variable is referred to as *discrete* if it takes a finite (or countable) number of possible values. For example, the random variable X in Eq. (2.40) was introduced to denote the outcomes of rolling a die and thus has totally six possible values (1–6). However, if a random variable has a value on a continuum (uncountable), it is known as *continuous*. Illustratively, the random variable Y in Eq. (2.41) is a continuous variable.

Generally, let X be a random variable with a probability function $\mathbb{P}$, the *cumulative distribution function* (CDF) of X, $F_X(x)$, is defined as follows,

$$F_X(x) = \mathbb{P}(X \le x), \quad \text{for } \forall x \in (-\infty, \infty) \tag{2.42}$$

Equation (2.42) implies that the CDF $F_X(x)$ equals the probability that X takes a value less than or equal to x. The function $F_X(x)$ has the following three properties,

(1) $F_X(x)$ is non-decreasing with respect to x;
(2) $\lim_{-\infty} F_X(x) = 0$;
(3) $\lim_{\infty} F_X(x) = 1$;

Property (1) holds by noting that for $x_1 < x_2$, the event space of $X \le x_1$ is contained by that of $X \le x_2$ and thus $F_X(x_2) < F_X(x_2)$. Properties (2) and (3) are due to the fact that the probability function $\mathbb{P}$ is strictly defined within $[0, 1]$.

Furthermore, with Eq. (2.42), if $x_1 < x_2$, then

$$\mathbb{P}(x_1 < X \le x_2) = \mathbb{P}(X \le x_2) - \mathbb{P}(X \le X_1) = F_X(x_2) - F_X(x_1) \tag{2.43}$$

If X is a discrete random variable with n possible values $\{x_1, x_2, \ldots, x_n\}$, its *probability mass function* (PMF), $p_X(x_i)$, is determined by

$$p_X(x_i) = \mathbb{P}(X = x_i) \tag{2.44}$$

for $i = 1, 2, \ldots, n$. With this, one has (1) $0 \leq p_X(x_i) \leq 1$, (2) $\sum_{i=1}^{n} p_X(x_i) = \sum_{i=1}^{n} \mathbb{P}(X = x_i) = 1$, and (3) $p_X(x) = 0$ if $x \notin \{x_1, x_2, \ldots, x_n\}$.

With Eqs. (2.42) and (2.44), it follows,

$$F_X(x) = \mathbb{P}(X \leq x) = \sum_{\forall x_i \leq x} p_X(x_i) \tag{2.45}$$

Example 2.9

Consider the post-hazard damage states of four buildings (failure or survival). Each building has an identical failure probability of 0.2, and the behaviour of each building is assumed to be independent of each other. Let random variable X represent the number of failed buildings.

(1) Find the CDF of X;
(2) Calculate $\mathbb{P}(1 < X \leq 3)$;
(3) Compare $\mathbb{P}(X > 2)$ and $\mathbb{P}(X \geq 2)$.

Solution

(1) Note that X takes a value from $\{0, 1, 2, 3, 4\}$. Thus,

$$
\begin{aligned}
\mathbb{P}(X = 0) &= C_4^0 0.2^0 (1 - 0.2)^4 = 0.4096 \\
\mathbb{P}(X = 1) &= C_4^1 0.2^1 (1 - 0.2)^3 = 0.4096 \\
\mathbb{P}(X = 2) &= C_4^2 0.2^2 (1 - 0.2)^2 = 0.1536 \\
\mathbb{P}(X = 3) &= C_4^3 0.2^3 (1 - 0.2)^1 = 0.0256 \\
\mathbb{P}(X = 4) &= C_4^4 0.2^4 (1 - 0.2)^0 = 0.0016
\end{aligned}
\tag{2.46}
$$

where C_n^k is the binomial coefficient (which equals to the coefficient of the monomial x^k in the expansion of $(1 + x)^n$, $C_n^k = \frac{n!}{k!(n-k)!}$). A quick check shows that $\sum_{i=0}^{4} \mathbb{P}(X = i) = 1$. Thus, the CDF of X is determined by

$$
F_X(x) = \begin{cases}
0, & x < 0 \\
0.4096, & 0 \leq x < 1 \\
0.8192, & 1 \leq x < 2 \\
0.9728, & 2 \leq x < 3 \\
0.9984, & 3 \leq x < 4 \\
1, & x \geq 4
\end{cases}
\tag{2.47}
$$

(2)

Method 1. $\mathbb{P}(1 < X \leq 3) = \mathbb{P}(X = 2) + \mathbb{P}(X = 3) = 0.1536 + 0.0256 = 0.1792$.

Method 2. $\mathbb{P}(1 < X \leq 3) = F_X(3) - F_X(1) = 0.9984 - 0.8192 = 0.1792$.

The two methods yield the same result.

(3) One can show that $\mathbb{P}(X > 2) = 1 - F_X(2) = 1 - 0.9728 = 0.0272$ and $\mathbb{P}(X \geq 2) = \mathbb{P}(X > 1) = 1 - F_X(1) = 1 - 0.8192 = 0.1808$. Thus, $\mathbb{P}(X \geq 2) > \mathbb{P}(X > 2)$.

Remark

The random variable X in Example 2.9 is referred to as a *binomial random variable*. Generally, we consider the case where, of n times of independent tests, the probability of survival for each trial is p and the failure probability is $1 - p$. Let X be the number of successful trials, then X follows a binomial distribution. $\mathbb{P}(X = k) = C_n^k p^k (1 - p)^{n-k}$ holds for $\forall k = 0, 1, 2, \ldots, n$. There are also other distribution types for discrete variables, as will be later discussed in Sect. 2.2.

Example 2.10

Consider the post-hazard damage states (failure or survival) of two building portfolios: A with four buildings and B with three buildings. All the buildings have an identical failure probability of p, and each building behaves independently of each other. Let q denote the probability that the damage ratio (i.e., number of damaged buildings to the total building number) of portfolio A is greater than that of B. Find the maximum value of q and the corresponding p.

Solution

We first summarize the possible scenarios of the two building portfolios as follows, with which the damage ratio of portfolio A is greater than that of B.

Number of damaged buildings in A, n_A	Number of damaged buildings in B, n_B
1	0
2	0,1
3	0,1,2
4	0,1,2

With this, we have

$$q = \sum_{i=1}^{4} \mathbb{P}(n_A = i)\mathbb{P}(n_B = 0) + \sum_{i=2}^{4} \mathbb{P}(n_A = i)\mathbb{P}(n_B = 1) + \sum_{i=3}^{4} \mathbb{P}(n_A = i)\mathbb{P}(n_B = 2)$$

$$= (1 - p)^3 \left[C_4^1 p(1 - p)^3 + C_4^2 p^2(1 - p)^2 + C_4^3 p^3(1 - p) + C_4^4 p^4 \right]$$

$$+ C_3^1 p(1 - p)^2 C_4^2 p^2(1 - p)^2 + C_3^1 p(1 - p)^2 C_4^3 p^3(1 - p)$$

$$+ C_3^2 p^2(1 - p) C_4^3 p^3(1 - p) + C_3^1 p(1 - p)^2 C_4^4 p^4 + C_3^2 p^2(1 - p) C_4^4 p^4$$

$$= (1 - p)^3 - (1 - p)^7 + 18p^3(1 - p)^4 + 12p^4(1 - p)^3 + 15p^5(1 - p)^2 + 3p^6(1 - p)$$

$$\tag{2.48}$$

Fig. 2.2 Relationship between p and q

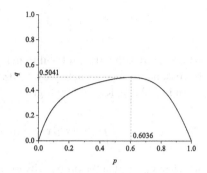

Figure 2.2 plots the relationship between p and q, from which it can be seen that the maximum value of q is 0.5041, achieved when $p = 0.6036$.

For a continuous random variable X, its *probability density function* (PDF), $f_X(x)$ for $x \in (-\infty, \infty)$, is defined by

$$\int_{\mathcal{D}} f_X(x)\mathrm{d}x = \mathbb{P}(X \in \mathcal{D}) \tag{2.49}$$

where $\mathcal{D}$ is a set of real numbers, $\mathcal{D} \subset (-\infty, \infty)$. Equation (2.49) implies that the probability that X lies within a set $\mathcal{D}$ equals the integration of the PDF over this set. Assigning $\mathcal{D} = (-\infty, \infty)$ in Eq. (2.49) yields

$$\int_{-\infty}^{\infty} f_X(x)\mathrm{d}x = \int_{-\infty}^{\infty} f_X(x)\mathrm{d}x = \mathbb{P}[X \in (-\infty, \infty)] = 1 \tag{2.50}$$

Remark

Recall Example 2.8. One can easily show that the updated PDF $f_{R'}(r)$ satisfies Eq. (2.50). For example, consider Eq. (2.39), which gives

$$\int_{-\infty}^{\infty} f_{R'}(r)\mathrm{d}r = \int_{-\infty}^{\infty} \frac{F_S(r)f_R(r)}{\int_{-\infty}^{\infty} F_S(\tau)f_R(\tau)\mathrm{d}\tau}\mathrm{d}r = \frac{\int_{-\infty}^{\infty} F_S(r)f_R(r)\mathrm{d}r}{\int_{-\infty}^{\infty} F_S(\tau)f_R(\tau)\mathrm{d}\tau} = 1 \tag{2.51}$$

If the set $\mathcal{D}$ is assigned to be $[a, b]$ in Eq. (2.49), then

$$\int_{a}^{b} f_X(x)\mathrm{d}x = \mathbb{P}(X \in [a, b]) \tag{2.52}$$

Specifically, if $b = a + \varepsilon$ in Eq. (2.52), where $\varepsilon \to 0$, then

$$P(X \in [a, a + \varepsilon]) = \int_a^{a+\varepsilon} f_X(x)dx \approx \varepsilon f_X(a) \qquad (2.53)$$

which implies that the probability that X lies within a small interval $[a, a + \varepsilon]$ can be reasonably approximately by $\varepsilon f_X(a)$.

In Eq. (2.52), letting $b = a$ gives

$$\int_a^a f_X(x)dx = P(a \le X \le a) = P(X = a) = 0 \qquad (2.54)$$

which suggests that the probability of a continuous random variable taking a particular value is zero. Thus, $F_X(x) = P(X \le x) = P(X < x)$ (this observation is different from the case of a discrete variable, c.f. Question (3) in Example 2.9).

Similar to Eq. (2.45), the relationship between the CDF and PDF of a continuous variable X is determined by considering Eq. (2.42), which gives

$$F_X(x) = P(X \le x) = P(-\infty \le X \le x) = \int_{-\infty}^x f_X(\tau)d\tau \qquad (2.55)$$

Taking the differential form of both sides of Eq. (2.55) further yields

$$f_X(x) = \frac{dF_X(x)}{dx} \qquad (2.56)$$

Example 2.11

If the PDF of random variable X is given as follows,

$$f_X(x) = \begin{cases} ax^2(1-x), & 0 \le x \le 1 \\ 0, & \text{otherwise} \end{cases} \qquad (2.57)$$

(1) Find the value of a;
(2) Derive the CDF of X, $F_X(x)$;
(3) Determine $P\left(\frac{1}{3} \le x \le \frac{1}{2}\right)$.

Solution

(1) With Eq. (2.50), the integration of $f_X(x)$ on $(-\infty, \infty)$ equals 1, which gives

$$\int_{-\infty}^{\infty} f_X(x)dx = \int_0^1 f_X(x)dx = a \int_0^1 x^2(1-x)dx$$

$$= a \left(\frac{1}{3}x^3 - \frac{1}{4}x^4\right)\Big|_0^1 = \frac{a}{12} = 1 \qquad (2.58)$$

$$\Rightarrow a = 12$$

(2) According to Eq. (2.55), $F_X(x) = \int_{-\infty}^{x} f_X(\tau)d\tau$. Thus, if $x < 0$, $F_X(x) = 0$ and if $x > 1$, $F_X(x) = 1$. For $x \in [0, 1]$,

$$
\begin{aligned}
F_X(x) &= \int_{-\infty}^{x} 12\tau^2(1-\tau)d\tau = \int_{0}^{x} 12\tau^2(1-\tau)d\tau \\
&= 12\left(\frac{1}{3}\tau^3 - \frac{1}{4}\tau^4\right)\Big|_{0}^{x} = 12\left(\frac{1}{3}x^3 - \frac{1}{4}x^4\right) \\
&= 4x^3 - 3x^4
\end{aligned}
\tag{2.59}
$$

Thus,

$$
F_X(x) = \begin{cases} 0, & x < 1 \\ 4x^3 - 3x^4, & 0 \le x \le 1 \\ 1, & x > 1 \end{cases}
\tag{2.60}
$$

(3) With the CDF of X obtained in (2),

$$
\mathbb{P}\left(\frac{1}{3} \le x \le \frac{1}{2}\right) = F_X\left(\frac{1}{2}\right) - F_X\left(\frac{1}{3}\right) = \frac{5}{16} - \frac{1}{9} = \frac{29}{144} = 0.201
\tag{2.61}
$$

Remark

The variable X in Example 2.11 is said to follow a Beta distribution (c.f. Sect. 2.2.7). There are also many other distribution types for continuous random variables, as will be further discussed in Sect. 2.2.

2.1.4 Mean and Variance of Random Variables

For a discrete random variable X with n possible values $\{x_1, x_2, \ldots, x_n\}$ and a PMF of $\{p_X(x_i)\}$, $i = 1, 2, \ldots, n$, the mean value (expectation) and variance of X, $\mathbb{E}(X)$ and $\mathbb{V}(X)$, are respectively

$$
\mathbb{E}(X) = \sum_{i=1}^{n} x_i p_X(x_i), \quad \mathbb{V}(X) = \sum_{i=1}^{n} (x_i - \mathbb{E}(X))^2 p_X(x_i)
\tag{2.62}
$$

Equation (2.62) indicates that the mean value of X is the *weighted* average of the possible values with a weight ratio of the corresponding probability mass $p_X(x_i)$. The variance of X can be treated as the mean value of $(X - \mathbb{E}(X))^2$, a function of X. More generally, if random variable Y is a function of X, denoted by $Y = g(X)$, then the mean value of Y is given by

$$\mathbb{E}(Y) = \mathbb{E}(g(X)) = \sum_{i=1}^{n} g(x_i) p_X(x_i) \tag{2.63}$$

For the continuous case, if the PDF of a continuous variable X is $f_X(x)$, then the mean value and variance of X are respectively determined by

$$\mathbb{E}(X) = \int_{-\infty}^{\infty} x f_X(x) \mathrm{d}x, \quad \mathbb{V}(X) = \int_{-\infty}^{\infty} (x - \mathbb{E}(X))^2 f_X(x) \mathrm{d}x \tag{2.64}$$

Equation (2.64) implies that the mean value of a continuous variable X, similar to the discrete case, equals the *weighted* average of the whole domain of X and the weight ratio is the PDF $F_x(X)$. The variance of X is the mean value of $(X - \mathbb{E}(X))^2$, which is a deterministic function of X. As before, the mean value of random variable Y, which is a function of X (i.e., $Y = g(X)$), is

$$\mathbb{E}(Y) = \mathbb{E}(g(X)) = \int_{-\infty}^{\infty} g(x) f_X(x) \mathrm{d}x \tag{2.65}$$

For both the discrete and continuous cases, the *standard deviation* of X is defined as the square root of the variance of X, which has the same unit as the mean value.

Importantly, we have the following three corollaries.

(1) For a random variable X and two constants a, b, $\mathbb{E}(aX + b) = a\mathbb{E}(X) + b$.

Proof For the discrete case, applying Eq. (2.63) gives

$$\mathbb{E}(aX + b) = \sum_{i=1}^{n}(ax_i + b)p_X(x_i) = a\sum_{i=1}^{n}x_i p_X(x_i) + b\sum_{i=1}^{n}p_X(x_i) = a\mathbb{E}(X) + b \tag{2.66}$$

For the continuous case, with Eq. (2.65), it follows,

$$\mathbb{E}(aX + b) = \int_{-\infty}^{\infty}(ax + b)f_X(x)\mathrm{d}x = a\int_{-\infty}^{\infty}x f_X(x)\mathrm{d}x + b\int_{-\infty}^{\infty}f_X(x)\mathrm{d}x = a\mathbb{E}(X) + b \tag{2.67}$$

with which the proof is complete. □

(2) $\mathbb{V}(X) = \mathbb{E}(X^2) - [\mathbb{E}(X)]^2$.

Proof For the discrete case, applying Eq. (2.63) yields

$$\mathbb{V}(X) = \mathbb{E}([X - \mathbb{E}(X)]^2) = \sum_{i=1}^{n}x_i^2 p_X(x_i) - 2\mathbb{E}(X)\sum_{i=1}^{n}x_i p_X(x_i) + \mathbb{E}^2(X)\sum_{i=1}^{n}p_X(x_i)$$
$$= \mathbb{E}(X^2) - 2\mathbb{E}^2(X) + \mathbb{E}^2(X) = \mathbb{E}(X^2) - [\mathbb{E}(X)]^2 \tag{2.68}$$

For the continuous case, with Eq. (2.65),

$$\mathbb{V}(X) = \mathbb{E}([X - \mathbb{E}(X)]^2) = \int_{-\infty}^{\infty} x^2 f_X(x) dx - 2\mathbb{E}(X) \int_{-\infty}^{\infty} x f_X(x) dx$$

$$+ \mathbb{E}^2(X) \int_{-\infty}^{\infty} f_X(x) dx \tag{2.69}$$

$$= \mathbb{E}(X^2) - 2\mathbb{E}^2(X) + \mathbb{E}^2(X) = \mathbb{E}(X^2) - [\mathbb{E}(X)]^2$$

which completes the proof. ☐

(3) For a random variable X and two constants a, b, $\mathbb{V}(aX + b) = a^2 \mathbb{V}(X)$.

Proof Note that $\mathbb{E}(aX + b) = a\mathbb{E}(X) + b$. For both cases of discrete and continuous variables,

$$\mathbb{V}(aX + b) = \mathbb{E}([aX + b - (a\mathbb{E}(X) + b)]^2) = \mathbb{E}([aX - a\mathbb{E}(X)]^2) = \mathbb{V}(aX)$$

$$= \mathbb{E}(a^2 X^2) - [\mathbb{E}(aX)]^2 = a^2 [\mathbb{E}(X^2) - [\mathbb{E}(X)]^2] = a^2 \mathbb{V}(X)$$

$$\tag{2.70}$$

which completes the proof. ☐

Example 2.12

For a (discrete) binomial variable X representing the number of successful tests out of n trials, find the mean value and variance of X given that the probability of success for each trial is p.

Solution

The variable X may take a value of $0, 1, 2, \ldots, n$, and

$$\mathbb{P}(X = k) = C_n^k p^k (1 - p)^{n-k}, \quad \forall k = 0, 1, 2, \ldots, n \tag{2.71}$$

Thus, according to Eq. (2.62), the mean value of X is estimated by

$$\mathbb{E}(X) = \sum_{k=0}^{n} k \cdot \mathbb{P}(X = k) = \sum_{k=0}^{n} k \cdot C_n^k p^k (1 - p)^{n-k} = \sum_{k=0}^{n} k \frac{n!}{k!(n-k)!} p^k (1 - p)^{n-k}$$

$$= np \sum_{k=1}^{n} \frac{(n-1)!}{(k-1)!((n-1)-(k-1))!} p^k (1 - p)^{n-1-(k-1)}$$

$$= np \sum_{k=0}^{n-1} C_{n-1}^k p^k (1 - p)^{n-1-k} = np$$

$$\tag{2.72}$$

Next, the variance of X, $\mathbb{V}(X)$, is calculated using Eq. (2.68). We first estimate $\mathbb{E}(X^2)$ as follows,

$$\mathbb{E}(X^2) = \sum_{k=0}^{n} k^2 \cdot \mathbb{P}(X=k) = \sum_{k=0}^{n} k^2 \cdot C_n^k p^k (1-p)^{n-k} = \sum_{k=0}^{n} k^2 \frac{n!}{k!(n-k)!} p^k (1-p)^{n-k}$$

$$= np \sum_{k=1}^{n} k \frac{(n-1)!}{(k-1)!((n-1)-(k-1))!} p^k (1-p)^{n-1-(k-1)}$$

$$= np \sum_{k=0}^{n-1} (k+1) C_{n-1}^k p^k (1-p)^{n-1-k}$$

$$= np \left[\sum_{k=0}^{n-1} k C_{n-1}^k p^k (1-p)^{n-1-k} + \sum_{k=0}^{n-1} C_{n-1}^k p^k (1-p)^{n-1-k} \right] = np[(n-1)p+1]$$

$$(2.73)$$

Thus,

$$\mathbb{V}(X) = \mathbb{E}(X^2) - \mathbb{E}^2(X) = np[(n-1)p+1] - (np)^2 = np(1-p) \qquad (2.74)$$

Example 2.13

For the random variable X in Example 2.11, find its mean value and variance.
Solution
The PDF of X is

$$f_X(x) = \begin{cases} 12x^2(1-x), & 0 \le x \le 1 \\ 0, & \text{otherwise} \end{cases} \qquad (2.75)$$

According to Eq. (2.64), the mean value of X is calculated as

$$\mathbb{E}(X) = \int_{-\infty}^{\infty} x f_X(x) \mathrm{d}x = \int_{0}^{1} x f_X(x) \mathrm{d}x = \int_{0}^{1} 12x^3(1-x) \mathrm{d}x$$

$$= \left(3x^4 - \frac{12}{5} x^5 \right) \Big|_{0}^{1} = \frac{3}{5} \qquad (2.76)$$

The variance of X is obtained using Eq. (2.69). The mean value of X^2 is first estimated as

$$\mathbb{E}(X^2) = \int_{0}^{1} x^2 f_X(x) \mathrm{d}x = \int_{0}^{1} 12x^4(1-x) \mathrm{d}x$$

$$= \left(\frac{12}{5} x^5 - 2x^6 \right) \Big|_{0}^{1} = \frac{2}{5} \qquad (2.77)$$

Thus,

$$\mathbb{V}(X) = \mathbb{E}(X^2) - \mathbb{E}^2(X) = \frac{2}{5} - \frac{9}{25} = \frac{1}{25} \qquad (2.78)$$

Example 2.14

Prove that

(1) (Markov's inequality) For a non-negative random variable X and a positive number a, $\mathbb{P}(X \geq a) \leq \frac{\mathbb{E}(X)}{a}$.

(2) (Chebyshev's inequality) For a random variable X with a mean value of μ and a standard deviation of σ, $\mathbb{P}(|X - \mu| \geq k) \leq \frac{\sigma^2}{k^2}$ holds for a positive number k.

Solution

(1) For the case of a discrete X with n possible values $\{x_1, x_2, \ldots, x_n\}$ in an ascending order and a PMF of $\{p_X(x_i)\}$, $i = 1, 2, \ldots, n$, suppose that $x_{k-1} < a \leq x_k$, with which

$$\mathbb{E}(X) = \sum_{i=1}^{n} x_i p_X(x_i) \geq \sum_{i=k}^{n} x_i p_X(x_i) \geq \sum_{i=k}^{n} a p_X(x_i) = a\mathbb{P}(X \geq a) \quad (2.79)$$

For the continuous case, if the PDF of X is $f_X(x)$,

$$\mathbb{E}(X) = \int_0^{\infty} x f_X(x)\mathrm{d}x \geq \int_a^{\infty} x f_X(x)\mathrm{d}x \geq \int_a^{\infty} a f_X(x)\mathrm{d}x = a\mathbb{P}(X \geq a) \quad (2.80)$$

Both Eqs. (2.79) and (2.80) yield $\mathbb{P}(X \geq a) \leq \frac{\mathbb{E}(X)}{a}$, which completes the proof.

(2) Based on the Markov's inequality, by noting that $\sigma^2 = \mathbb{E}[(X - \mu)^2]$ by definition, we have

$$\frac{\sigma^2}{k^2} = \frac{\mathbb{E}[(X - \mu)^2]}{k^2} \geq \mathbb{P}((X - \mu)^2 \geq k^2) = \mathbb{P}(|X - \mu| \geq k) \quad (2.81)$$

which completes the proof.

It is noticed that both the mean value and variance of a random variable, say, X, are closely related to its moments. Generally, the expectation of X^n is called the nth moment ($n = 1, 2, \ldots$) of X, calculated as follows,

$$\mathbb{E}(X^n) = \begin{cases} \sum_{i=1}^{n} x_i^n p_X(x_i), & \text{discrete} \\ \int_{-\infty}^{\infty} x^n f_X(x)\mathrm{d}x, & \text{continuous} \end{cases} \quad (2.82)$$

Clearly, the mean value of X is exactly the first-order moment, and the variance of X is the second-order moment subtracting the square of the first-order moment. The higher-order moments of a random variable are also representative of its probabilistic characteristics. For instance, the *skewness* of a variable, which is closely related to the variable's third-order moment, can be used to measure the asymmetry of the probability distribution [33]. Moreover, for most cases, the moments and the probability distribution of a random variable can uniquely determine each other, as will be further discussed in Sect. 2.1.6.

2.1.5 Independent Random Variables

We first extend the definition of CDF to two variables (c.f. Eq. (2.42)). For two random variables X and Y, their *joint CDF*, $F_{X,Y}(x, y)$, is defined by

$$F_{X,Y}(x, y) = \mathbb{P}(X \leq x \cap Y \leq y), \quad \forall x, y \in (-\infty, \infty) \tag{2.83}$$

With Eq. (2.83), the marginal CDFs of X and Y are respectively

$$F_X(x) = \mathbb{P}(X \leq x) = \mathbb{P}(X \leq x \cap Y \leq \infty) = F_{X,Y}(x, \infty) \tag{2.84}$$
$$F_Y(y) = \mathbb{P}(Y \leq y) = \mathbb{P}(X \leq \infty \cap Y \leq y) = F_{X,Y}(\infty, y) \tag{2.85}$$

Remark

The joint CDF in Eq. (2.83) can be further extended for the case of n random variables, say, X_1 through X_n, as follows,

$$F_{X_1, X_2, \ldots, X_n}(x_1, x_2, \ldots, x_n) = \mathbb{P}(X_1 \leq x_1 \cap X_2 \leq x_2 \cap \cdots \cap X_n \leq x_n),$$
$$\forall x_1, x_2, \ldots, x_n \in (-\infty, \infty) \tag{2.86}$$

With this, the marginal CDF of X_i, $F_{X_i}(x)$, can be obtained by assigning $x_i = x$ and $x_j = \infty$ in Eq. (2.86) for $\forall j \neq i$.

If both X and Y are discrete, their *joint probability mass function* is given by

$$p_{X,Y}(x_i, y_j) = \mathbb{P}(X = x_i \cap Y = y_j) \tag{2.87}$$

where $\{x_i\}$ and $\{y_j\}$ are the two sets from which X and Y take values. With Eq. (2.87), the PMFs of X and Y are respectively

$$p_X(x_i) = \sum_{\forall j} p_{X,Y}(x_i, y_j); \quad p_Y(y_j) = \sum_{\forall i} p_{X,Y}(x_i, y_j) \tag{2.88}$$

For the case where both X and Y are continuous, their *joint probability density function* of X and Y, $f_{X,Y}(x, y)$, satisfies that for two sets A and B,

$$\mathbb{P}(x \in A \cap y \in B) = \int_A \int_B f_{X,Y}(x, y) \mathrm{d}y \mathrm{d}x \tag{2.89}$$

Similar to Eq. (2.88), the PDFs of X and Y, $f_X(x)$ and $f_Y(y)$, can be obtained based on $f_{X,Y}(x, y)$ as follows.

$$\int_A f_X(x)dx = \mathbb{P}(x \in A) = \mathbb{P}(x \in A \cap y \in (-\infty, \infty)) = \int_A \int_{-\infty}^{\infty} f_{X,Y}(x, y)dy dx$$

$$\Rightarrow f_X(x) = \int_{-\infty}^{\infty} f_{X,Y}(x, y)dy$$

$$(2.90)$$

$$\int_B f_Y(y)dy = \mathbb{P}(y \in B) = \mathbb{P}(x \in (-\infty, \infty) \cap y \in B) = \int_B \int_{-\infty}^{\infty} f_{X,Y}(x, y)dx dy$$

$$\Rightarrow f_Y(y) = \int_{-\infty}^{\infty} f_{X,Y}(x, y)dx$$

$$(2.91)$$

Recall Eq. (2.56), which gives a relationship between the PDF and CDF of a random variable. This relationship is extended herein for the joint CDF and joint PDF of two variables (and for multiple variables in a similar manner). Note that according to Eqs. (2.83) and (2.89)

$$F_{X,Y}(x, y) = \mathbb{P}(X \le x \cap Y \le y) = \int_{-\infty}^{x} \int_{-\infty}^{y} f(x_1, y_1)dy_1 dx_1 \qquad (2.92)$$

Differentiating both sides of Eq. (2.92) twice (with respect to x and y) yields

$$f_{X,Y}(x, y) = \frac{\partial^2}{\partial x \partial y} F_{X,Y}(x, y) \qquad (2.93)$$

If random variable Z is a function of X and Y, denoted by $Z = g(X, Y)$, then, similar to Eqs. (2.63) and (2.65), the expectation of Z is given by

$$\mathbb{E}(Z) = \mathbb{E}(g(X, Y)) = \begin{cases} \displaystyle\sum_{\forall i} \sum_{\forall j} g(x_i, y_j) p_{X,Y}(x_i, y_j), & \text{discrete } X \text{ and } Y \\[2ex] \displaystyle\int_{-\infty}^{\infty} \int_{-\infty}^{\infty} g(x, y) f_X(x) f_Y(y)dx dy, & \text{continuous } X \text{ and } Y \end{cases}$$

$$(2.94)$$

An application of Eq. (2.94) is that when $Z = g(X, Y) = X + Y$,

$$\mathbb{E}(Z) = \mathbb{E}(X) + \mathbb{E}(Y) \qquad (2.95)$$

The proof of Eq. (2.95) is as follows.

Proof For the discrete case,

$$\mathbb{E}(X+Y) = \sum_{\forall i} \sum_{\forall j} (x_i + y_j) p_{X,Y}(x_i, y_j)$$

$$= \sum_{\forall i} \sum_{\forall j} x_i p_{X,Y}(x_i, y_j) + \sum_{\forall i} \sum_{\forall j} y_j p_{X,Y}(x_i, y_j) \qquad (2.96)$$

$$= \sum_{\forall i} x_i p_X(x_i) + \sum_{\forall j} y_j p_Y(y_j)$$

$$= \mathbb{E}(X) + \mathbb{E}(Y)$$

For the case of continuous X and Y,

$$\mathbb{E}(X+Y) = \int_{-\infty}^{\infty} \int_{-\infty}^{\infty} (x+y) f_X(x) f_Y(y) \mathrm{d}x \mathrm{d}y$$

$$= \int_{-\infty}^{\infty} \int_{-\infty}^{\infty} x f_X(x) f_Y(y) \mathrm{d}x \mathrm{d}y + \int_{-\infty}^{\infty} \int_{-\infty}^{\infty} y f_X(x) f_Y(y) \mathrm{d}x \mathrm{d}y \quad (2.97)$$

$$= \int_{-\infty}^{\infty} x f_X(x) \mathrm{d}x + \int_{-\infty}^{\infty} y f_Y(y) \mathrm{d}y = \mathbb{E}(X) + \mathbb{E}(Y)$$

which completes the proof.   □

Remark

We can further generalize Eq. (2.95) to the case of n random variables $X_1, X_2, \ldots, X_n$, which states that

$$\mathbb{E}\left(\sum_{i=1}^{n} X_i\right) = \sum_{i=1}^{n} \mathbb{E}(X_i) \qquad (2.98)$$

The proof of Eq. (2.98) is straightforward based on Eq. (2.95). Furthermore, a similar relationship is that $\mathbb{E}\left(\sum_{i=1}^{n} a_i X_i\right) = \sum_{i=1}^{n} a_i \mathbb{E}(X_i)$, where a_1 through a_n are constants. This is because $\mathbb{E}(aX) = a\mathbb{E}(X)$ (c.f. Eq. (2.67)).

Recall that we have earlier discussed the independence between two events (c.f. Eq. (2.17)). Now we consider the dependence of two random variables X and Y. If the following equation holds for $\forall x, y \in (-\infty, \infty)$, then X and Y are deemed as *independent random variables*.

$$\mathbb{P}(X \le x \cap Y \le y) = \mathbb{P}(X \le x) \cdot \mathbb{P}(Y \le y) \qquad (2.99)$$

Equation (2.99) is equivalent to

$$F_{X,Y}(x, y) = F_X(x) \cdot F_Y(y) \qquad (2.100)$$

For the discrete case, Eq. (2.100) reduces to

$$p_{X,Y}(x_i, y_j) = p_X(x_i)p_Y(y_j) \tag{2.101}$$

For the continuous case, with Eq. (2.100), the condition of independence reduces to

$$f_{X,Y}(x, y) = f_X(x) \cdot f_Y(y) \tag{2.102}$$

Example 2.15

For two independent random variables X and Y, and two functions h and g, show that $\mathbb{E}[h(X)g(Y)] = \mathbb{E}[h(X)] \cdot \mathbb{E}[g(Y)]$.

Solution

For the discrete case,

$$\mathbb{E}[h(X)g(Y)] = \sum_i \sum_j h(x_i)g(y_j)p_{X,Y}(x_i, y_j) = \sum_i \sum_j h(x_i)g(y_j)p_X(x_i)p_Y(y_j)$$

$$= \sum_i h(x_i)p_X(x_i) \cdot \sum_j g(y_j)p_Y(y_j) = \mathbb{E}[h(X)] \cdot \mathbb{E}[g(Y)] \tag{2.103}$$

For the continuous case,

$$\mathbb{E}[h(X)g(Y)] = \int \int h(x)g(y)f_{X,Y}(x, y)\mathrm{d}x\mathrm{d}y = \int \int h(x)g(y)f_X(x)f_Y(y)\mathrm{d}x\mathrm{d}y$$

$$= \int h(x)f_X(x)\mathrm{d}x \int g(x)f_Y(y)\mathrm{d}y = \mathbb{E}[h(X)] \cdot \mathbb{E}[g(Y)] \tag{2.104}$$

which completes the proof.

Specifically, if $h(X) = X$ and $g(Y) = Y$, then $\mathbb{E}(XY) = \mathbb{E}(X) \cdot \mathbb{E}(Y)$.

Example 2.16

For two continuous variables X and Y, their joint PDF is given as follows,

$$f_{X,Y}(x, y) = \frac{a \exp(-x^2)}{y^2 + 1}, \quad x, y \in (-\infty, \infty) \tag{2.105}$$

(1) Find the value of a; (2) Determine the marginal PDFs for X and Y respectively; (3) Show that X and Y are independent.

Solution

(1) The integral of $f_{X,Y}(x, y)$ on $(-\infty, \infty) \times (-\infty, \infty)$ equals 1, with which

$$\int_{-\infty}^{\infty} \int_{-\infty}^{\infty} f_{X,Y}(x, y) = \int_{-\infty}^{\infty} \int_{-\infty}^{\infty} \frac{a \exp(-x^2)}{y^2 + 1} = a\pi^{3/2} = 1$$

$$\Rightarrow a = \frac{1}{\pi^{3/2}} \tag{2.106}$$

(2) Using Eqs. (2.90) and (2.91),

$$f_X(x) = \int_{-\infty}^{\infty} f_{X,Y}(x, y)dy = \int_{-\infty}^{\infty} \frac{\exp(-x^2)}{\pi^{3/2}(y^2 + 1)}dy$$

$$= \exp(-x^2) \int_{-\infty}^{\infty} \frac{1}{\pi^{3/2}(y^2 + 1)}dy = \frac{1}{\sqrt{\pi}}\exp(-x^2) \qquad (2.107)$$

and

$$f_Y(y) = \int_{-\infty}^{\infty} f_{X,Y}(x, y)dx = \int_{-\infty}^{\infty} \frac{\exp(-x^2)}{\pi^{3/2}(y^2 + 1)}dx$$

$$= \frac{1}{\pi^{3/2}(y^2 + 1)} \int_{-\infty}^{\infty} \exp(-x^2)dx = \frac{1}{\pi(y^2 + 1)} \qquad (2.108)$$

(3) By noting that

$$f_X(x) \cdot f_Y(y) = \frac{1}{\sqrt{\pi}}\exp(-x^2) \cdot \frac{1}{\pi(y^2 + 1)} = f_{X,Y}(x, y) \qquad (2.109)$$

it is concluded that X and Y are independent.

Similar to Eq. (2.95), for two independent variables X and Y, we can show that

$$\mathbb{V}(X + Y) = \mathbb{V}(X) + \mathbb{V}(Y) \qquad (2.110)$$

Proof Due to the independence between X and Y, $\mathbb{E}(XY) = \mathbb{E}(X)\mathbb{E}(Y)$. Thus,

$$\begin{aligned}
\mathbb{V}(X + Y) &= \mathbb{E}[(X + Y)^2] - \mathbb{E}^2(X + Y) \\
&= \mathbb{E}(X^2) + \mathbb{E}(Y^2) + 2\mathbb{E}(XY) - [\mathbb{E}^2(X) + \mathbb{E}^2(Y) + 2\mathbb{E}(X)\mathbb{E}(Y)] \\
&= [\mathbb{E}(X^2) - \mathbb{E}^2(X)] + [\mathbb{E}(Y^2) - \mathbb{E}^2(Y)] \\
&= \mathbb{V}(X) + \mathbb{V}(Y)
\end{aligned}$$

$$(2.111)$$

which completes the proof.   □

Remark

Equation (2.110) can be further extended to the case of n *mutually independent* random variables $X_1, X_2, \ldots, X_n$, that is,

$$\mathbb{V}\left(\sum_{i=1}^{n} X_i\right) = \sum_{i=1}^{n} \mathbb{V}(X_i) \qquad (2.112)$$

A further generalization of Eq. (2.112) gives $\mathbb{V}\left(\sum_{i=1}^{n} a_i X_i\right) = \sum_{i=1}^{n} a_i^2 \mathbb{V}(X_i)$ for mutually independent $X_1, X_2, \ldots, X_n$, where a_1 through a_n are constants. This is because $\mathbb{V}(aX) = a^2\mathbb{V}(X)$ (c.f. Eq. (2.70)).

Recall the binomial variable X in Example 2.12, which is the number of successful tests out of n trials. Now we introduce a variable X_i for the ith trial, which takes a value of 1 if the trial is successful and 0 otherwise. With this, $X = \sum_{i=1}^{n} X_i$. Find the mean value and variance of X.

Solution

First, the mean value and variance of X_i are calculated with Eq. (2.62) as follows,

$$\mathbb{E}(X_i) = 1 \times p + 0 \times (1 - p) = p \tag{2.113}$$

$$\mathbb{V}(X_i) = \mathbb{E}(X_i^2) - \mathbb{E}^2(X_i) = \mathbb{E}(X_i) - \mathbb{E}^2(X_i) = p - p^2 = p(1 - p) \tag{2.114}$$

With this, the mean value of X is obtained according to Eq. (2.98), with which

$$\mathbb{E}(X) = \mathbb{E}\left(\sum_{i=1}^{n} X_i\right) = \sum_{i=1}^{n} \mathbb{E}(X_i) = \sum_{i=1}^{n} p = np \tag{2.115}$$

Next, by noting the mutual independence between each X_i, with Eq. (2.112),

$$\mathbb{V}(X) = \mathbb{V}\left(\sum_{i=1}^{n} X_i\right) = \sum_{i=1}^{n} \mathbb{V}(X_i) = \sum_{i=1}^{n} p(1 - p) = np(1 - p) \tag{2.116}$$

which is consistent with the results from Example 2.12.

For the case of dependent variables, their joint probabilistic behaviour will be later discussed in Sect. 2.3.

2.1.6 Moment Generating Function and Characteristic Function

For a random variable X, its *moment generating function* (MGF), $\psi_X(\tau)$, is defined as follows [27],

$$\psi_X(\tau) = \mathbb{E}(\exp(\tau X)) \tag{2.117}$$

For a discrete random variable X with n possible values $\{x_1, x_2, \ldots, x_n\}$ and a PMF of $\{p_X(x_i)\}$, $i = 1, 2, \ldots, n$, Eq. (2.117) becomes

$$\psi_X(\tau) = \sum_{i=1}^{n} \exp(x_i \tau) p_X(x_i) \tag{2.118}$$

Likewise, for the continuous case, if the PDF of a continuous variable X is $f_X(x)$, then

$$\psi_X(\tau) = \int_{-\infty}^{\infty} \exp(x\tau) f_X(x) \mathrm{d}x \tag{2.119}$$

Remark

Equation (2.119) implies that the MGF $\psi_X(\tau)$ is a Laplace transform of the PDF $f_X(x)$ if the variable X is defined on $[0, \infty)$. Generally, for a function $f(x)$ defined for all real numbers $x \geq 0$, its Laplace transform, $\mathcal{L}\{f\}(\tau)$, is defined by

$$\mathcal{L}\{f\}(\tau) = \int_0^{\infty} f(x) \exp(-\tau x) \mathrm{d}x \tag{2.120}$$

With Eq. (2.120), it can be shown that $\psi_X(\tau) = \mathcal{L}\{f_X\}(-\tau)$ for a non-negative random variable X.

The function $\psi_X(\tau)$ is called the moment generating function because it "generates" the moments for the variable X by differentiating $\psi_X(\tau)$. That is, the nth derivative of $\psi_X(\tau)$ evaluated at $\tau = 0$ equals $\mathbb{E}(X^n)$, or,

$$\psi_X^{(n)}(0) = \mathbb{E}(X^n), \quad n = 1, 2, 3 \ldots \tag{2.121}$$

The proof is as follows.

Proof First, note that $\frac{\mathrm{d}^n}{\mathrm{d}\tau} \exp(\tau x) = x^n \exp(\tau x)$ holds for a real number or a random variable x. With this, for $n = 1, 2, 3, \ldots$, it follows,

$$
\begin{aligned}
\psi_X^{(n)}(0) &= \frac{\mathrm{d}^n}{\mathrm{d}\tau^n} \psi_X(\tau) \bigg|_{\tau=0} = \frac{\mathrm{d}^n}{\mathrm{d}\tau^n} \mathbb{E}(\exp(\tau X)) \bigg|_{\tau=0} \\
&= \mathbb{E}\left(\frac{\mathrm{d}^n}{\mathrm{d}\tau^n} \exp(\tau X) \right) \bigg|_{\tau=0} = \mathbb{E}\left(X^n \exp(\tau X) \right) \big|_{\tau=0} \\
&= \mathbb{E}(X^n)
\end{aligned}
\tag{2.122}
$$

which completes the proof. $\square$

For two independent variables X and Y with their MGFs being $\psi_X(\tau)$ and $\psi_Y(\tau)$ respectively, the MGF of $X + Y$, $\psi_{X+Y}(\tau)$, equals $\psi_X(\tau) \cdot \psi_Y(\tau)$. This can be shown by noting that

$$\psi_{X+Y}(\tau) = \mathbb{E}\{\exp[\tau(X+Y)]\} = \mathbb{E}\{\exp[\tau X]\} \cdot \mathbb{E}\{\exp[\tau Y]\}$$
$$= \psi_X(\tau) \cdot \psi_Y(\tau) \tag{2.123}$$

Example 2.18

For the binomial variable X in Example 2.12, (1) derive its MGF; (2) calculate the mean value and variance of X based on the MGF.

Solution

(1) Notice that the PMF of X takes the form of

$$\mathbb{P}(X = k) = C_n^k p^k (1-p)^{n-k}, \quad \forall k = 0, 1, 2, \ldots, n \tag{2.124}$$

Thus, with Eq. (2.118), the MGF of X is determined by

$$\psi_X(\tau) = \sum_{k=0}^{n} \exp(k\tau)\mathbb{P}(X = k) = \sum_{i=0}^{n} \exp(k\tau) C_n^k p^k (1-p)^{n-k}$$
$$= \sum_{i=0}^{n} C_n^k (e^\tau p)^k (1-p)^{n-k} = (pe^\tau + 1 - p)^n \tag{2.125}$$

(2) The first- and second-order moments of X are respectively

$$\mathbb{E}(X) = \psi_X'(0) = n(1 - p + pe^\tau)pe^\tau\big|_{\tau=0} = np \tag{2.126}$$

and

$$\mathbb{E}^2(X) = \psi_X''(0) = npe^\tau(1 - p + pe^\tau)^{n-2}(npe^\tau + 1 - p)\big|_{\tau=0} = np(np + 1 - p) \tag{2.127}$$

Thus,

$$\mathbb{V}(X) = \mathbb{E}^2(X) - \mathbb{E}^2(X) = np(1-p) \tag{2.128}$$

which is consistent with the results in Example 2.12.

*Example 2.19

[*Readers may refer to Sect. 2.2.4 for relevant background information.*]

For two independent random variables X_1 and X_2 that follow a normal distribution, if the mean value and standard deviation of X_1 are μ_1 and σ_1 respectively, then the MGF of X_1 is $\psi_{X_1}(\tau) = \exp\left(\mu_1\tau + \frac{1}{2}\sigma_1^2\tau^2\right)$. Similarly, if the mean and standard deviation of X_2 are μ_2 and σ_2, then $\psi_{X_2}(\tau) = \exp\left(\mu_2\tau + \frac{1}{2}\sigma_2^2\tau^2\right)$. Show that $X_1 + X_2$ also follows a normal distribution with a mean value of $\mu_1 + \mu_2$ and a standard deviation of $\sqrt{\sigma_1^2 + \sigma_2^2}$.

Solution

The MGF of $X_1 + X_2$ is estimated by

$$\psi_{X_1+X_2}(\tau) = \exp\left(\mu_1\tau + \frac{1}{2}\sigma_1^2\tau^2\right) \cdot \exp\left(\mu_2\tau + \frac{1}{2}\sigma_2^2\tau^2\right)$$

$$= \exp\left[(\mu_1+\mu_1)\tau + \frac{1}{2}\left(\sqrt{\sigma_1^2+\sigma_2^2}\right)^2\tau^2\right] \tag{2.129}$$

Thus, it can be seen that $X_1 + X_2$ also follows a normal distribution with a mean value of $\mu_1 + \mu_2$ and a standard deviation of $\sqrt{\sigma_1^2 + \sigma_2^2}$.

Based on the MGF, the characteristic function (CF) of X, $\psi_{iX}(\tau)$, is defined by [3]

$$\psi_{iX}(\tau) = \psi_X(i\tau) = \mathbb{E}(\exp(i\tau X)) \tag{2.130}$$

where $i = \sqrt{-1}$ is the imaginary unit.

Similar to Eqs. (2.118) and (2.119), when X is discrete with a PMF of $p_X(x_i)$, $i = 1, 2, \ldots, n$,

$$\psi_{iX}(\tau) = \sum_{i=1}^{n} \exp(ix_i\tau)p_X(x_i) \tag{2.131}$$

For a continuous X whose PDF is $f_X(x)$,

$$\psi_{iX}(\tau) = \int_{-\infty}^{\infty} \exp(ix\tau)f_X(x)dx \tag{2.132}$$

Equation (2.132) implies that $\psi_{iX}(\tau)$ and $f_X(x)$ is a Fourier transform pair for a continuous variable. Thus,

$$f_X(x) = \frac{1}{2\pi}\int_{-\infty}^{\infty} \exp(-\tau ix)\psi_{iX}(\tau)d\tau \tag{2.133}$$

With Eq. (2.133), it can be seen that the PDF of a random variable can be uniquely determined provided that the CF of the variable is known.

Example 2.20

For a random variable X whose moment generating function is $\psi_X(\tau) = \exp\left(\mu\tau + \frac{1}{2}\sigma^2\tau^2\right)$, determine its PDF, $f_X(x)$.

Solution

First, the CF of X is $\psi_{iX}(\tau) = \exp\left(\mu\tau i - \frac{1}{2}\sigma^2\tau^2\right)$. Thus, $f_X(x)$ is determined uniquely as follows according to Eq. (2.133).

$$
\begin{aligned}
f_X(x) &= \frac{1}{2\pi} \int_{-\infty}^{\infty} \exp(-\tau i x) \exp\left(\mu\tau i - \frac{1}{2}\sigma^2\tau^2\right) d\tau \\
&= \frac{1}{2\pi} \int_{-\infty}^{\infty} \exp(i(\mu\tau - x\tau)) \exp\left(-\frac{1}{2}\sigma^2\tau^2\right) d\tau \\
&= \frac{1}{2\pi} \int_{-\infty}^{\infty} \cos(\mu\tau - x\tau) \exp\left(-\frac{1}{2}\sigma^2\tau^2\right) d\tau \\
&= \frac{1}{\sqrt{2\pi}\sigma} \exp\left\{-\frac{(x-\mu)^2}{2\sigma^2}\right\}
\end{aligned}
\tag{2.134}
$$

2.2 Selected Probability Distribution Types

In this section, we discuss some probability distribution types for both discrete and continuous random variables. These distribution types are chosen on the basis that they are frequently-used in practical engineering to model the probabilistic behaviour of random quantities.

2.2.1 Bernoulli, Binomial and Geometric Distributions

2.2.1.1 Bernoulli Distribution

Consider an experiment with two possible outcomes: success or failure. We introduce a (discrete) random variable X which equals 1 if the outcome is "success" and 0 otherwise. If the probability of success is p ($0 \leq p \leq 1$), the failure probability is correspondingly $1 - p$. With this, the variable X has a PMF as follows,

$$
\mathbb{P}(X = 1) = p; \quad \mathbb{P}(X = 0) = 1 - p
\tag{2.135}
$$

In such a case, X is deemed as a Bernoulli random variable. The mean value and variance of X are respectively p and $p(1 - p)$ (c.f. Eq. (2.113)). The MGF of X is

$$
\psi_X(\tau) = \exp(\tau) \cdot p + \exp(\tau \times 0) \cdot (1 - p) = e^\tau p + 1 - p
\tag{2.136}
$$

2.2.1.2 Binomial Distribution

Now we repeat the aforementioned experiment in Sect. 2.2.1.1 for n times. If the outcome of each trial is not affected by the other trials, then the number of successful tests, X, is a binomial random variable. The PMF, mean value and variance of X have been discussed earlier (c.f. Examples 2.9, 2.12, 2.17 and 2.18), and are summarized herein for better readability. The PMF of X is

$$\mathbb{P}(X = k) = C_n^k p^k (1 - p)^{n-k}, \quad \forall k = 0, 1, 2, \ldots, n \qquad (2.137)$$

The mean value and variance of X are respectively np and $np(1 - p)$. The MGF of X is $\psi_X(\tau) = (e^\tau + 1 - p)^n$.

2.2.1.3 Geometric Distribution

Reconsider the experiment in Sect. 2.2.1.1 which has a success probability of p and a failure probability of $1 - p$. Now we perform the experiment repeatedly and stop until a successful trial occurs. Let random variable X denote the number of trials. In such a case, X is a geometric random variable. The PMF of X is

$$\mathbb{P}(X = k) = (1 - p)^{k-1} p, \quad k = 1, 2, \ldots \qquad (2.138)$$

It is easy to verify that $\sum_{k=1}^{\infty} \mathbb{P}(X = k) = \sum_{k=1}^{\infty} (1 - p)^{k-1} p = p \sum_{k=1}^{\infty} (1 - p)^{k-1} = 1$. Correspondingly, the CDF of X is

$$F_X(k) = \mathbb{P}(X \leq k) = 1 - \mathbb{P}(X > k) = 1 - (1 - p)^k, \quad k = 1, 2, \ldots \qquad (2.139)$$

The mean value and variance of the geometric variable X can be estimated according to Eq. (2.62) as follows.

$$
\begin{aligned}
\mathbb{E}(X) &= \sum_{k=1}^{\infty} k(1 - p)^{k-1} p = \sum_{k=0}^{\infty} (k + 1)(1 - p)^k p \\
&= (1 - p) \left[\sum_{k=1}^{\infty} k(1 - p)^{k-1} p + \sum_{k=1}^{\infty} (1 - p)^{k-1} p + \frac{p}{1 - p} \right] \\
&= (1 - p) \left(\mathbb{E}(X) + 1 + \frac{p}{1 - p} \right) \\
&\Rightarrow \mathbb{E}(X) = \frac{1}{p}
\end{aligned}
\qquad (2.140)
$$

$$\mathbb{E}(X^2) = \sum_{k=1}^{\infty} k^2(1-p)^{k-1}p = \sum_{k=0}^{\infty}(k+1)^2(1-p)^k p$$

$$= \sum_{k=1}^{\infty} k^2(1-p)^k p + \sum_{k=0}^{\infty}(1-p)^k p + 2\sum_{k=1}^{\infty}k(1-p)^k p \qquad (2.141)$$

$$= \mathbb{E}(X^2)(1-p) + 1 + \frac{2-2p}{p}$$

$$\Rightarrow \mathbb{E}(X^2) = \frac{2-p}{p^2}$$

and thus,

$$\mathbb{V}(X) = \mathbb{E}(X^2) - \mathbb{E}^2(X) = \frac{2-p}{p^2} - \frac{1}{p^2} = \frac{1-p}{p^2} \qquad (2.142)$$

The MGF of X is, with Eq. (2.118), obtained as

$$\psi_X(\tau) = \sum_{k=1}^{\infty} \exp(k\tau)p(1-p)^{k-1} = \sum_{k=1}^{\infty}[\exp(\tau)(1-p)]^k \frac{p}{1-p}$$

$$= \frac{pe^\tau[1-(e^\tau(1-p))^\infty]}{1-e^\tau(1-p)} \qquad (2.143)$$

which converges when $e^\tau(1-p) < 1$, or equivalently, $\tau < -\ln(1-p)$. With this condition,

$$\psi_X(\tau) = \frac{pe^\tau}{1-e^\tau(1-p)} \qquad (2.144)$$

Example 2.21

Consider a building subjected to wind loads. The building will fail (collapse) if the wind speed exceeds 50 m/s, and the annual probability of wind speed exceeding 50 m/s is $p = 0.1$.

(1) Find the return period of the wind loads exceeding 50 m/s, $\bar{\tau}$. The *return period* is defined as the mean value (average) of the time between two successive events that are statistically independent [21].
(2) Find the CDF of the service life (in years).
(3) Find the probability that the maximum wind speed within the return period exceeds 50 m/s.

Solution

(1) Let random variable X denote the year in which the wind speed first exceeds 50 m/s (i.e., the service life). It is easy to see that X is a geometric variable, and $\mathbb{E}(X) = \frac{1}{p} = 10$ (c.f. Eq. (2.140)). Thus, $\bar{\tau} = \mathbb{E}(X) = 10$ years.
(2) According to Eq. (2.139)

$$F_X(k) = 1 - (1 - p)^k = 1 - 0.9^k, \quad k = 1, 2, \ldots \tag{2.145}$$

(3) The probability that the maximum wind speed within the return period exceeds 50 m/s is estimated as

$$\mathbb{P}(X \le \bar{t}) = F_X(\bar{t}) = 1 - 0.9^{10} = 0.6513. \tag{2.146}$$

Remark

In Question (3) of Example 2.21, as $p \to 0$ (i.e., the return period $1/p$ is large enough), $\mathbb{P}(X \le \bar{t}) = F_X(\bar{t}) = 1 - (1 - p)^{1/p} = 1 - e^{-1} = 0.6321$, implying that the probability of structural failure is approximately 63.21% if the designed service life equals the return period.

2.2.2   Uniform Distribution

If the PDF of a continuous random variable X defined in $[a, b]$ can be described as follows, then X is deemed to follow a *uniform distribution*.

$$f_X(x) = \mathbb{I}(x \in [a, b]) \frac{1}{b - a} \tag{2.147}$$

where $\mathbb{I}(\bullet)$ is an indicative function, which returns 1 if the event in the bracket is true and 0 otherwise. The mean value and variance of X are $a + b$ and $\frac{(b-a)^2}{12}$, respectively. With Eq. (2.147), the CDF of X is given by

$$F_X(x) = \begin{cases} 0, & x < a \\ \dfrac{x - a}{b - a}, & a \le x < b \\ 1, & x \ge b \end{cases} \tag{2.148}$$

The MGF of X is given by

$$\psi_X(\tau) = \int_a^b \exp(x\tau) \cdot f_X(x) \mathrm{d}x = \int_a^b \frac{\exp(x\tau)}{b - a} \mathrm{d}x = \frac{1}{\tau(b - a)} \exp(x\tau) \Big|_a^b = \frac{e^{b\tau} - e^{a\tau}}{\tau(b - a)} \tag{2.149}$$

Remark

For an arbitrary continuous random variable, its CDF follows a uniform distribution within [0, 1]. This fact is basis for the "inverse transformation method" [27] that can be used to generate a realization of a random variable with a known CDF, as will be discussed in Sect. 3.2.1.

2.2.3 Poisson and Exponential Distributions

For a time period of $[0, T]$, discrete events may occur randomly within this period. The Poisson process is frequently employed to model the random occurrence of a sequence of events during a time period of interest. Dividing $[0, T]$ into n identical sections, the duration of each interval, $\Delta t = \frac{T}{n}$, becomes short enough if n is sufficiently large. We assume that,

(1) The probability of occurrence of one event during each small interval is identically p.
(2) At most one event possibly occurs during each interval.
(3) The event occurrence within each interval is independent of those of other intervals.

With these assumptions, we let random variable X represent the number of events that occur within $[0, T]$. For an arbitrary non-negative integer k, the probability of "$X = k$" can be evaluated by

$$\mathbb{P}(X = k) = C_n^k p^k (1 - p)^{n-k} \tag{2.150}$$

As n is large enough,

$$\begin{aligned} \mathbb{P}(X = k) &= \lim_{n \to \infty} \frac{n(n-1)\dots(n-k+1)}{k!} p^k (1-p)^{n-k} \\ &= \lim_{n \to \infty} \frac{1}{k!} (np)^k (1-p)^{n-k} = \frac{(np)^k}{k!} \exp(-np) \end{aligned} \tag{2.151}$$

Let $\lambda T = np$ (λ is referred to as the "occurrence rate"), and Eq. (2.151) becomes

$$\mathbb{P}(X = k) = \frac{(\lambda T)^k}{k!} \exp(-\lambda T), \quad k = 0, 1, 2, \dots \tag{2.152}$$

Equation (2.152) gives the PMF of the variable X. We call X a *Poisson random variable*. The mean value and variance of X are both equal to λT. The MGF of X is

$$\psi_X(\tau) = \sum_{k=0}^{\infty} \frac{(\lambda T)^k}{k!} \exp(-\lambda T) \cdot \exp(k\tau) = \exp(-\lambda T) \sum_{k=0}^{\infty} \frac{(\lambda T e^\tau)^k}{k!} \quad (2.153)$$
$$= \exp(\lambda T e^\tau - \lambda T) = \exp\left[\lambda T(e^\tau - 1)\right]$$

Equation (2.150) implies that the Poisson distribution is a special case of the binomial distribution (c.f. Sect. 2.2.1.2) with the number of trials approaching infinity. Recall that in Example 2.13, the mean and variance of a binomial variable X were obtained as np and $np(1 - p)$, respectively. This further yields a mean value and variance of $np = \lambda T$ and $np(1 - p) = \lambda T(1 - \lambda T/n) = \lambda T$ for a Poisson variable X with $n \to \infty$. Moreover, since n is large enough, the moment generating function of the binomial distribution (c.f. Eq. (2.125)) becomes

$$\psi_X(\tau) = \lim_{n \to \infty}(pe^\tau + 1 - p)^n = \lim_{n \to \infty} \exp\{n \ln[1 + p(e^\tau - 1)]\}$$
$$= \lim_{n \to \infty} \exp\left[np(e^\tau - 1)\right] = \exp\left[\lambda T(e^\tau - 1)\right] \quad (2.154)$$

which is the same as Eq. (2.153).

Recall that in Example 2.17, the binomial variable was treated as the sum of Bernoulli variables. This implies that one can also use the sum of Bernoulli variables to approximate a Poisson variable. With Eq. (2.150), we introduce a Bernoulli variable X_i to represent the event occurrence within the ith interval ($i = 1, 2, \ldots, n$), which equals 1 if an event occurs and 0 otherwise. With this, $\mathbb{P}(X_i = 1) = \lambda \frac{T}{n}$, and it follows, according to the definition of Poisson variables, that

$$X = \sum_{i=1}^{n} X_i \quad (2.155)$$

where X is the Poisson random variable in Eq. (2.150). Since the MGF of X_i equals $\psi_{X_i}(\tau) = e^\tau \lambda \frac{T}{n} + 1 - \lambda \frac{T}{n}$ (c.f. Eq. (2.136)), according to Eq. (2.155), the MGF of X is obtained as

$$\psi_X(\tau) = \lim_{n \to \infty} \prod_{i=1}^{n} \psi_{X_i}(\tau) = \lim_{n \to \infty} \left[1 + (e^\tau - 1)\lambda\frac{T}{n}\right]^n \quad (2.156)$$
$$= \exp\left[\lambda T(e^\tau - 1)\right]$$

yielding the same result as in Eq. (2.153).

Remark

The poisson process has been widely employed to model the random occurrence of discrete events. For example, it can be used to describe the occurrence of tropical cyclones for a specific region of interest [19, 35]. Figure 2.3 examines the cyclone (known as hurricane at the Atlantic Basin) occurrence for Miami-Dade County, US, an area that has suffer significantly from historical hurricanes. Figure 2.3a shows the historical tracks of hurricanes

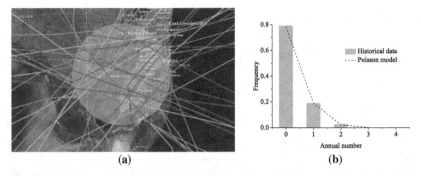

Fig. 2.3 Poisson distribution for the annual hurricane number of Miami-Dade County, US

from 1901 to 2010 [19], which was originally reproduced from the US National Hurricane Center website [24]. Figure 2.3b shows the comparison between the recorded and theoretical (Poisson) probability distribution of the annual hurricane events [35]. Totally 27 hurricanes occurred from 1901 to 2010, with which $\lambda = 0.245$/year. The consistency of the recorded and theoretical results in Fig. 2.3b indicates that the hurricane occurrence can be well described by a Poisson random process.

Example 2.22

With the information observed from Fig. 2.3, calculate the probability that at most one hurricane occurs affecting Miami-Dade County in a future year.

Solution

Using Eq. (2.152), since $\lambda = 0.245$/year, the probability that at most one hurricane occurs in one year ($T = 1$ year) is estimated by

$$\mathbb{P}(X = 0) + \mathbb{P}(X = 1) = \frac{0.245^0}{0!} \exp(-0.245) + \frac{0.245^1}{1!} \exp(-0.245) = 97.45\% \tag{2.157}$$

Example 2.23

If two Poisson random variables X_1 and X_2 are independent and have mean values of ν_1 and ν_2 respectively, show that $X_1 + X_2$ also follows a Poisson distribution with a mean value of $\nu_1 + \nu_2$.

Solution

We can complete the proof with two methods.

Method 1. Note that the PMFs of X_1 and X_2, with Eq. (2.152), are respectively

$$\mathbb{P}(X_1 = k) = \frac{\nu_1^k}{k!} \exp(-\nu_1), \quad \mathbb{P}(X_2 = k) = \frac{\nu_2^k}{k!} \exp(-\nu_2) \tag{2.158}$$

Thus,

$$\mathbb{P}(X_1 + X_2 = k) = \sum_{l=0}^{k} \mathbb{P}(X_1 = l)\mathbb{P}(X_2 = k - l) = \sum_{l=0}^{k} \frac{\nu_1^l}{l!} \exp(-\nu_1) \frac{\nu_2^{k-l}}{(k-l)!} \exp(-\nu_2)$$

$$= \frac{\exp(-\nu_1 - \nu_2)}{k!} \sum_{l=0}^{k} \frac{k! \nu_1^l \nu_2^{k-l}}{l!(k-l)!} = \frac{\exp(-\nu_1 - \nu_2)}{k!} (\nu_1 + \nu_2)^k$$

$$(2.159)$$

with which it is concluded that $X_1 + X_2$ follows a Poisson distribution and its mean value
is $\nu_1 + \nu_2$.

Method 2. Consider the MGFs of X_1 and X_2, $\psi_{X_1}(\tau)$ and $\psi_{X_2}(\tau)$, respectively. With
Eq. (2.153),

$$\psi_{X_1}(\tau) = \exp\left[\nu_1(e^\tau - 1)\right], \quad \psi_{X_2}(\tau) = \exp\left[\nu_2(e^\tau - 1)\right] \tag{2.160}$$

Since X_1 and X_2 are independent, the MGF of $X_1 + X_2$, $\psi_{X_1+X_2}(\tau)$, equals

$$\psi_{X_1+X_2}(\tau) = \psi_{X_1}(\tau) \cdot \psi_{X_2}(\tau) = \exp\left[(\nu_1 + \nu_2)(e^\tau - 1)\right] \tag{2.161}$$

Thus, $X_1 + X_2$ is a Poisson random variable with a mean value of $\nu_1 + \nu_2$.

If T is large enough, let T_1 denote the arriving time of the first event. Clearly, T_1
is a continuous random variable defined in $[0, T]$. Let $F_{T_1}(t)$ be the CDF of T_1. By
definition, $1 - F_{T_1}(t)$ equals the probability that no event occurs before time t (i.e.,
the probability of $T_1 > t$). With this,

$$F_{T_1}(t) = 1 - \lim_{n \to \infty} (1 - \lambda \Delta t)^{tn/T} = 1 - \exp(-\lambda t), \quad t \geq 0 \tag{2.162}$$

With Eq. (2.162), T_1 is said to follow an *exponential distribution*. Furthermore, the
PDF of T_1 can be obtained from Eq. (2.162) as follows,

$$f_{T_1}(t) = \lambda \exp(-\lambda t), \quad t \geq 0 \tag{2.163}$$

The mean value and variance of T_1 are $\frac{1}{\lambda}$ and $\frac{1}{\lambda^2}$, respectively, because

$$\mathbb{E}(T_1) = \int_0^\infty t\lambda \exp(-\lambda t)\mathrm{d}t = -\frac{e^{-\lambda t}(1 + \lambda t)}{\lambda} \bigg|_0^\infty = \frac{1}{\lambda} \tag{2.164}$$

and

$$\mathbb{E}(T_1^2) = \int_0^\infty t^2 \lambda \exp(-\lambda t)dt = -\frac{e^{-\lambda t}(2 + 2\lambda t + \lambda^2 t^2)}{\lambda^2}\bigg|_0^\infty = \frac{2}{\lambda^2} \tag{2.165}$$

$$\Rightarrow \mathbb{V}(T_1) = \mathbb{E}(T_1^2) - \mathbb{E}^2(T_1) = \frac{1}{\lambda^2}$$

The MGF of T_1 is obtained as

$$\psi_{T_1}(\tau) = \int_0^\infty \exp(t\tau)\lambda \exp(-\lambda t) = \lambda \frac{\lambda}{\tau - \lambda} \exp[t(\tau - \lambda)]\bigg|_0^\infty$$

$$= \frac{\lambda}{\lambda - \tau}, \text{ conditional on } \tau < \lambda \tag{2.166}$$

Example 2.24

[*Readers may refer to Sect. 2.3.2 for relevant background information.*]

For two independent and exponentially distributed random variables X_1 and X_2, if their mean values are $\frac{1}{\lambda_1}$ and $\frac{1}{\lambda_2}$ respectively, find the probability that $X_1 > X_2$.

Solution

The probability of $X_1 > X_2$ is estimated, using the law of total probability (c.f. Eq. (2.23)), as follows,

$$\mathbb{P}(X_1 > X_2) = \int_0^\infty \mathbb{P}(X_1 > X_2 | X_2 = t) f_{X_2}(t)dt = \int_0^\infty [1 - \mathbb{P}(X_1 \le X_2 | X_2 = t)] f_{X_2}(t)dt$$

$$= \int_0^\infty \exp(-\lambda_1 t)\lambda_2 \exp(-\lambda_2 t)dt$$

$$= -\frac{\lambda_2}{\lambda_1 + \lambda_2} \exp(-(\lambda_1 + \lambda_2)t)\bigg|_0^\infty = \frac{\lambda_2}{\lambda_1 + \lambda_2} \tag{2.167}$$

An interesting property of the exponential distribution is that it can be used to model a "memoryless process". This property is explained by the fact that

$$\mathbb{P}(T_1 > x + y | T_1 > x) = \mathbb{P}(T_1 > y), \forall x, y \ge 0 \tag{2.168}$$

Furthermore, with Eq. (2.15), since $\mathbb{P}(T_1 > x + y | T_1 > x) = \frac{\mathbb{P}(T_1 > x + y \cap T_1 > x)}{\mathbb{P}(T_1 > x)} = \frac{\mathbb{P}(T_1 > x + y)}{\mathbb{P}(T_1 > x)}$, Eq. (2.168) is equivalent to

$$\mathbb{P}(T_1 > x + y) = \mathbb{P}(T_1 > x) \cdot \mathbb{P}(T_1 > y), \forall x, y \ge 0 \tag{2.169}$$

The proof of Eq. (2.169) is as follows.

Proof With the CDF of T_1 in Eq. (2.162),

$$\mathbb{P}(T_1 > x + y) = 1 - \mathbb{P}(T_1 \le x + y) = 1 - F_{T_1}(x + y)$$
$$= \exp(-\lambda(x + y)) = \exp(-\lambda x) \cdot \exp(-\lambda x) \qquad (2.170)$$
$$= (1 - F_{T_1}(x)) \cdot (1 - F_{T_1}(y)) = \mathbb{P}(T_1 > x) \cdot \mathbb{P}(T_1 > y)$$

The memoryless property of the exponential distribution (c.f. Eq. (2.168)) indicates that the time interval of two adjacent successive Poisson events follows an exponential distribution. With this, for a Poisson process with an occurrence rate of λ, the *return period* is simply $\frac{1}{\lambda}$ (i.e., the mean value of T_1).

Remark

For the case where the time interval between two successive events does not follow an exponential but an arbitrary distribution, the Poisson process is generalized as a "renewal process" [27]. This will be later discussed in Chap. 3.

Now we consider a more general case of the Poisson process where the occurrence rate λ is not a constant but varies with time, denoted by $\lambda(t)$ (i.e., on average $\lambda(t) \cdot \Delta t$ events occur within a time duration of Δt at time t, $\Delta t \to 0$). With this, Eq. (2.152) becomes

$$\mathbb{P}(X = k) = \frac{\left(\int_0^T \lambda(t)dt\right)^k \cdot \exp\left(-\int_0^T \lambda(t)dt\right)}{k!}, \quad k = 0, 1, 2, \ldots \qquad (2.171)$$

The proof of Eq. (2.171) is as follows.

Proof Similar to Eq. (2.155), we introduce a Bernoulli variable X_i to represent the event occurrence within the ith interval as before ($i = 1, 2, \ldots, n$), which equals 1 if an event occurs and 0 otherwise. With this, $\mathbb{P}(X_i = 1) = \lambda_i \frac{T}{n}$, and the MGF of the variable X in Eq. (2.171) is

$$\psi_X(\tau) = \lim_{n \to \infty} \prod_{i=1}^{n} \psi_{X_i}(\tau) = \lim_{n \to \infty} \exp\left\{\ln\left[1 + (e^{\tau} - 1)\lambda_i \frac{T}{n}\right]\right\}$$
$$= \lim_{n \to \infty} \exp\left\{(e^{\tau} - 1)\lambda_i \frac{T}{n}\right\} = \exp\left[(e^{\tau} - 1)\int_0^T \lambda(t)dt\right] \qquad (2.172)$$

Thus, X in Eq. (2.171) is a Poisson random variable with a mean value of $\int_0^T \lambda(t)dt$, which implies the validity of Eq. (2.171). $\square$

Remark

Recall that we have used the Poisson process to model the occurrence randomness of hurricane events (c.f. Fig. 2.3). Figure 2.4 presents the yearly occurrence rate of hurricanes at the Atlantic Basin [22, 24]. Due to the potential impacts of climate change [2, 9], the future hurricane occurrence rate may change with time (reflecting the non-stationarity in

Fig. 2.4 Hurricane frequency from HURDAT database and the fitted prediction for the Atlantic Basin: 1850–2100

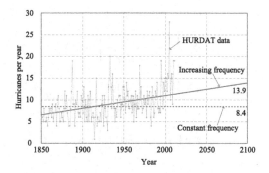

the Poisson process). It was predicted [22] that the annual occurrence rate of hurricanes at the Atlantic basin, which is on average 8.4/year currently, will increase to 13.9/year in year 2100. This is, as a result, indicative of a greater hurricane risk to the coastal areas with an increasing trend of occurrence rate.

Example 2.25

A structure is subjected to a sequence of hazardous events whose occurrence is modelled by a non-stationary Poisson process [18]. At time t, the time-variant occurrence rate is $\lambda(t)$, and the probability of survival is $\eta(t)$ conditional on the occurrence of one hazardous event ($\eta(t)$ may decrease with time due to the potential impact of structural deterioration and/or the increasing trend of hazard magnitude). Estimate the structure's probability of survival for a reference period of $[0, T]$.

Solution

The probability of structural survival within $[0, T]$ can be assessed by two approaches.

Method 1. First consider the structural survival associated with one hazardous event. The occurrence time of a single event has a PDF of $f_1(t) = \frac{\lambda(t)}{\int_0^T \lambda(t)\mathrm{d}t}$. With this, using the law of total probability (c.f. Eq. (2.23)), the survival probability conditional on the occurrence of one event is

$$\mathbb{P}(\text{survival}|\text{single event}) = \mathbb{P}_1 = \int_0^T \eta(t)f_1(t)\mathrm{d}t = \frac{\int_0^T \eta(t)\lambda(t)\mathrm{d}t}{\int_0^T \lambda(t)\mathrm{d}t} \qquad (2.173)$$

In the presence of totally k events, the survival probability equals $(\mathbb{P}_1)^k$. Furthermore, with Eq. (2.171), using the law of total probability again gives the unconditional survival probability within $[0, T]$ as follows,

$\mathbb{P}$(survival within $[0, T]$)

$$= \sum_{k=0}^{\infty} \left\{ \frac{\int_0^T \eta(t)\lambda(t)dt}{\int_0^T \lambda(t)dt} \right\}^k \cdot \mathbb{P}(X = k)$$

$$= \sum_{k=0}^{\infty} \left\{ \frac{\int_0^T \eta(t)\lambda(t)dt}{\int_0^T \lambda(t)dt} \cdot \int_0^T \lambda(t)dt \right\}^k \frac{\exp\left(-\int_0^T \lambda(t)dt\right)}{k!}$$

$$= \exp\left(\int_0^T \eta(t)\lambda(t)dt\right) \cdot \exp\left(-\int_0^T \lambda(t)\right) = \exp\left(\int_0^T [\eta(t) - 1]\lambda(t)dt\right)$$

(2.174)

Method 2. We divide the service period of $[0, T]$ into n identical sections, and let n be sufficiently large so that the duration of each interval, $\Delta t = \frac{T}{n}$, is small enough. For the ith interval, $[t_{i-1}, t_i]$, the probability of structural survive is given by

$$\mathbb{P}(i) = \eta(t_i) \cdot \underbrace{[\lambda(t_i)\Delta t]}_{\text{prob. of occurrence}} + \underbrace{[1 - \lambda(t_i)\Delta t]}_{\text{prob. of no occurrence}} = 1 + \lambda(t_i)[\eta(t_i) - 1]\Delta t$$

(2.175)

With this,

$$\mathbb{P}(\text{survival within } [0, T]) = \prod_{i=1}^{n} \mathbb{P}(i) = \prod_{i=1}^{n} [1 + \lambda(t_i)[\eta(t_i) - 1]\Delta t]$$

$$= \exp\left\{ \sum_{i=1}^{n} \ln\left[(1 + \lambda(t_i)[\eta(t_i) - 1]\Delta t)\right] \right\}$$

$$\approx \exp\left\{ \sum_{i=1}^{n} \lambda(t_i)[\eta(t_i) - 1]\Delta t \right\} = \exp\left(\int_0^T [\eta(t) - 1]\lambda(t)dt\right)$$

(2.176)

2.2.4 Normal Distribution

For a normal distributed random variable X with parameters μ and σ, its PDF can be expressed as

$$f_X(x) = \frac{1}{\sqrt{2\pi}\sigma} \exp\left\{ -\frac{(x - \mu)^2}{2\sigma^2} \right\}, \quad x \in (-\infty, \infty)$$

(2.177)

The MGF of X is $\psi_X(\tau) = \exp\left(\mu\tau + \frac{1}{2}\sigma^2\tau^2\right)$ (c.f. Example 2.20). Based on this,

$$\mathbb{E}(X) = \psi_X'(0) = \exp\left(\mu\tau + \frac{1}{2}\sigma^2\tau^2\right)(\mu + \sigma^2\tau)\bigg|_{\tau=0} = \mu$$

$$\mathbb{E}(X^2) = \psi_X''(0) = \exp\left(\mu\tau + \frac{1}{2}\sigma^2\tau^2\right)[(\mu + \sigma^2\tau)^2 + \sigma^2]\bigg|_{\tau=0} = \mu^2 + \sigma^2$$

$$(2.178)$$

implying that the mean and variance of X are μ and σ^2, respectively. Specially, if $\mu = 0$ and $\sigma = 1$, X follows a *standard normal distribution*, whose PDF and CDF can be written exclusively as $\phi(x)$ and $\Phi(x)$. With this, the PDF and CDF of X in Eq. (2.177) are

$$f_X(x) = \frac{1}{\sigma}\phi\left(\frac{x-\mu}{\sigma}\right); \quad F_X(x) = \Phi\left(\frac{x-\mu}{\sigma}\right) \qquad (2.179)$$

For an arbitrary real number x, the CDF $\Phi(x)$ satisfies

$$\Phi(-x) + \Phi(x) = 1 \qquad (2.180)$$

due to the symmetry of the standard normal distribution about the y axis. Furthermore, it is easy to verify that $F_X(x) + F_X(2\mu - x) = 1$ also holds.

The CDF of X is closely related to the error function, $\mathrm{erf}(x)$, defined as $\mathrm{erf}(x) = \frac{2}{\sqrt{\pi}}\int_0^x \exp(-z^2)dz$. Mathematically, it follows,

$$\Phi(x) = \frac{1}{2}\left(1 + \mathrm{erf}\frac{x}{\sqrt{2}}\right); \quad \mathrm{erf}(x) = 2\Phi(\sqrt{2}x) - 1 \qquad (2.181)$$

With this, the CDF of X can be alternatively expressed as $F_X(x) = \frac{1}{2}\left(1 + \mathrm{erf}\frac{x-\mu}{\sqrt{2}\sigma}\right)$. Equation (2.181) provides us an insight into the relationship between the error function and the CDF of a normal variable.

Remark

If X is a normal variable with a mean value of 0 and a standard deviation of σ, then the CDF of $|X|$ is $F_{|X|}(x) = \mathrm{erf}\left(\frac{x}{\sqrt{2}\sigma}\right)$ for $x \geq 0$. In such a case, $|X|$ is deemed to follow a *half-normal distribution* [15]. The PDF of $|X|$ is

$$f_{|X|}(x) = \frac{\sqrt{2}}{\sqrt{\pi}\sigma}\exp\left(-\frac{x^2}{2\sigma^2}\right), \quad x \geq 0 \qquad (2.182)$$

The MGF of $|X|$ is

$$\psi_{|X|}(\tau) = \int_0^\infty \exp(x\tau) \frac{\sqrt{2}}{\sqrt{\pi}\sigma} \exp\left(-\frac{x^2}{2\sigma^2}\right) dx = \int_0^\infty \frac{2}{\sqrt{2\pi}\sigma} \exp\left(-\frac{(x - \tau\sigma^2)^2}{2\sigma^2} + \frac{\tau^2\sigma^2}{2}\right) dx$$

$$= 2\exp\left(\frac{1}{2}\tau^2\sigma^2\right)\left[1 - \Phi\left(-\frac{\tau\sigma^2}{\sigma}\right)\right] = 2\exp\left(\frac{1}{2}\tau^2\sigma^2\right)[1 - \Phi(-\tau\sigma)]$$

$$= 2\exp\left(\frac{1}{2}\tau^2\sigma^2\right)\Phi(\tau\sigma)$$

$$(2.183)$$

The mean value and variance of $|X|$ are $\sigma\sqrt{\frac{2}{\pi}}$ and $\sigma^2\left(1 - \frac{2}{\pi}\right)$, respectively, by using $\mathbb{E}(|X|) = \psi'_{|X|}(0)$ and $\mathbb{V}(|X|) = \psi''_{|X|}(0) - \mathbb{E}^2(|X|)$.

Example 2.26

The Fick's second law provides a powerful tool to describe the diffusion process of chloride in a concrete structure (usually in the concrete cover) that is subjected to marine environment [17, 32], taking a form of $\frac{\partial C(x,t)}{\partial t} = D \cdot \frac{\partial^2 C(x,t)}{\partial x^2}$, where $C(x, t)$ is the chloride concentration at a distance x from the concrete surface at time t, and D is the diffusion coefficient. Verify that the solution of $C(x, t)$ is

$$C(x, t) = C_s \cdot \left[1 - \text{erf}\left(\frac{x}{2\sqrt{Dt}}\right)\right] = C_s \cdot \left[2 - 2\Phi\left(\frac{x}{\sqrt{2Dt}}\right)\right] \qquad (2.184)$$

where C_s is the surface chloride concentration.

Solution

First, the boundary conditions of $C(x, t)$ are satisfied since $C(0, t) = C_s$ and $C(x, 0) = 0$.
Next, with Eq. (2.184),

$$\frac{\partial C(x, t)}{\partial t} = C_s\phi\left(\frac{x}{\sqrt{2Dt}}\right)\frac{x}{\sqrt{2D}}t^{-\frac{3}{2}}$$

$$\frac{\partial C(x, t)}{\partial x} = -2C_s\phi\left(\frac{x}{\sqrt{2Dt}}\right)\cdot\frac{1}{\sqrt{2Dt}} \qquad (2.185)$$

$$\frac{\partial^2 C(x, t)}{\partial x^2} = C_s\phi\left(\frac{x}{\sqrt{2Dt}}\right)\frac{1}{\sqrt{2Dt}}\frac{x}{Dt}$$

As such, it is easy to see that $\frac{\partial C(x,t)}{\partial t} = D \cdot \frac{\partial^2 C(x,t)}{\partial x^2}$ is satisfied.

Example 2.27

Recall the two random variables X and Y in Example 2.16. (1) Find the joint CDF of X and Y; (2) Find the CDF of X.

Solution

(1) According to Eq. (2.92),

$$F_{X,Y}(x, y) = \int_{-\infty}^{x} \int_{-\infty}^{y} \frac{\exp(-x_1^2)}{\pi^{3/2}(y_1^2 + 1)} dy_1 dx_1$$

$$= \frac{1}{\pi^{3/2}} \int_{-\infty}^{x} \exp(-x_1^2) dx_1 \int_{-\infty}^{y} \frac{1}{y_1^2 + 1} dy_1 \qquad (2.186)$$

$$= \frac{1}{\pi} \Phi(\sqrt{2}x) \left(\frac{\pi}{2} + \arctan y \right)$$

(2) With Eq. (2.84),

$$F_X(x) = F_{X,Y}(x, \infty) = \Phi(\sqrt{2}x) \qquad (2.187)$$

Example 2.28

Recall the two random variables X_1 and X_2 in Example 2.24. Estimate the probability of $X_1 > X_2^2$.

Solution

The probability of $X_1 > X_2^2$ is obtained by

$$P(X_1 > X_2^2) = \int_0^{\infty} P(X_1 > X_2^2 | X_2 = t) f_{X_2}(t) dt = \int_0^{\infty} [1 - P(X_1 \le X_2^2 | X_2 = t)] f_{X_2}(t) dt$$

$$= \int_0^{\infty} \exp(-\lambda_1 t^2) \lambda_2 \exp(-\lambda_2 t) dt$$

$$= \frac{\lambda_2}{2} \sqrt{\frac{\pi}{\lambda_1}} \exp\left(\frac{\lambda_2^2}{4\lambda_1} \right) \left[2\Phi\left(\frac{\lambda_2 + 2\lambda_1 t}{\sqrt{2\lambda_1}} \right) - 1 \right] \Big|_0^{\infty}$$

$$= \lambda_2 \sqrt{\frac{\pi}{\lambda_1}} \exp\left(\frac{\lambda_2^2}{4\lambda_1} \right) \left[1 - \Phi\left(\frac{\lambda_2}{\sqrt{2\lambda_1}} \right) \right]$$

$$(2.188)$$

Example 2.29

For a half-normal variable X whose PDF is as in Eq. (2.182), derive its nth moment, $\mathbb{E}(|X|^n)$ for $n = 1, 2, \ldots$.

Hint. The Gamma function $\Gamma(\bullet)$ may be used, which is defined as $\Gamma(x) = \int_0^{\infty} \tau^{x-1} \exp(-\tau) d\tau$.

Solution

The nth moment of $|X|$ is obtained as

$$\mathbb{E}(|X|^n) = \int_0^{\infty} x^n \frac{\sqrt{2}}{\sqrt{\pi}\sigma} \exp\left(-\frac{x^2}{2\sigma^2} \right) dx \qquad (2.189)$$

Let $t = \frac{x^2}{2\sigma^2}$, with which $x = \sqrt{2}\sigma\sqrt{t}$ and $dx = \frac{\sigma}{\sqrt{2t}} dt$. Thus, Eq. (2.189) becomes

$$\mathbb{E}(|X|^n) = \int_0^\infty \frac{\sqrt{2}}{\sqrt{\pi}\sigma} \exp(-t)(\sqrt{2t}\sigma)^n \frac{\sigma}{\sqrt{2t}} dt$$

$$= \frac{1}{\sqrt{\pi}}(\sqrt{2}\sigma)^n \int_0^\infty t^{\frac{1}{2}(n+1)-1} \exp(-t) dt \qquad (2.190)$$

$$= \frac{1}{\sqrt{\pi}}(\sqrt{2}\sigma)^n \Gamma\left(\frac{1}{2}(n+1)\right)$$

Assigning $n = 1$ and 2 in Eq. (2.190) respectively yields $\mathbb{E}(|X|) = \frac{\sqrt{2}}{\sqrt{\pi}}\sigma$ and $\mathbb{E}(|X|^2) = \sigma^2$ and thus $\mathbb{V}(|X|) = \sigma^2\left(1 - \frac{2}{\pi}\right)$.

An important theorem related to the normal distribution is the *central limit theorem*, which states that for n independent and identically distributed random variables X_1 through X_n with a mean value of μ and a variance of σ^2 (that is independent of n), the distribution of $\frac{X_1+X_2+\cdots X_n-n\mu}{\sigma\sqrt{n}}$ approaches a standard normal distribution if $n \to \infty$, i.e.,

$$\mathbb{P}\left(\frac{X_1 + X_2 + \cdots X_n - n\mu}{\sigma\sqrt{n}} \le x\right) \to \Phi(x) \qquad (2.191)$$

The proof is as follows.

Proof Consider the moment generating function of $\frac{X_1+X_2+\cdots X_n-n\mu}{\sigma\sqrt{n}}$, which is given by

$$\psi(\tau) = \mathbb{E}\left\{\exp\left[\tau\left(\frac{X_1 + X_2 + \cdots X_n - n\mu}{\sigma\sqrt{n}}\right)\right]\right\} = \left(\mathbb{E}\left[\exp\left(\frac{\tau X}{\sqrt{n}}\right)\right]\right)^n \qquad (2.192)$$

where $X = \frac{X_1-\mu}{\sigma}$. Since n is large enough, using the Taylor series expansion, one has

$$\exp\left(\frac{\tau X}{\sqrt{n}}\right) \approx 1 + \frac{\tau X}{\sqrt{n}} + \frac{\tau^2 X^2}{2n} \qquad (2.193)$$

By noting that $\mathbb{E}(X) = 0$ and $\mathbb{E}(X^2) = 1$, it follows,

$$\mathbb{E}\left[\exp\left(\frac{\tau X}{\sqrt{n}}\right)\right] \approx 1 + \frac{\tau\mathbb{E}(X)}{\sqrt{n}} + \frac{\tau^2\mathbb{E}(X^2)}{2n} = 1 + \frac{\tau^2}{2n} \qquad (2.194)$$

Thus,

$$\lim_{n\to\infty}\left(\mathbb{E}\left[\exp\left(\frac{\tau X}{\sqrt{n}}\right)\right]\right)^n = \lim_{n\to\infty}\left(1 + \frac{\tau^2}{2n}\right)^n = \exp\left(\frac{\tau^2}{2}\right) \qquad (2.195)$$

implying that $\frac{X_1+X_2+\cdots X_n-n\mu}{\sigma\sqrt{n}}$ follows a normal distribution with a mean value of 0 and a standard deviation of 1. $\qquad\square$

A more general statement of the central limit theorem is given as follows.

Theorem 2.1 (Lyapunov Theorem) *(e.g., Theorem 27.3 in [3]) For a sequence of independent random variables $X_1, X_2, \ldots, X_n$, let μ_i and σ_i denote the mean value and standard deviation of X_i respectively for $i = 1, 2, \ldots, n$. Defining $S_n = \left(\sum_{i=1}^n \sigma_i^2 \right)^{1/2}$, the sum of X_i, $\sum_{i=1}^n X_i$, converges to a normal distribution as n is large enough if the sequence $X_1, X_2, \ldots, X_n$ satisfies the Lyapunov condition, i.e., there exists a positive number δ such that $\lim_{n \to \infty} \frac{1}{S_n^{2+\delta}} \sum_{i=1}^n \mathbb{E}\left[|X_i - \mu_i|^{2+\delta} \right] = 0$.*

Example 2.30

Consider an engineered building portfolio with n buildings. In the presence of a hazardous event, we use a Bernoulli random variable, $Y_i \in \{0, 1\}$, to denote the post-hazard state of the ith building. $Y_i = 1$ if the ith building fails and 0 otherwise. Suppose that each Y_i is independent of each other and $\mathbb{P}(Y_i = 1) = p_i$ for $i = 1, 2, \ldots, n$. The number of failed buildings due to the hazardous event is $\sum_{i=1}^n Y_i$. Show that $\sum_{i=1}^n Y_i$ follows a normal distribution.

Solution

We use the Lyapunov Theorem to show the normality of the sum of Y_i. Note that

$$\lim_{n \to \infty} \frac{1}{S_n^4} \sum_{i=1}^n \mathbb{E}\left[|Y_i - \mu_i|^4 \right] = \lim_{n \to \infty} \frac{\sum_{i=1}^n p_i(1 - p_i)(1 + 3p_i^2 - 3p_i)}{\left[\sum_{i=1}^n p_i(1 - p_i) \right]^2} \quad (2.196)$$

Since $1 + 3p_i^2 - 3p_i = 1 - 3p_i(1 - p_i) < 1$, it follows,

$$0 \le \lim_{n \to \infty} \frac{1}{S_n^4} \sum_{i=1}^n \mathbb{E}\left[|Y_i - \mu_i|^4 \right] < \lim_{n \to \infty} \frac{\sum_{i=1}^n p_i(1 - p_i)}{\left[\sum_{i=1}^n p_i(1 - p_i) \right]^2} = \lim_{n \to \infty} \frac{1}{\sum_{i=1}^n p_i(1 - p_i)} = 0 \quad (2.197)$$

Thus, $\lim_{n \to \infty} \frac{1}{S_n^4} \sum_{i=1}^n \mathbb{E}\left[|Y_i - \mu_i|^4 \right] = 0$, suggesting that $\{Y_1, Y_2, \ldots, Y_n\}$ satisfies the Lyapunov condition with $\delta = 2$. This observation supports the normality of $\sum_{i=1}^n Y_i$ as n is large enough.

Example 2.31

This example is to show that the sum of n independent and identically distributed random variables as n is sufficiently large is not necessarily a normal variable. Consider a community that is subjected to tropical cyclone damages [36]. The damage loss conditional on the occurrence of one cyclone event is D, with a mean value of μ_D and a standard deviation of σ_D. The cyclone occurrence is modelled as a stationary Poisson process with a mean occurrence rate of λ. Assume that the post-hazard damaged buildings are restored to the pre-damage state before the occurrence of the next cyclone event. Consider the cumulative cyclone damage costs for a reference period of $[0, T]$.

We divide the time period $[0, T]$ into n identical sections, where n is sufficiently large. The probability of occurrence of one cyclone event during each time interval is identically

$\lambda T/n$, with which we introduce a Bernoulli variable B_i to represent the cyclone occurrence for the ith interval, $\mathbb{P}(B_i = 1) = \mathbb{P}(\text{occurrence}) = \lambda T/n$ and $\mathbb{P}(B_i = 0) = 1 - \lambda T/n$. Let $\widetilde{D}_i$ denote the cyclone damage loss associated with the ith time interval. Clearly, $\widetilde{D}_i = B_i \cdot D$. The cumulative cyclone damage costs is $\sum_{i=1}^{n} \widetilde{D}_i$. Show that while each $\widetilde{D}_i$ is independent and identically distributed, $\sum_{i=1}^{n} \widetilde{D}_i$ does not satisfy the Lyapunov condition and thus does not necessarily follow a normal distribution.

Solution

Note that $\mathbb{E}(\widetilde{D}_i) = \frac{\lambda T}{n}\mu_D$ for $i = 1, 2, \ldots, n$. For an arbitrary positive δ,

$$
\begin{aligned}
\lim_{n \to \infty} \sum_{i=1}^{n} \mathbb{E}\left[|\widetilde{D}_i - \mathbb{E}(\widetilde{D}_i)|^{2+\delta}\right] &= \lim_{n \to \infty} n\mathbb{E}\left[\left|B_i \cdot D - \frac{\lambda T}{n}\mu_D\right|^{2+\delta}\right] \\
&= \lim_{n \to \infty} n\mathbb{E}\left[\left|D - \frac{\lambda T}{n}\mu_D\right|^{2+\delta}\right] \cdot \frac{\lambda T}{n} + \lim_{n \to \infty} n\mathbb{E}\left[\left(\frac{\lambda T}{n}\mu_D\right)^{2+\delta}\right] \cdot \left(1 - \frac{\lambda T}{n}\right) \\
&= \lambda T \mathbb{E}\left(D^{2+\delta}\right)
\end{aligned}
$$

$$(2.198)$$

and

$$
\begin{aligned}
S_n^2 = \lim_{n \to \infty} \mathbb{V}\left[\sum_{i=1}^{n} \widetilde{D}_i\right] &= \lim_{n \to \infty} \sum_{i=1}^{n} \mathbb{V}[B_i \cdot D] = \lim_{n \to \infty} \sum_{i=1}^{n} \left\{\mathbb{E}\left[B_i^2 \cdot D^2\right] - (\mathbb{E}[B_i \cdot D])^2\right\} \\
&= n\left\{\frac{\lambda T}{n}(\mu_D^2 + \sigma_D^2) - \left(\frac{\lambda T}{n}\mu_D\right)^2\right\} = \lambda T(\mu_D^2 + \sigma_D^2)
\end{aligned}
$$

$$(2.199)$$

Thus,

$$
\lim_{n \to \infty} \frac{1}{S_n^{2+\delta}} \sum_{i=1}^{n} \mathbb{E}\left[|\widetilde{D}_i - \mathbb{E}(\widetilde{D}_i)|^{2+\delta}\right] = \frac{\mathbb{E}\left(D^{2+\delta}\right)}{(\lambda T)^{0.5\delta}(\mu_D^2 + \sigma_D^2)^{1+0.5\delta}} \neq 0 \quad (2.200)
$$

indicating that the sequence $\{\widetilde{D}_i\}$ does not satisfy the Lyapunov condition.

The normal distribution is closely related to the uniform distribution (c.f. Eq. (2.147)) and the exponential distribution (c.f. Eq. (2.162)). Consider the vector $\mathbf{z}$ in Fig. 2.5, where the square of the magnitude of $\mathbf{z}$, R^2, follows an exponential distribution with a mean value of $2\sigma^2$ (i.e., $\lambda = \frac{1}{2\sigma^2}$ in Eq. (2.162)). The angle between $\mathbf{z}$ and the x-axis, Θ, is uniformly distributed within $[0, 2\pi]$. Let $X = R\cos\Theta$ and $Y = R\sin\Theta$. It can be shown that X and Y are statistically independent and identically normally distributed, with a mean value of 0 and a standard deviation of σ [21, 27]. Based on this, it is straightforward to see that both $X + \mu$ and $Y + \mu$ are normal variables with a mean value of μ and a standard deviation of σ.

Fig. 2.5 Graphical illustration of generating normal random variables

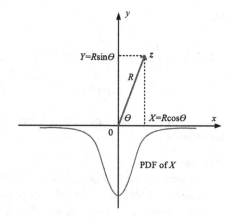

2.2.5 Lognormal Distribution

If the natural logarithm of a random variable X (rather than X itself) has a normal distribution, X is deemed to follow a *lognormal distribution*, whose CDF and PDF are respectively given by

$$F_X(x) = \Phi\left(\frac{\ln x - \lambda}{\varepsilon}\right), \quad x > 0 \tag{2.201}$$

and

$$f_X(x) = \frac{\mathrm{d}F_X(x)}{\mathrm{d}x} = \frac{1}{\sqrt{2\pi}x\varepsilon}\exp\left[-\frac{1}{2}\left(\frac{\ln x - \lambda}{\varepsilon}\right)^2\right], \quad x > 0 \tag{2.202}$$

The items λ and ε in Eqs. (2.201) and (2.202) are the mean value and standard deviation of $\ln X$ respectively, which can be determined uniquely with the mean value and variance of X, $\mathbb{E}(X)$ and $\mathbb{V}(X)$, according to

$$\mathbb{E}(X) = \exp\left(\lambda + \frac{1}{2}\varepsilon^2\right); \quad \mathbb{V}(X) = \mathbb{E}^2(X)[\exp(\varepsilon^2) - 1] \tag{2.203}$$

Equation (2.203) holds by noting that

$$\mathbb{E}(X) = \int_0^\infty x f_X(x)\mathrm{d}x = \int_0^\infty \frac{1}{\sqrt{2\pi}\varepsilon}\exp\left[-\frac{1}{2}\left(\frac{\ln x - \lambda}{\varepsilon}\right)^2\right]\mathrm{d}x$$

$$\underset{\ln x = t}{=\!=\!=\!=} \int_{-\infty}^\infty \frac{1}{\sqrt{2\pi}\varepsilon}\exp\left[-\frac{1}{2}\left(\frac{t - \lambda}{\varepsilon}\right)^2\right] \cdot \exp(t)\mathrm{d}t \tag{2.204}$$

$$= \psi(1) = \exp\left(\lambda + \frac{1}{2}\varepsilon^2\right)$$

and

$$\mathbb{E}(X^2) = \int_0^\infty x^2 f_X(x)\mathrm{d}x \xrightarrow{\ln x = t} \int_{-\infty}^\infty \frac{1}{\sqrt{2\pi}\varepsilon} \exp\left[-\frac{1}{2}\left(\frac{t-\lambda}{\varepsilon}\right)^2\right] \cdot \exp(2t)\mathrm{d}t$$

$$= \psi(2) = \exp\left(2\lambda + 2\varepsilon^2\right) \Rightarrow \mathbb{V}(X) = \mathbb{E}^2(X)[\exp(\varepsilon^2) - 1]$$

(2.205)

where $\psi(\tau)$ is the MGF of a normal variable with a mean value of λ and a standard deviation of ε, i.e., $\psi(\tau) = \exp\left(\lambda\tau + \frac{1}{2}\varepsilon^2\tau^2\right)$.

Remark

A generalized form of Eqs. (2.204) and (2.205) is that $\mathbb{E}(X^n) = \psi(n) = \exp\left(\lambda n + \frac{1}{2}\varepsilon^2 n^2\right)$. However, the MGF of lognormal distribution does not exist because the lognormal distribution cannot be uniquely determined by its moments [12].

According to Eq. (2.203), the coefficient of variation (COV, defined as the ratio of standard deviation to mean value of a random variable) of X, v_X, is $v_X = \sqrt{\exp(\varepsilon^2) - 1}$. Furthermore, $\varepsilon = \sqrt{\ln(v_X^2 + 1)} \approx v_X$ if the COV of X is small enough.

The median of X, med_X, is obtained by assigning $F_X(\mathrm{med}_X) = \frac{1}{2}$, which yields $\mathrm{med}_X = \exp(\lambda)$. The median value is smaller than the mean value, indicating a positive skewness [7] of the lognormal distribution.

Remark

In practical engineering, the lognormal distribution is widely utilized to model the probabilistic behaviour of a random variable that is by definition the product of several random variables, e.g., the structural resistance (c.f. Fig. 1.3) [8, 25, 37]. Moreover, the fragility curves of an engineered structure (or a component from the structure) are typically described by the CDF of a lognormal distribution [20, 26].

For example, according to the ASHTO LRFD Bridge Design Specifications, the bridge capacity R is determined as the product of several parameters as follows [25, 37],

$$R = R_\mathrm{n} \cdot M \cdot F \cdot P$$

(2.206)

where R_n is the nominal resistance, and the other three parameters M, F, P are representative of the uncertainties arising from material, fabrication and structural modelling respectively. With this, one may model R as a lognormal variable under the assumption of statistical independence between M, F and P. Further details can be found in Sect. 4.4.1.

2.2.6 Rayleigh Distribution

A random variable X follows a *Rayleigh distribution* if its PDF takes the form of

$$f_X(x) = \frac{x}{\sigma^2} \exp\left(-\frac{x^2}{2\sigma^2}\right); \quad x \geq 0 \tag{2.207}$$

where σ is the scale parameter. Correspondingly, the CDF of X is

$$F_X(x) = 1 - \exp\left(-\frac{x^2}{2\sigma^2}\right); \quad x \geq 0 \tag{2.208}$$

The mean value and variance of X are $\sigma\sqrt{\frac{\pi}{2}}$ and $\frac{4-\pi}{2}\sigma^2$, respectively. More generally, the nth moment of the Rayleigh variable X, $\mathbb{E}(X^n)$, is obtained as follows by referring to Example 2.29.

$$\begin{aligned}
\mathbb{E}(X^n) &= \int_0^\infty x^n \frac{x}{\sigma^2} \exp\left(-\frac{x^2}{2\sigma^2}\right) dx = \frac{\sqrt{\pi}}{\sqrt{2}\sigma} \int_0^\infty \frac{\sqrt{2}x^{n+1}}{\sqrt{\pi}\sigma} \exp\left(-\frac{x^2}{2\sigma^2}\right) dx \\
&= \frac{\sqrt{\pi}}{\sqrt{2}\sigma} \cdot \frac{1}{\sqrt{\pi}} (\sqrt{2}\sigma)^{n+1} \Gamma\left(\frac{1}{2}(n+2)\right) = (\sqrt{2}\sigma)^n \Gamma\left(\frac{1}{2}n + 1\right)
\end{aligned} \tag{2.209}$$

The MGF of X is obtained as

$$\begin{aligned}
\psi_X(\tau) &= \int_0^\infty \frac{x}{\sigma^2} \exp\left(-\frac{x^2}{2\sigma^2} + x\tau\right) dx \\
&= \int_0^\infty \left(\frac{x - \tau\sigma^2}{\sigma^2} + \tau\right) \exp\left(-\frac{(x - \tau\sigma^2)^2}{2\sigma^2} + \frac{\tau^2\sigma^2}{2}\right) dx \\
&= \exp\left(\frac{\tau^2\sigma^2}{2}\right) \left[\int_0^\infty \frac{x - \tau\sigma^2}{\sigma^2} \exp\left(-\frac{(x - \tau\sigma^2)^2}{2\sigma^2}\right) dx \right. \\
&\qquad \left. + \sqrt{2\pi}\sigma\tau \int_0^\infty \frac{1}{\sqrt{2\pi}\sigma} \exp\left(-\frac{(x - \tau\sigma^2)^2}{2\sigma^2}\right) dx\right] \\
&= \exp\left(\frac{\tau^2\sigma^2}{2}\right) \left[\exp\left(-\frac{\tau^2\sigma^2}{2}\right) + \sqrt{2\pi}\sigma\tau(1 - \Phi(-\tau\sigma))\right] \\
&= 1 + \sqrt{2\pi}\tau\sigma\Phi(\tau\sigma) \exp\left(\frac{\tau^2\sigma^2}{2}\right)
\end{aligned} \tag{2.210}$$

The Rayleigh distribution is closely related to the normal distribution (c.f. Sect. 2.2.4). Consider two orthogonal and independent vectors $\mathbf{U}$ and $\mathbf{V}$. If the magnitude of both vectors follow a normal distribution with a mean value of 0 and a standard deviation of σ, then the magnitude of $\mathbf{Z} = \mathbf{U} + \mathbf{V}$ follows a Rayleigh distribution. The CDF of $Z = |\mathbf{Z}|$ can be derived by recalling Fig. 2.5. We let $|\mathbf{U}| = R\cos\Theta$ and $|\mathbf{V}| = R\sin\Theta$, with which $Z^2 = |\mathbf{U}|^2 + |\mathbf{V}|^2 = R^2$. Thus,

$$\mathbb{P}(Z \leq z) = \mathbb{P}(R^2 \leq z) = F_R(z^2) = 1 - \exp\left(-\frac{z^2}{2\sigma^2}\right); \quad z \geq 0 \qquad (2.211)$$

consistent with Eq. (2.208), where $F_R(\bullet)$ is the CDF of R (c.f. Eq. (2.162)).

If the magnitudes of the two vectors $\mathbf{U}$ and $\mathbf{V}$ are fully correlated, then it is easy to see that Z follows a half-normal distribution with a mean value of $\frac{2\sigma}{\sqrt{\pi}}$ (c.f. Eq. (2.182)).

2.2.7 Gamma and Beta Distributions

If a random variable X follows a *Gamma distribution* with a shape parameter of $a > 0$ and a scale parameter of $b > 0$, its PDF takes the form of

$$f_X(x) = \frac{(x/b)^{a-1}}{b\Gamma(a)} \exp(-x/b); \quad x > 0 \qquad (2.212)$$

where $\Gamma(\cdot)$ is the Gamma function (c.f. Example 2.29).

Remark

The Gamma function $\Gamma(x)$ is a generalization of the factorial of an integer, i.e., $\Gamma(x) = (x-1)!$ holds for positive integers x. Moreover,

$$\Gamma(x + 1) = x\Gamma(x) \qquad (2.213)$$

for $x > 0$.

Proof

$$\begin{aligned}
\Gamma(x + 1) &= \int_0^\infty \tau^x \exp(-\tau)\mathrm{d}\tau = \int_0^\infty -\tau^x \mathrm{d}(\exp(-\tau)) \\
&= -\tau^x \exp(-\tau)\Big|_0^\infty + \int_0^\infty \exp(-\tau)x\tau^{x-1}\mathrm{d}(x) = x\Gamma(x)
\end{aligned} \qquad (2.214)$$

Example 2.32

Given that $\Gamma\left(\frac{1}{2}\right) = \sqrt{\pi}$, show that $\Gamma\left(\frac{1}{2} + n\right) = \frac{(2n)!}{4^n n!}\sqrt{\pi}$ for $n = 0, 1, 2, \ldots$.

Solution

We prove inductively. First, for the case of $n = 0$, the statement is true (as given). Now we assume that the statement holds for $n = k$ (i.e., $\Gamma\left(\frac{1}{2} + k\right) = \frac{(2k)!}{4^k k!}\sqrt{\pi}$) and consider the case of $n = k + 1$. With Eq. (2.213),

$$\Gamma\left(k + 1 + \frac{1}{2}\right) = \left(k + \frac{1}{2}\right)\Gamma\left(k + \frac{1}{2}\right) = \left(k + \frac{1}{2}\right)\frac{(2k)!}{4^k k!}\sqrt{\pi} = \frac{(2k + 2)!}{4^{k+1}(k + 1)!}\sqrt{\pi}$$

$$(2.215)$$

which complete the proof.

The MGF of X is

$$
\begin{aligned}
\psi_X(\tau) &= \int_0^\infty \frac{(x/b)^{a-1}}{b\Gamma(a)}\exp(-x/b + \tau x)\mathrm{d}x \\
&\underset{-x/b+\tau x=-x/b'}{=\!=\!=\!=\!=} \int_0^\infty \frac{(x/b')^{a-1}\cdot(b'/b)^{a-1}}{b'\Gamma(a)\cdot b/b'}\exp(-x/b')\mathrm{d}x \qquad (2.216)\\
&= \frac{(b'/b)^{a-1}}{b/b'} = (b'/b)^a = \left(\frac{1}{1 - b\tau}\right)^a
\end{aligned}
$$

Thus, the mean and variance of X are ab and ab^2 respectively, with their derivation presented in the following.

$$
\begin{aligned}
\mathbb{E}(X) &= \psi_X'(0) = ab(1 - b\tau)^{-a-1}\big|_{\tau=0} = ab \\
\mathbb{E}(X^2) &= \psi_X''(0) = ab^2(a + 1)(1 - b\tau)^{-a-2}\big|_{\tau=0} = ab^2(a + 1) \Rightarrow \mathbb{V}(X) = ab^2
\end{aligned}
$$

$$(2.217)$$

Remark

A generalized form of Eq. (2.217) is that for $n = 1, 2, \ldots$, $\mathbb{E}(X^n) = b^n\prod_{i=0}^{n-1}(a + i)$.
Proof We first consider the nth derivative of the MGF, $\psi_X(\tau)$. Using the method of induction, it can be shown that $\psi_X(\tau)^{(n)} = b^n\prod_{i=0}^{n-1}(a + i)(1 - b\tau)^{-a-n}$. With this,

$$\mathbb{E}(X^n) = \psi_X^{(n)}(0) = b^n\prod_{i=0}^{n-1}(a + i) \qquad (2.218)$$

which completes the proof.   □

Based on Eq. (2.212), the *generalized Gamma distribution*, with three parameters a, d and p, has a PDF taking the form of [31]

$$f_X(x) = \frac{(p/a^d)}{\Gamma\left(\frac{d}{p}\right)}x^{d-1}\exp\left(-(x/a)^p\right), \quad x \geq 0 \qquad (2.219)$$

Table 2.1 Relationship between the generalized Gamma distribution and some other distributions

Case no.	Parameter values	Distribution	Reference
1	$p = 1$	Gamma	Sect. 2.2.7
2	$d = 1, p = 2$	Half-normal	Eq. (2.182)
3	$d = 2, p = 2$	Rayleigh	Sect. 2.2.6
4	$d = p$	Weibull	Sect. 2.2.8

For a random variable whose PDF is as in Eq. (2.219), the nth order moment is estimated as

$$\mathbb{E}(X^n) = \int_0^\infty x^n \frac{(p/a^d)}{\Gamma\left(\frac{d}{p}\right)} x^{d-1} \exp\left(-(x/a)^p\right) dx$$

$$\overset{(x/a)^p = t}{=\!=\!=\!=} \frac{(p/a^d)}{\Gamma\left(\frac{d}{p}\right)} \int_0^\infty a^{d-1+n} t^{\frac{d+n-1}{p}} \exp(-t) \frac{a}{p} t^{\frac{1}{p}-1} dt \qquad (2.220)$$

$$= a^n \frac{\Gamma\left(\frac{d+n}{p}\right)}{\Gamma\left(\frac{d}{p}\right)}, \quad n = 0, 1, 2, \ldots$$

When the three parameters a, d and p take some specific values, the generalized Gamma distribution simply becomes some related distributions, as summarized in Table 2.1.

An important property of the Gamma distribution is that for two independent Gamma variables X_1 (with shape and scale parameters of a_1 and b) and X_2 (with a_2 and b), $X_1 + X_2$ also follows a Gamma distribution with shape and scale parameters of $a_1 + a_2$ and b. The Gamma distribution of $X_1 + X_2$ is explained by the fact that the MGF of $X_1 + X_2$ is

$$\psi_{X_1+X_2}(\tau) = \psi_{X_1}(\tau) \cdot \psi_{X_2}(\tau) = \left(\frac{1}{1-b\tau}\right)^{a_1} \cdot \left(\frac{1}{1-b\tau}\right)^{a_2} = \left(\frac{1}{1-b\tau}\right)^{a_1+a_2}$$
$$(2.221)$$

which is exactly the MGF of a Gamma variable with shape and scale parameters of $a_1 + a_2$ and b respectively.

Remark

For n statistically independent and identically exponentially distributed variables $X_1, X_2, \ldots, X_n$ with a mean value of $\frac{1}{\lambda}$, their sum, $\sum_{i=1}^n X_i$, follows a Gamma distribution. This is verified by noting that the MGF of X_i is $\frac{\lambda}{\lambda-\tau}$ and that of $\sum_{i=1}^n X_i$ equals $\left(\frac{\lambda}{\lambda-\tau}\right)^n$, which is exactly the MGF of a Gamma distribution with a shape parameter of n and a scale parameter of $\frac{1}{\lambda}$. Furthermore, when the shape parameter (n) is sufficiently large, a Gamma distribution approaches to a normal distribution with the central limit theorem (c.f. Sect. 2.2.4). This can also be verified through considering the MGF of $\frac{1}{n}\sum_{i=1}^{n}X_i$, $\left(\frac{n\lambda}{n\lambda-\tau}\right)^n$. Note that

$$
\begin{aligned}
\lim_{n\to\infty}\left(\frac{n\lambda}{n\lambda-\tau}\right)^n &= \lim_{n\to\infty}\left(1-\frac{\tau}{n\lambda}\right)^{-n} = \lim_{n\to\infty}\exp\left[-n\ln\left(1-\frac{\tau}{n\lambda}\right)\right] \\
&\approx \exp\left[-n\left(-\frac{\tau}{n\lambda}-\frac{\tau^2}{2n^2\lambda^2}\right)\right] = \exp\left(\frac{\tau}{\lambda}+\frac{\tau^2}{2n\lambda^2}\right)
\end{aligned}
\tag{2.222}
$$

which is exactly the MGF of a normal distribution with a mean value of $\frac{1}{\lambda}$ and a variance of $\frac{1}{n\lambda^2}$.

For random variables X_1 and X_2 as in Eq. (2.221), let $X = \frac{X_1}{X_1+X_2}$, then X follows a *Beta distribution* and is independent of $X_1 + X_2$. Clearly, X is defined strictly within [0, 1]. The PDF of X takes a form of

$$
f_X(x) = \frac{\Gamma(\eta+\zeta)}{\Gamma(\eta)\Gamma(\zeta)}x^{\eta-1}(1-x)^{\zeta-1}, \quad 0 \le x \le 1
\tag{2.223}
$$

where $\Gamma(\bullet)$ is the Gamma function, and η and ζ are two shape parameters. The mean value and COV of X are $\frac{\eta}{\eta+\zeta}$ and $\sqrt{\frac{\zeta}{\eta(\eta+\zeta+1)}}$ respectively. In fact, the nth moment of X is estimated by

$$
\begin{aligned}
\mathbb{E}(X^n) &= \int_0^1 x^n \frac{\Gamma(\eta+\zeta)}{\Gamma(\eta)\Gamma(\zeta)}x^{\eta-1}(1-x)^{\zeta-1}\mathrm{d}x \\
&= \frac{\Gamma(\eta+n)\Gamma(\eta+\xi)}{\Gamma(\eta)\Gamma(\eta+\xi+n)}\cdot\int_0^1 \frac{\Gamma(\eta+\zeta+n)}{\Gamma(\eta+n)\Gamma(\zeta)}x^{\eta+n-1}(1-x)^{\zeta-1}\mathrm{d}x \\
&= \frac{\Gamma(\eta+n)\Gamma(\eta+\xi)}{\Gamma(\eta)\Gamma(\eta+\xi+n)}
\end{aligned}
\tag{2.224}
$$

If $\mathbb{E}(X)$ and $\mathbb{V}(X)$ are known, the two parameters η and ζ are determined by

$$
\eta = \frac{\mathbb{E}^2(X)-\mathbb{E}^3(X)}{\mathbb{V}(X)} - \mathbb{E}(X), \quad \zeta = \left(\frac{\mathbb{E}(X)-\mathbb{E}^2(X)}{\mathbb{V}(X)}-1\right)\cdot(1-\mathbb{E}(X))
\tag{2.225}
$$

2.2.8 Extreme Value Distributions

Let $\{X_1, X_2, \ldots, X_n\}$ be a sequence of independent and identically distributed random variables with a CDF of F. The maxima of the sequence is denoted by M_n, whose CDF is given by

$$F_{M_n}(x) = \mathbb{P}(M_n \leq x) = \mathbb{P}(X_1 \leq x \cap X_2 \leq x \cap \cdots X_n \leq x) = F^n(x) \quad (2.226)$$

For the case where F is unknown, the following theorem gives an asymptotic estimate of F_{M_n} [10].

Theorem 2.2 (The Fisher–Tippett–Gnedenko theorem) *Suppose that there exists constants $a_n > 0$, $b_n \in (-\infty, \infty)$, and $n > 1$ such that*

$$\mathbb{P}[(M_n - b_n)/a_n \leq x] = F^n(a_n x + b_n) \rightarrow G(x) \quad (2.227)$$

for $n \rightarrow \infty$ and non-degenerate G, then $G(x) \propto \exp\left[-(1 + \xi x)^{-1/\xi}\right]$, where ξ is dependent on the tail shape of the distribution. The cases of $\xi = 0$, $\xi > 0$ and $\xi < 0$ correspond to the Gumbel, Fréchet and Weibull families respectively.

Depending on the specific value of ξ, $G(x)$ takes a standard form of one of the following three types.

$$\text{Type 1}\quad G(x) = \exp\left[-\exp(-x)\right], x \in (-\infty, \infty) \quad (2.228)$$

$$\text{Type 2}\quad G(x) = \begin{cases} 0, & x < 0 \\ \exp(-x^{-\alpha}), & x \geq 0, \alpha > 0 \end{cases} \quad (2.229)$$

and

$$\text{Type 3}\quad G(x) = \begin{cases} \exp(-(-x)^\alpha), & x < 0 \\ 1, & x \geq 0, \alpha > 0 \end{cases} \quad (2.230)$$

2.2.8.1 Gumbel Distribution

If a random variable X follows an *Extreme type I distribution* (Gumbel distribution) with location and scale parameters of μ and σ respectively, then its CDF is as follows, which is of the same type of Eq. (2.228).

$$F_X(x) = \exp\left[-\exp\left(-\frac{x - \mu}{\sigma}\right)\right], x \in (-\infty, \infty) \quad (2.231)$$

With Eq. (2.231), the PDF of X is

$$f_X(x) = \frac{1}{\sigma}\exp\left(-\frac{x - \mu}{\sigma}\right) \cdot \exp\left[-\exp\left(-\frac{x - \mu}{\sigma}\right)\right], x \in (-\infty, \infty) \quad (2.232)$$

The MGF of X is

$$\psi_X(\tau) = \int_{-\infty}^{\infty} \frac{1}{\sigma} \exp\left(-\frac{x-\mu}{\sigma}\right) \cdot \exp\left[-\exp\left(-\frac{x-\mu}{\sigma}\right)\right] \exp(\tau x) dx$$

$$\underset{\exp(-\frac{x-\mu}{\sigma})=t}{=\!=\!=\!=\!=} \int_0^{\infty} \frac{t}{\sigma} \exp(-t) \exp(\tau\mu - \sigma\tau \ln t) \frac{\sigma}{t} dt \qquad (2.233)$$

$$= \exp(\mu\tau) \int_0^{\infty} \exp(-t) t^{(1-\sigma\tau)-1} = \exp(\mu\tau)\Gamma(1-\sigma\tau)$$

Using the MGF, the mean value of X, $\mathbb{E}(X)$, is obtained as

$$\mathbb{E}(X) = \psi_X'(0) = \exp(\mu\tau)\left[\mu\Gamma(1-\sigma\tau) - \sigma\Gamma'(1-\sigma\tau)\right]\big|_{\tau=0} = \mu + \gamma\sigma \quad (2.234)$$

where γ is the Euler–Mascheroni constant,

$$\gamma = -\Gamma'(1) = -\int_0^{\infty} \exp(-t)\ln t\, dt \approx 0.5772 \qquad (2.235)$$

Moreover, since

$$\mathbb{E}(X^2) = \psi_X''(0) = \exp(\mu\tau)\left[\sigma^2\Gamma''(1-\sigma\tau) - 2\mu\sigma\Gamma'(1-\sigma\tau) + \mu^2\Gamma(1-\sigma\tau)\right]\big|_{\tau=0}$$
$$= \sigma^2\Gamma''(1) - 2\mu\sigma\Gamma'(1) + \mu^2 \qquad (2.236)$$

The variance of X is estimated as

$$\mathbb{V}(X) = \mathbb{E}(X^2) - \mathbb{E}^2(X) = \sigma^2\left\{\Gamma''(1) - [\Gamma'(1)]^2\right\} = \frac{\pi^2}{6}\sigma^2 \qquad (2.237)$$

Example 2.33

For n statistically independent and identically exponentially distributed variables X_1 through X_n, show that their maximum value follows an Extreme Type I distribution.
Solution
Suppose that the CDF of X_i is $F_X(x) = 1 - \exp(-\lambda x)$ for $x \geq 0$ (c.f. Eq. (2.162)). Then according to Eq. (2.226), the maxima of $\{X_i\}$ has a CDF of

$$F_{\max\{X_i\}}(x) = [F_X(x)]^n = \left[1 - \exp(-\lambda x)\right]^n = \exp\left(n\ln\left[1 - \exp(-\lambda x)\right]\right)$$

$$\approx \exp\left(-\exp(\ln n)\cdot\exp(-\lambda x)\right) = \exp\left[-\exp\left(-\frac{x - \frac{1}{\lambda}\ln n}{\frac{1}{\lambda}}\right)\right] \qquad (2.238)$$

As such, the maxima of $\{X_i\}$ follows an Extreme Type I distribution with a location parameter of $\frac{1}{\lambda}\ln n$ and a scale parameter of $\frac{1}{\lambda}$.

2.2.8.2 Fréchet Distribution

If a random variable X follows an *Extreme Type II distribution* (Fréchet distribution), then its CDF takes the form of

$$F_X(x) = \exp\left[-\left(\frac{x}{\varepsilon}\right)^{-k}\right], \quad x \geq 0 \tag{2.239}$$

in which ε is scale parameter and k is shape parameter. The PDF of X is

$$f_X(x) = k\varepsilon^k x^{-k-1} \exp\left[-\left(\frac{x}{\varepsilon}\right)^{-k}\right], \quad x \geq 0 \tag{2.240}$$

Furthermore, the nth moment of X is estimated as

$$
\begin{aligned}
\mathbb{E}(X^n) &= \int_0^\infty x^n f_X(x)\mathrm{d}x = \int_0^\infty k\varepsilon^k x^{-k-1+n} \exp\left[-\left(\frac{x}{\varepsilon}\right)^{-k}\right] \\
&\xlongequal{(x/\varepsilon)^{-k}=t} \int_0^\infty k\varepsilon^k \left(\varepsilon t^{-\frac{1}{k}}\right)^{n-1-k} \exp(-t)\frac{\varepsilon}{k}t^{-1-\frac{1}{k}}\mathrm{d}t \\
&= \varepsilon^n \Gamma\left(1 - \frac{n}{k}\right)
\end{aligned}
\tag{2.241}
$$

which holds when $n < k$. Specifically, the mean value and variance of X, $\mathbb{E}(X)$ and $\mathbb{V}(X)$, are respectively

$$
\begin{aligned}
\mathbb{E}(X) &= \varepsilon \cdot \Gamma\left(1 - \frac{1}{k}\right) \text{ if } k > 1 \\
\mathbb{V}(X) &= \varepsilon^2\left[\Gamma\left(1 - \frac{2}{k}\right) - \Gamma^2\left(1 - \frac{1}{k}\right)\right] \text{ if } k > 2
\end{aligned}
\tag{2.242}
$$

Remark

The maximum peak ground acceleration (PGA) within a reference period (say, one year) follows an Extreme Type II distribution [6], as briefly illustrated in the following.

The occurrence of earthquake event is modelled by a Poisson process, with an occurrence rate of λ. Within a reference period of $[0, T]$, the CDF of the maximum PGA, denoted by A_m, is obtained as follows using the law of total probability,

$$F_{A_m}(a) = \mathbb{P}(A_m \leq a) = \sum_{k=0}^\infty \mathbb{P}(X = k) \cdot [1 - \mathbb{P}_c(A > a)]^k \tag{2.243}$$

where X is the number of earthquake events within the considered reference period, and $\mathbb{P}_c(A > a)$ is the probability that the PGA, A, exceeds a conditional on the occurrence of an earthquake event.

$$\mathbb{P}_c(A > a) = \int_R \int_M \mathbb{I}(A > a|r_0, m_0)\, f_R(r_0) f_M(m_0) dm_0 dr \qquad (2.244)$$

where $f_R(r_0)$ and $f_M(m_0)$ are the PDFs of the focal distance R and the earthquake magnitude M, respectively. The CDF of M, F_M, takes a form of

$$F_M(x) = 1 - \exp(-\beta(x - M_0)) \qquad (2.245)$$

where β is a constant and M_0 is the minimum magnitude that is considered. Furthermore, using an earthquake attenuation model, the relationship between A and M, R is

$$A = \gamma_1 \exp(\gamma_2 M) R^{-\gamma_3} \qquad (2.246)$$

where γ_1, γ_2 and γ_3 are three constants. Thus,

$$
\begin{aligned}
\mathbb{P}_c(A > a) &= \int_R \int_M \mathbb{I}\left(\gamma_1 \exp(\gamma_2 m_0) r_0^{-\gamma_3} > a\right) f_R(r_0) f_M(m_0) dm_0 dr_0 \\
&= \int_R \int_M \mathbb{I}\left(m_0 > \frac{r_0^{\gamma_3}}{\gamma_2} \ln\left(\frac{a}{\gamma_1}\right)\right) f_R(r_0) f_M(m_0) dm_0 dr_0 \\
&= \int_R \exp\left[-\beta\left(\frac{r_0^{\gamma_3}}{\gamma_2} \ln\left(\frac{a}{\gamma_1}\right) - M_0\right)\right] f_R(r_0) dr_0 \\
&\propto a^{-\frac{\beta}{\gamma_2}}
\end{aligned}
\qquad (2.247)
$$

With this, we let $\mathbb{P}_c(A > a) = \eta a^{-\frac{\beta}{\gamma_2}}$, where η is a parameter that is independent of a. Thus, Eq. (2.243) becomes

$$
\begin{aligned}
F_{A_m}(a) &= \sum_{k=0}^{\infty} \mathbb{P}(X = k) \cdot \left[1 - \eta a^{-\frac{\beta}{\gamma_2}}\right]^k = \sum_{k=0}^{\infty} \frac{(\lambda T)^k}{k!} \exp(-\lambda T) \cdot \left[1 - \eta a^{-\frac{\beta}{\gamma_2}}\right]^k \\
&= \exp(-\lambda T) \cdot \exp\left[\lambda T\left(1 - \eta a^{-\frac{\beta}{\gamma_2}}\right)\right] = \exp\left(-\lambda T \eta a^{-\frac{\beta}{\gamma_2}}\right)
\end{aligned}
\qquad (2.248)
$$

Equation (2.248) indicates that the maximum PGA follows an Extreme Type II distribution.

2.2.8.3 Weibull Distribution

The *Extreme Type III distribution* is closely related to the *Weibull distribution*. For a random variable X following a Weibull distribution, its two-parameter CDF takes the form of

$$F_X(x) = 1 - \exp\left[-\left(\frac{x}{u}\right)^{\alpha}\right], \quad x \geq 0 \qquad (2.249)$$

in which u and α are the scale and shape parameters, respectively. With this, the PDF of X is

$$f_X(x) = \frac{\alpha}{u}\left(\frac{x}{u}\right)^{\alpha-1}\exp\left[-\left(\frac{x}{u}\right)^{\alpha}\right], \quad x \geq 0 \qquad (2.250)$$

The nth moment of X is obtained as

$$
\begin{aligned}
\mathbb{E}(X^n) &= \int_0^\infty x^n \frac{\alpha}{u}\left(\frac{x}{u}\right)^{\alpha-1}\exp\left[-\left(\frac{x}{u}\right)^{\alpha}\right]dx \\
&\underset{(x/u)^\alpha=t}{=\!=\!=\!=} \frac{\alpha}{u}\int_0^\infty u^n t^{\frac{n}{\alpha}} \cdot t^{\frac{\alpha-1}{\alpha}}\exp(-t)\frac{u}{\alpha}t^{\frac{1}{\alpha}-1}dt \qquad (2.251) \\
&= u^n \int_0^\infty t^{1+\frac{n}{\alpha}-1}\exp(-t)dt = u^n\Gamma\left(1+\frac{n}{\alpha}\right)
\end{aligned}
$$

Thus, the mean and variance of X are given by

$$\mathbb{E}(X) = u\cdot\Gamma\left(1+\frac{1}{\alpha}\right), \quad \mathbb{V}(X) = u^2\cdot\left[\Gamma\left(1+\frac{2}{\alpha}\right)-\Gamma^2\left(1+\frac{1}{\alpha}\right)\right] \quad (2.252)$$

Remark

The Weibull distribution is by nature the Extreme Type III distribution for minimum values. This is briefly illustrated by considering a sequence of positive variables $\{X_1, X_2, \ldots, X_n\}$, whose minima is denoted by $X_{\min}$. The CDF of $X_{\min}$ is by definition determined as follows,

$$
\begin{aligned}
F_{X_{\min}}(x) &= \mathbb{P}\left(\min_{i=1}^n\{X_i\} \leq x\right) = 1 - \mathbb{P}\left(\min_{i=1}^n\{X_i\} > x\right) \\
&= 1 - \mathbb{P}\left(\bigcap_{i=1}^n X_i > x\right) = 1 - \mathbb{P}\left(\bigcap_{i=1}^n -X_i < -x\right)
\end{aligned}
\qquad (2.253)
$$

Suppose that the maxima of the sequence $\{-X_i\}$ follows an Extreme Type III distribution whose CDF takes a standard form of $\exp(-(-x)^\alpha)$ (c.f. Eq. (2.230)). With this,

$$F_{X_{\min}}(x) = 1 - \exp(-(-(-x))^\alpha) = 1 - \exp(-x^\alpha) \qquad (2.254)$$

As such, it can be seen that $X_{\min}$ follows a Weibull distribution.

Example 2.34

For a Weibull random variable X whose PDF is as in Eq. (2.250), derive its MGF.

Solution

First, the nth moment of X is estimated according to Eq. (2.251). With this,

$$\psi_X(\tau) = \mathbb{E}\left(e^{X\tau}\right) = \mathbb{E}\left(\sum_{i=0}^{\infty} \frac{(X\tau)^i}{i!}\right) = \sum_{i=0}^{\infty} \frac{\tau^i}{i!}\mathbb{E}\left(X^i\right) = \sum_{i=0}^{\infty} \frac{(u\tau)^i}{i!}\Gamma\left(1 + \frac{i}{\alpha}\right)$$

(2.255)

Example 2.35

For a non-negative random variable X which follows a Fréchet distribution, show that $\frac{1}{X}$ follows a Weibull distribution.

Solution

Let $Y = \frac{1}{X}$. Since X is a Fréchet random variable, its CDF takes the form of $\mathbb{P}(X \leq x) = F_X(x) = \exp\left[-\left(\frac{x}{\varepsilon}\right)^{-k}\right]$ (c.f. Eq. (2.239)). Thus, the CDF of Y is

$$F_Y(y) = \mathbb{P}(Y \leq y) = \mathbb{P}\left(\frac{1}{X} \leq y\right) = \mathbb{P}\left(X \geq \frac{1}{y}\right) = 1 - \mathbb{P}\left(X \leq \frac{1}{y}\right)$$

$$= 1 - \exp\left[-(y\varepsilon)^k\right] = 1 - \exp\left[-\left(\frac{y}{\frac{1}{\varepsilon}}\right)^k\right]$$

(2.256)

Thus, $Y = \frac{1}{X}$ follows a Weibull distribution with a shape parameter of k and a scale parameter of $\frac{1}{\varepsilon}$.

2.3 Joint Probabilistic Behaviour of Random Variables

Recall that in Sect. 2.1.5, we have discussed the joint CDF of two (or multiple) independent variables. In this section, the joint probabilistic behaviour of dependent variables will be addressed.

2.3.1 Joint Distribution of Dependent Variables

For two random variables X and Y, their *covariance*, $\mathbb{C}(X, Y)$, is defined by

$$\mathbb{C}(X, Y) = \mathbb{E}[(X - \mathbb{E}(X))(Y - \mathbb{E}(Y))] = \mathbb{E}(XY) - \mathbb{E}(X) \cdot \mathbb{E}(Y) \quad (2.257)$$

Clearly, if X and Y are independent, $\mathbb{E}(XY) - \mathbb{E}(X) \cdot \mathbb{E}(Y) = 0$ (c.f. Example 2.15) and thus $\mathbb{C}(X, Y) = 0$. We call X and Y *uncorrelated random variables* when $\mathbb{C}(X, Y) = 0$.

Remark

It is emphasized that two random variables being uncorrelated does not necessarily mean that they are *independent*. A counterexample is that, if X is a uniform variable in $\left[-\frac{1}{2}, \frac{1}{2}\right]$, then the covariance between $Y = X^2$ and $Z = \sin(2\pi X)$ is zero, by noting that $\mathbb{E}(Z) = \int_{-\frac{1}{2}}^{\frac{1}{2}} \sin(2\pi x)dx = 0$, and thus

$$\mathbb{E}(YZ) - \mathbb{E}(Y) \cdot \mathbb{E}(Z) = \int_{-\frac{1}{2}}^{\frac{1}{2}} x^2 \sin(2\pi x)dx = 0 \qquad (2.258)$$

However, Y and Z are not independent, by examining, for example, that $\mathbb{P}\left(Y \leq \frac{1}{9}\right) = \mathbb{P}\left(-\frac{1}{3} \leq x \leq \frac{1}{3}\right) = \frac{2}{3}$, $\mathbb{P}\left(Z \leq \frac{\sqrt{2}}{2}\right) = \mathbb{P}\left(-\frac{1}{2} \leq x \leq \frac{1}{8}\right) = \frac{5}{8}$, $\mathbb{P}\left(Y \leq \frac{1}{9} \cap Z \leq \frac{\sqrt{2}}{2}\right) = \mathbb{P}\left(-\frac{1}{3} \leq x \leq \frac{1}{8}\right) = \frac{11}{24}$, and thus $\mathbb{P}\left(Y \leq \frac{1}{9} \cap Z \leq \frac{\sqrt{2}}{2}\right) \neq \mathbb{P}\left(Y \leq \frac{1}{9}\right) \cdot \mathbb{P}\left(Z \leq \frac{\sqrt{2}}{2}\right)$.

In Eq. (2.257), assigning $X = Y$ yields $\mathbb{C}(X, X) = \mathbb{V}(X)$, implying that the covariance is an extension of the variance for different variables. It indeed provides a quantitative measure for the association between two random variables. The following properties of covariance hold.

(1) For two variables X and Y and a constant a,

$$\mathbb{C}(aX, Y) = a\mathbb{C}(X, Y) \qquad (2.259)$$

Proof

$$\mathbb{C}(aX, Y) = \mathbb{E}(aXY) - \mathbb{E}(aX) \cdot \mathbb{E}(Y) = a[\mathbb{E}(XY) - \mathbb{E}(X) \cdot \mathbb{E}(Y)] = a\mathbb{C}(X, Y) \qquad (2.260)$$

(2) For random variables $X_1, X_2, \ldots, X_m, Y_1, Y_2, \ldots, Y_n$,

$$\mathbb{C}\left(\sum_{i=1}^{m} X_i, \sum_{j=1}^{n} Y_j\right) = \sum_{i=1}^{m}\sum_{j=1}^{n} \mathbb{C}(X_i, Y_j) \qquad (2.261)$$

Proof We first show that for random variables X, Y and Z, $\mathbb{C}(X + Y, Z) = \mathbb{C}(X, Z) + \mathbb{C}(Y, Z)$. This is because

$$\mathbb{C}(X + Y, Z) = \mathbb{E}[(X + Y)Z] - \mathbb{E}(X + Y)\mathbb{E}(Z)$$
$$= \mathbb{E}(XZ) + \mathbb{E}(YZ) - \mathbb{E}(X)\mathbb{E}(Z) - \mathbb{E}(Y)\mathbb{E}(Z) = \mathbb{C}(X, Z) + \mathbb{C}(Y, Z) \qquad (2.262)$$

Thus,

$$\mathbb{C}\left(\sum_{i=1}^{m} X_i, \sum_{j=1}^{n} Y_j\right) = \sum_{i=1}^{m} \mathbb{C}\left(X_i, \sum_{j=1}^{n} Y_j\right) = \sum_{i=1}^{m}\left[\sum_{j=1}^{n} \mathbb{C}(X_i, Y_j)\right] = \sum_{i=1}^{m}\sum_{j=1}^{n} \mathbb{C}(X_i, Y_j)$$

$$(2.263)$$

(3) For random variables $X_1, X_2, \ldots, X_n$,

$$\mathbb{V}\left(\sum_{i=1}^{n} X_i\right) = \sum_{i=1}^{n} \mathbb{V}(X_i) + 2\sum_{i=1}^{n}\sum_{j<i} \mathbb{C}(X_i, X_j) \qquad (2.264)$$

Proof According to Eq. (2.261),

$$\mathbb{V}\left(\sum_{i=1}^{n} X_i\right) = \mathbb{C}\left(\sum_{i=1}^{n} X_i, \sum_{j=1}^{n} X_j\right) = \sum_{i=1}^{n}\sum_{j=1}^{n} \mathbb{C}(X_i, X_j)$$

$$= \sum_{i=1}^{n} \mathbb{C}(X_i, X_i) + \sum_{i=1}^{n}\sum_{j\neq i} \mathbb{C}(X_i, X_j) = \sum_{i=1}^{n} \mathbb{V}(X_i) + 2\sum_{i=1}^{n}\sum_{j<i} \mathbb{C}(X_i, X_j)$$

$$(2.265)$$

Remark

(1) With Eq. (2.261), for a sequence of statistically independent and identically distributed random variables $X_1, X_2, \ldots, X_n$,

$$\mathbb{C}\left(\frac{1}{n}\sum_{i=1}^{n} X_i, X_i - \frac{1}{n}\sum_{i=1}^{n} X_i\right) = 0 \qquad (2.266)$$

holds for $i = 1, 2, \ldots, n$. This is because

$$\mathbb{C}\left(\frac{1}{n}\sum_{i=1}^{n} X_i, X_i - \frac{1}{n}\sum_{i=1}^{n} X_i\right) = \mathbb{C}\left(\frac{1}{n}\sum_{j=1}^{n} X_j, X_i\right) - \mathbb{C}\left(\frac{1}{n}\sum_{j=1}^{n} X_j, \frac{1}{n}\sum_{j=1}^{n} X_j\right)$$

$$= \frac{1}{n}\sum_{j=1}^{n} \mathbb{C}\left(X_j, X_i\right) - \frac{1}{n^2}\sum_{j=1}^{n}\sum_{k=1}^{n} \mathbb{C}\left(X_j, X_k\right)$$

$$= \frac{1}{n}\mathbb{C}\left(X_i, X_i\right) - \frac{1}{n^2}\sum_{j=1}^{n} \mathbb{C}\left(X_j, X_j\right)$$

$$= \frac{1}{n}\mathbb{V}\left(X_i\right) - \frac{n}{n^2}\mathbb{V}\left(X_j\right) = 0$$

$$(2.267)$$

Equation (2.266) implies that the *sample mean* of $\{X_i\}$, $\frac{1}{n}\sum_{i=1}^{n} X_i$, and the difference between each X_i and the sample mean are uncorrelated.

(2) In Eq. (2.264), if each X_i is independent of others (i.e., $\mathbb{C}(X_i, X_j) = 0$ if $i \neq j$), then $\mathbb{V}\left(\sum_{i=1}^{n} X_i\right) = \sum_{i=1}^{n} \mathbb{V}(X_i)$, which is consistent with Eq. (2.112).

Example 2.36

For two continuous variables X and Y, their joint PDF is given as follows,

$$f_{X,Y}(x, y) = a(x + y)\exp(-x - y), \quad x \geq 0, y \geq 0 \tag{2.268}$$

(1) Find the value of a; (2) Find the covariance of X and Y; (3) Comment on the dependence between X and Y.

Solution

(1) Similar to Example 2.16, the integral of $f_{X,Y}(x, y)$ on $(-\infty, \infty) \times (-\infty, \infty)$ equals 1, with which

$$\int_{-\infty}^{\infty}\int_{-\infty}^{\infty} f_{X,Y}(x, y) = \int_{0}^{\infty}\int_{0}^{\infty} a(x + y)\exp(-x - y) = 2a = 1 \Rightarrow a = \frac{1}{2} \tag{2.269}$$

Thus, the joint PDF of X and Y is

$$f_{X,Y}(x, y) = \frac{1}{2}(x + y)\exp(-x - y), \quad x \geq 0, y \geq 0 \tag{2.270}$$

(2) To calculate the covariance between X and Y, we first find the PDFs of X and Y. According to Eqs. (2.90) and (2.91),

$$f_X(x) = \int_{-\infty}^{\infty} f_{X,Y}(x, y)\mathrm{d}y = \int_{0}^{\infty} \frac{1}{2}(x + y)\exp(-x - y)\mathrm{d}y = \frac{1}{2}(x + 1)\exp(-x)$$
$$f_Y(y) = \int_{-\infty}^{\infty} f_{X,Y}(x, y)\mathrm{d}x = \int_{0}^{\infty} \frac{1}{2}(x + y)\exp(-x - y)\mathrm{d}x = \frac{1}{2}(y + 1)\exp(-y)$$
$$\tag{2.271}$$

Thus,

$$\mathbb{E}(X) = \int_{0}^{\infty} \frac{x}{2}(x + 1)\exp(-x)\mathrm{d}x = \frac{3}{2}, \quad \mathbb{E}(Y) = \mathbb{E}(X) = \frac{3}{2}$$
$$\mathbb{E}(XY) = \int_{0}^{\infty}\int_{0}^{\infty} \frac{xy}{2}(x + y)\exp(-x - y)\mathrm{d}x\mathrm{d}y = 2 \tag{2.272}$$

According to Eq. (2.257), $\mathbb{C}(X, Y) = \mathbb{E}(XY) - \mathbb{E}(X)\mathbb{E}(Y) = 2 - \frac{9}{4} = -\frac{1}{4}$.
(3) Since $\mathbb{E}(XY) - \mathbb{E}(X)\mathbb{E}(Y) \neq 0$, X and Y are not independent.

A normalized form of the covariance between two random variables X and Y is called *linear correlation coefficient* (or the *Pearson's correlation coefficient*), denoted by $\rho(X, Y)$, which is calculated by

$$\rho(X, Y) = \frac{\mathbb{C}(X, Y)}{\sqrt{\mathbb{V}(X) \cdot \mathbb{V}(Y)}} \tag{2.273}$$

It can be shown that $\rho(X, Y)$ strictly varies within $[-1, 1]$ (see Example 2.37). The cases of $\rho(X, Y) > 0, = 0$ and < 0 correspond to X and Y being positively correlated, uncorrelated, and negatively correlated, respectively.

Example 2.37

For two random variables X and Y and their correlation coefficient $\rho(X, Y)$ defined in Eq. (2.273), show that $\rho(X, Y) \in [-1, 1]$.

Solution

Consider function $f(t) = \mathbb{E}\left\{([X - \mathbb{E}(X)]t + [Y - \mathbb{E}(Y)])^2\right\}$ with respect to a real number t. Obviously, $f(t) \geq 0$, i.e.,

$$f(t) = \mathbb{E}\left\{[X - \mathbb{E}(X)]^2 t^2 + [Y - \mathbb{E}(Y)]^2 + 2[X - \mathbb{E}(X)][Y - \mathbb{E}(Y)]t\right\}$$

$$= \mathbb{E}\{[X - \mathbb{E}(X)]^2\}t^2 + \mathbb{E}\{[Y - \mathbb{E}(Y)]^2\} + 2\mathbb{E}\{[X - \mathbb{E}(X)][Y - \mathbb{E}(Y)]\}t$$

$$= \mathbb{V}(X)t^2 + 2\mathbb{C}(X, Y)t + \mathbb{V}(Y) \geq 0$$

$$(2.274)$$

Thus,

$$[2\mathbb{C}(X, Y)]^2 - 4\mathbb{V}(X)\mathbb{V}(Y) \leq 0 \Rightarrow \rho(X, Y)^2 \leq 1 \Rightarrow \rho(X, Y) \in [-1, 1] \quad (2.275)$$

which completes the proof.

Example 2.38

Recall the two random variables X and Y in Example 2.36. Find the linear correlation coefficient between the two variables.

Solution

In Example 2.36, we have obtained that $\mathbb{C}(X, Y) = -\frac{1}{4}$. Now we estimate the variances of X and Y, respectively.

$$\mathbb{E}(X^2) = \int_0^\infty \frac{x^2}{2}(x + 1)\exp(-x)dx = 4 \Rightarrow \mathbb{V}(X) = 4 - \frac{9}{4} = \frac{7}{4}$$

$$\mathbb{E}(Y^2) = 4 \Rightarrow \mathbb{V}(Y) = \frac{7}{4}$$

$$(2.276)$$

Thus, $\rho(X, Y) = \frac{\mathbb{C}(X,Y)}{\sqrt{\mathbb{V}(X)\cdot\mathbb{V}(Y)}} = \frac{-\frac{1}{4}}{\frac{7}{4}} = -\frac{1}{7}$.

For two continuous random variables Y_1 and Y_2 with an unknown joint PDF of $f_{Y_1,Y_2}(y_1, y_2)$, if they are functions of random variables X_1 and X_2 with a joint PDF of $f_{X_1,X_2}(x_1, x_2)$, i.e., $Y_1 = h(X_1, X_2)$ and $Y_2 = g(X_1, X_2)$, one can derive $f_{Y_1,Y_2}(y_1, y_2)$ based on $f_{X_1,X_2}(x_1, x_2)$ according to

$$f_{Y_1,Y_2}(y_1, y_2) = |J|^{-1} f_{X_1,X_2}(x_1, x_2) \quad (2.277)$$

where x_1 and x_2 are expressed in terms of y_1 and y_2, and J is the determinant of a Jacobian matrix taking a form of

$$J = J(x_1, x_2) = \begin{vmatrix} \frac{\partial y_1}{\partial x_1} & \frac{\partial y_1}{\partial x_2} \\ \frac{\partial y_2}{\partial x_1} & \frac{\partial y_2}{\partial x_2} \end{vmatrix} \qquad (2.278)$$

Equation (2.277) holds based on the following two conditions: (1) x_1 and x_2 can be uniquely determined given y_1 and y_2 by jointly solving the equations $y_1 - h(x_1, x_2) = 0$ and $y_2 - g(x_1, x_2) = 0$; (2) the functions h and g are partially differentiable and $\frac{\partial y_1}{\partial x_1} \cdot \frac{\partial y_2}{\partial x_2} \neq \frac{\partial y_1}{\partial x_2} \cdot \frac{\partial y_2}{\partial x_1}$ for $\forall(x_1, x_2)$ (so that $J \neq 0$). The absolute value of J in Eq. (2.277) guarantees that $f_{Y_1, Y_2}(y_1, y_2)$ is always non-negative.

Example 2.39

For two independent normal variables X and Y, show that $X + Y$ is independent of $X - Y$ if both X and Y have a common variance.

Solution

Let μ_1, μ_2 denote the mean values of X and Y respectively, and σ the standard deviation of X and Y. The joint PDF of $U = X + Y$ and $V = X - Y$ is estimated according to Eq. (2.277). First, note that the determinant of the Jacobian matrix from (X, Y) to (U, V) is

$$J = \begin{vmatrix} \frac{\partial u}{\partial x} & \frac{\partial u}{\partial y} \\ \frac{\partial v}{\partial x} & \frac{\partial v}{\partial y} \end{vmatrix} = \begin{vmatrix} 1 & 1 \\ 1 & -1 \end{vmatrix} = -2 \qquad (2.279)$$

Thus,

$$
\begin{aligned}
f_{U,V}(u, v) &= \frac{1}{|J|} f_{X,Y}(x(u, v), y(u, v)) = \frac{1}{2} f_X(x(u, v)) f_Y(y(u, v)) \\
&= \frac{1}{2} \frac{1}{\sqrt{2\pi}\sigma} \exp\left\{ -\frac{\left(\frac{u+v}{2} - \mu_1\right)^2}{2\sigma^2} \right\} \frac{1}{\sqrt{2\pi}\sigma} \exp\left\{ -\frac{\left(\frac{u-v}{2} - \mu_2\right)^2}{2\sigma^2} \right\} \\
&= \frac{1}{2} \frac{1}{\sqrt{2\pi}\sigma} \cdot \frac{1}{\sqrt{2\pi}\sigma} \exp\left\{ -\frac{(u + v - 2\mu_1)^2 + (u - v - 2\mu_2)^2}{8\sigma^2} \right\} \\
&= \frac{1}{\sqrt{2\pi} \cdot \sqrt{2}\sigma} \cdot \frac{1}{\sqrt{2\pi} \cdot \sqrt{2}\sigma} \exp\left\{ -\frac{[u - (\mu_1 + \mu_2)]^2 + [v - (\mu_1 - \mu_2)]^2}{2(\sqrt{2}\sigma)^2} \right\} \\
&= \frac{1}{\sqrt{2\pi} \cdot \sqrt{2}\sigma} \exp\left\{ -\frac{[u - (\mu_1 + \mu_2)]^2}{2(\sqrt{2}\sigma)^2} \right\} \cdot \frac{1}{\sqrt{2\pi} \cdot \sqrt{2}\sigma} \exp\left\{ -\frac{[v - (\mu_1 - \mu_2)]^2}{2(\sqrt{2}\sigma)^2} \right\}
\end{aligned}
$$
$$(2.280)$$

which implies that (1) U follows a normal distribution with a mean value of $\mu_1 + \mu_2$ and a standard deviation of $\sqrt{2}\sigma$, (2) V also follows a normal distribution with a mean value of $\mu_1 - \mu_2$ and a standard deviation of $\sqrt{2}\sigma$, and (3) $U + V$ and $U - V$ are independent of each other.

Example 2.40

Recall that in Fig. 2.5, it was claimed that $X = R \cos \Theta$ and $Y = R \sin \Theta$ are statistically independent and identically normally distributed, with a mean value of 0 and a standard deviation of σ. Prove this statement.

Solution

Consider the joint PDF of X and Y, $f_{X,Y}(x, y)$, which is derived by using $X = \sqrt{L} \cos \Theta$ and $Y = \sqrt{L} \sin \Theta$, where $L = R^2$. The determinant of the Jacobian matrix of the transformation from (X, Y) to (L, Θ) is given by

$$J = \begin{vmatrix} \dfrac{\partial x}{\partial l} & \dfrac{\partial x}{\partial \theta} \\ \dfrac{\partial y}{\partial l} & \dfrac{\partial y}{\partial \theta} \end{vmatrix} = \begin{vmatrix} \dfrac{\cos \theta}{2\sqrt{l}} & -\sqrt{l} \sin \theta \\ \dfrac{\sin \theta}{2\sqrt{l}} & \sqrt{l} \cos \theta \end{vmatrix} = \frac{1}{2} \qquad (2.281)$$

With this,

$$f_{X,Y}(x, y) = \frac{1}{|J|} f_{L,\Theta}(l(x, y), \theta(x, y)) = 2 f_L(l(x, y)) \cdot f_\Theta(\theta(x, y))$$

$$= 2 \cdot \frac{1}{2\sigma^2} \exp\left(-\frac{l(x, y)}{2\sigma^2}\right) \cdot \frac{1}{2\pi} = \frac{1}{\left(\sqrt{2\pi}\sigma\right)^2} \exp\left(-\frac{x^2 + y^2}{2\sigma^2}\right)$$

$$= \frac{1}{\sqrt{2\pi}\sigma} \exp\left(-\frac{x^2}{2\sigma^2}\right) \cdot \frac{1}{\sqrt{2\pi}\sigma} \exp\left(-\frac{y^2}{2\sigma^2}\right)$$

$$(2.282)$$

Thus, it can be seen that X and Y are independent and normally distributed with a standard deviation of σ.

Example 2.41

Recall that in Sect. 2.2.7, we have claimed that for two independent Gamma variables X_1 (with shape and scale parameters of a_1 and b) and X_2 (with a_2 and b) (c.f. Eq. (2.221)), $X = \frac{X_1}{X_1 + X_2}$ follows a Beta distribution and is independent of $X_1 + X_2$. Prove this statement. *Hint. To find the PDF of X, one may let $Y = X_1 + X_2$, and calculate the joint PDF of X and Y first.*

Solution

We find the PDF of X by introducing $Y = X_1 + X_2$ and calculating the joint PDF of X and Y. As before, the determinant of the Jacobian matrix of the transformation from (X_1, X_2) to (X, Y) is given by

$$J = \begin{vmatrix} \dfrac{\partial x}{\partial x_1} & \dfrac{\partial x}{\partial x_2} \\ \dfrac{\partial y}{\partial x_1} & \dfrac{\partial y}{\partial x_2} \end{vmatrix} = \begin{vmatrix} \dfrac{x_2}{(x_1 + x_2)^2} & \dfrac{-x_1}{(x_1 + x_2)^2} \\ 1 & 1 \end{vmatrix} = \frac{1}{x_1 + x_2} \qquad (2.283)$$

Also, given x and y, x_1 and x_2 are uniquely determined by $x_1 = xy$ and $x_2 = y - xy$. Thus, the joint PDF of X and Y is obtained as

$$
\begin{aligned}
f_{X,Y}(x, y) &= \frac{1}{|J|} f_{X_1, X_2}(x_1(x, y), x_2(x, y)) = (x_1 + x_2) \cdot f_{X_1}[x_1(x, y)] \cdot f_{X_2}[x_2(x, y)] \\
&= (x_1 + x_2) \cdot \frac{(x_1/b)^{a_1 - 1}}{b\Gamma(a_1)} \exp(-x_1/b) \cdot \frac{(x_2/b)^{a_2 - 1}}{b\Gamma(a_2)} \exp(-x_2/b) \\
&= y \cdot \frac{(xy)^{a_1 - 1}}{b^{a_1}\Gamma(a_1)} \exp(-xy/b) \cdot \frac{(y - xy)^{a_2 - 1}}{b^{a_2}\Gamma(a_2)} \exp(-y/b + xy/b) \\
&= \frac{\Gamma(a_1 + a_2)x^{a_1 - 1}(1 - x)^{a_2 - 1}}{\Gamma(a_1)\Gamma(a_2)} \cdot \frac{(y/b)^{a_1 + a_2 - 1}}{b\Gamma(a_1 + a_2)} \exp(-y/b)
\end{aligned}
$$

$$(2.284)$$

which indicates that (1) X follows a Beta distribution with two shape parameters of a_1 and a_2 (c.f. Eq. (2.223)), (2) Y follows a Gamma distribution as expected, and (3) X and Y are independent.

Example 2.42

For a normal variable X with a mean value of μ and a standard deviation of σ, find the PDF of $Y = \exp(-X)$.

Solution

Method 1. We first find the CDF of Y by definition:

$$
\begin{aligned}
F_Y(y) &= \mathbb{P}(Y \le y) = \mathbb{P}(\exp(-X) \le y) = \mathbb{P}(X \ge -\ln y) \\
&= 1 - \mathbb{P}(X < -\ln y) = 1 - \Phi\left(\frac{-\ln y - \mu}{\sigma}\right) = \Phi\left(\frac{\ln y + \mu}{\sigma}\right)
\end{aligned}
$$

$$(2.285)$$

Thus,

$$
f_Y(y) = F_Y'(y) = \frac{1}{y\sigma} \phi\left(\frac{\ln y + \mu}{\sigma}\right) \tag{2.286}
$$

Method 2. We consider Eq. (2.277), which is simply

$$
f_Y(y) = |J|^{-1} f_X(x) \tag{2.287}
$$

in the presence of a single random variable, where $J = \frac{dy}{dx}$. With this, $x = -\ln y$, $J = -\exp(-x) = -y$, and

$$
f_Y(y) = \frac{1}{|J|} \cdot \frac{1}{\sigma} \phi\left(\frac{x - \mu}{\sigma}\right) = \frac{1}{y\sigma} \phi\left(\frac{-\ln y - \mu}{\sigma}\right) = \frac{1}{y\sigma} \phi\left(\frac{\ln y + \mu}{\sigma}\right) \tag{2.288}
$$

We can further extend Eq. (2.277) to an n-dimensional case. Suppose that continuous random variables $Y_1, Y_2, \ldots, Y_n$ are functions of $X_1, X_2, \ldots, X_n$, denoted by $Y_i = h_i(X_1, X_2, \ldots, X_n)$ for $i = 1, 2, \ldots, n$, if (1) $x_1, x_2, \ldots, x_n$ can be uniquely solved by $y_i = h_i(x_1, x_2, \ldots, x_n)$ $(i = 1, 2, \ldots, n)$ simultaneously with known $y_1, y_2, \ldots, y_n$; (2) the functions h_i's are partially differentiable and the following determinant

$$J = J(x_1, x_2, \ldots, x_n) = \begin{vmatrix} \frac{\partial y_1}{\partial x_1} & \frac{\partial y_1}{\partial x_2} & \cdots & \frac{\partial y_1}{\partial x_n} \\ \frac{\partial y_2}{\partial x_1} & \frac{\partial y_2}{\partial x_2} & \cdots & \frac{\partial y_2}{\partial x_n} \\ \vdots & \vdots & \ddots & \vdots \\ \frac{\partial y_n}{\partial x_1} & \frac{\partial y_n}{\partial x_2} & \cdots & \frac{\partial y_n}{\partial x_n} \end{vmatrix} \qquad (2.289)$$

is non-zero (i.e., $J(x_1, x_2, \ldots, x_n) \neq 0$), then the joint PDF of $Y_1, Y_2, \ldots, Y_n$, $f_{Y_1, Y_2, \ldots, Y_n}(y_1, y_2, \ldots, y_n)$, is obtained by

$$f_{Y_1, Y_2, \ldots, Y_n}(y_1, y_2, \ldots, y_n) = |J|^{-1} f_{X_1, X_2, \ldots, X_n}(x_1, x_2, \ldots, x_n) \qquad (2.290)$$

2.3.2 Conditional Distribution

Recall that in Sect. 2.1.2, we have discussed the conditional probability of an event on the occurrence of another event, which is representative of the dependence between the two random events. The concept of conditional probability can be actually extended to random variables. In fact, different random variables may be dependent on each other and accordingly their probabilistic behaviours will be affected mutually. In many practical cases, we are interested in calculating the probability associated with one random variable conditional on the realization (specific value) of another. For example, consider two steel bars, A and B, produced by the same factory whose tensile strengths are positively correlated and identically distributed (say, following a normal distribution with a mean value of 235 MPa and a COV of 0.2). A testing shows that the strength of A is 270 MPa. This information can be used to help us form a better understanding of the strength of bar B: that is, provided the testing result of A, the probability that the strength of B being greater than 235 MPa increases to be greater than $\frac{1}{2}$ (otherwise this probability equals $\frac{1}{2}$). Clearly, the probability behaviours of the two strengths affect each other.

For a discrete random variable X with m possible values $\{x_1, x_2, \ldots, x_m\}$ and a PMF of $\{p_X(x_i)\}$, $i = 1, 2, \ldots, m$, and a discrete variable Y with n possible values $\{y_1, y_2, \ldots, y_n\}$ and a PMF of $\{p_Y(x_i)\}$, $i = 1, 2, \ldots, n$, the conditional PMF of X on $Y = y_j$ is defined as

$$p_{X|Y}(x_i|y_j) = \mathbb{P}(X = x_i|Y = y_j) = \frac{\mathbb{P}(X = x_i \cap Y = y_j)}{\mathbb{P}(Y = y_j)} = \frac{p_{X,Y}(x_i, y_j)}{p_Y(y_j)}$$
(2.291)

where $p_{X,Y}$ is the joint PMF of X and Y. With this, the conditional CDF of X on $Y = y_j$, similar to Eq. (2.45), is

$$F_{X|Y}(x|Y = y_j) = \sum_{\forall x_i \leq x} \mathbb{P}(X = x_i|Y = y_j) = \sum_{\forall x_i \leq x} p_{X|Y}(x_i|y_j)$$
(2.292)

The conditional mean value and variance of X given that $Y = y_j$ are estimated as follows (c.f. Eq. (2.62)).

$$\mathbb{E}(X|Y = y_j) = \sum_{i=1}^{m} x_i p_{X|Y}(x_i|y_j), \quad \mathbb{V}(X|Y = y_j) = \sum_{i=1}^{m} [x_i - \mathbb{E}(X|Y = y_j)]^2 p_X(x_i|y_j)$$
(2.293)

Remark

Note that in Eq. (2.291), if X and Y are independent, then $p_{X,Y}(x_i, y_j) = p_X(x_i)p_Y(y_j)$ according to Eq. (2.101), and further, $p_{X|Y}(x_i|y_j) = \frac{p_{X,Y}(x_i,y_j)}{p_Y(y_j)} = p_X(x_i)$, which implies that the conditional PMF of X on Y equals the PMF of X itself. This observation is similar to the case of independent random events (c.f. Eq. (2.18)).

Example 2.43

For two independent Poisson variables X and Y with mean values of λ_1 and λ_2 respectively, find the conditional mean value and variance of X on $X + Y = n$, where n is a non-negative integer.

Solution

First, note that $X + Y$ is also a Poisson variable with a mean value of $\lambda_1 + \lambda_2$ (c.f. Example 2.23). Thus, according to Eq. (2.152), for $k = 0, 1, 2, \ldots,$

$$\mathbb{P}(X = k) = \frac{\lambda_1^k}{k!} \exp(-\lambda_1), \quad \mathbb{P}(Y = k) = \frac{\lambda_2^k}{k!} \exp(-\lambda_2),$$

$$\mathbb{P}(X + Y = k) = \frac{(\lambda_1 + \lambda_2)^k}{k!} \exp[-(\lambda_1 + \lambda_2)]$$
(2.294)

Now we consider the conditional PMF of X on $X + Y = n$, which is estimated by

$$\mathbb{P}(X = k | X + Y = n) = \frac{\mathbb{P}(X = k \cap X + Y = n)}{\mathbb{P}(X + Y = n)} = \frac{\mathbb{P}(X = k) \cdot \mathbb{P}(Y = n - k)}{\mathbb{P}(X + Y = n)}$$

$$= \frac{\frac{\lambda_1^k}{k!} \exp(-\lambda_1) \cdot \frac{\lambda_2^{n-k}}{(n-k)!} \exp(-\lambda_2)}{\frac{(\lambda_1 + \lambda_2)^n}{n!} \exp[-(\lambda_1 + \lambda_2)]}$$

$$= \frac{n! \lambda_1^k \cdot \lambda_2^{n-k}}{k!(n-k)!(\lambda_1 + \lambda_2)^n} = \frac{n!}{k!(n-k)!} \left(\frac{\lambda_1}{\lambda_1 + \lambda_2} \right)^k \cdot \left(\frac{\lambda_2}{\lambda_1 + \lambda_2} \right)^{n-k}$$

$$(2.295)$$

Thus, the conditional X on $X + Y = n$ is a binomial random variable (c.f. Eq. (2.137)). The mean value and variance of X conditional on $X + Y = n$ are $\frac{n\lambda_1}{\lambda_1 + \lambda_2}$ and $\frac{n\lambda_1\lambda_2}{(\lambda_1 + \lambda_2)^2}$ respectively.

Now we extend Eq. (2.291) to the continuous case. For two continuous random variables X and Y with a joint PDF of $f_{X,Y}(x, y)$, the conditional PDF of X on $Y = y$, $f_{X|Y}(x|y)$, is estimated by

$$\lim_{dx \to 0} f_{X|Y}(x|y) dx = \lim_{dx,dy \to 0} \mathbb{P}(x \le X \le x + dx | y \le Y \le y + dy)$$

$$= \lim_{dx,dy \to 0} \frac{\mathbb{P}(x \le X \le x + dx \cap y \le Y \le y + dy)}{\mathbb{P}(y \le Y \le y + dy)}$$

$$= \lim_{dx,dy \to 0} \frac{f_{X,Y}(x, y) dx dy}{f_Y(y) dy} \qquad (2.296)$$

$$= \lim_{dx \to 0} \frac{f_{X,Y}(x, y) dx}{f_Y(y)}$$

$$\Rightarrow f_{X|Y}(x|y) = \frac{f_{X,Y}(x, y)}{f_Y(y)}$$

Similar to Eqs. (2.64) and (2.293), the conditional mean value and variance of X on $Y = y$ are respectively

$$\mathbb{E}(X|Y = y) = \int_{-\infty}^{\infty} x f_{X|Y}(x|y) dx, \quad \mathbb{V}(X|Y = y) = \int_{-\infty}^{\infty} [x - \mathbb{E}(X|Y = y)]^2 f_{X|Y}(x|y) dx$$

$$(2.297)$$

Remark

With Eqs. (2.293) and (2.297), similar to Eqs. (2.68) and (2.69), $\mathbb{V}(X|Y = y) = \mathbb{E}(X^2|Y = y) - [\mathbb{E}(X|Y = y)]^2$. The proof for the continuous case is presented as follows.

$$\mathbb{V}(X|Y = y) = \int_{-\infty}^{\infty} [x - \mathbb{E}(X|Y = y)]^2 f_{X|Y}(x|y)\mathrm{d}x$$

$$= \int_{-\infty}^{\infty} x^2 f_{X|Y}(x|y)\mathrm{d}x + [\mathbb{E}(X|Y = y)]^2$$

$$\int_{-\infty}^{\infty} f_{X|Y}(x|y)\mathrm{d}x - 2\mathbb{E}(X|Y = y) \int_{-\infty}^{\infty} x f_{X|Y}(x|y)\mathrm{d}x \qquad (2.298)$$

$$= \mathbb{E}(X^2|Y = y) + [\mathbb{E}(X|Y = y)]^2 - 2[\mathbb{E}(X|Y = y)]^2$$

$$= \mathbb{E}(X^2|Y = y) - [\mathbb{E}(X|Y = y)]^2$$

Example 2.44

Recall the two random variables X and Y in Example 2.36. Calculate the conditional mean value and variance of X on $Y = y$.

Solution

According to Eq. (2.296),

$$f_{X|Y}(x|y) = \frac{f_{X,Y}(x, y)}{f_Y(y)} = \frac{\frac{1}{2}(x + y)\exp(-x - y)}{\frac{1}{2}(y + 1)\exp(-y)} = \frac{(x + y)\exp(-x)}{y + 1} \qquad (2.299)$$

Thus,

$$\mathbb{E}(X|Y = y) = \int_{-\infty}^{\infty} x f_{X|Y}(x|y)\mathrm{d}x = \int_{0}^{\infty} \frac{x(x + y)\exp(-x)}{y + 1}\mathrm{d}x = \frac{2 + y}{1 + y} \qquad (2.300)$$

Further, since

$$\mathbb{E}(X^2|Y = y) = \int_{-\infty}^{\infty} x^2 f_{X|Y}(x|y)\mathrm{d}x = \int_{0}^{\infty} \frac{x^2(x + y)\exp(-x)}{y + 1}\mathrm{d}x = \frac{2(3 + y)}{1 + y}$$
$$(2.301)$$

the conditional variance of X on $Y = y$ is estimated by

$$\mathbb{V}(X|Y = y) = \mathbb{E}(X^2|Y = y) - [\mathbb{E}(X|Y = y)]^2 = \frac{2(3 + y)}{1 + y} - \left(\frac{2 + y}{1 + y}\right)^2 = \frac{y^2 + 4y + 2}{(1 + y)^2}$$
$$(2.302)$$

Example 2.45

For two independent continuous variables X and Y with PDFs of $f_X(x)$ and $f_Y(y)$ respectively, show that the PDF of $Z = X + Y$ is

$$f_Z(z) = \int_{-\infty}^{\infty} f_X(z - y) f_Y(y)\mathrm{d}y \qquad (2.303)$$

Solution

We first consider the CDF of Z, which is estimated by

$$F_Z(z) = \mathbb{P}(X + Y \leq z) = \int_{-\infty}^{\infty} \mathbb{P}(X + y \leq z | Y = y) f_Y(y) \mathrm{d}y = \int_{-\infty}^{\infty} F_X(z - y) f_Y(y) \mathrm{d}y$$

$$(2.304)$$

where $F_X(x)$ is the CDF of X. Taking the differential form of both sides with respect to z gives

$$f_Z(z) = \frac{\mathrm{d}F_Z(z)}{\mathrm{d}z} = \int_{-\infty}^{\infty} f_X(z - y) f_Y(y) \mathrm{d}y \qquad (2.305)$$

Let $\mathbb{E}(X|Y)$ denote a function of Y which equals $\mathbb{E}(X|Y = y)$ when $Y = y$. Clearly, $\mathbb{E}(X|Y)$ itself is a random variable with uncertainty arising from Y. With the law of total probability (c.f. Eq. (2.23)), the law of total expectation or total variance holds, as presented in the following.

Theorem 2.3 (Law of total expectation and law of total variance (e.g., [5])) *For two random variables X and Y, if the mean of X, $\mathbb{E}(X)$, is defined, then*

$$\mathbb{E}(X) = \mathbb{E}(\mathbb{E}(X|Y)) \qquad (2.306)$$

and

$$\mathbb{V}(X) = \mathbb{E}[\mathbb{V}(X|Y)] + \mathbb{V}[\mathbb{E}(X|Y)] \qquad (2.307)$$

The proof of Eqs. (2.306) and (2.307) for the case where X and Y are continuous variables is presented in the following.

Proof We first show that for a scalar variable X and a random vector Y, $\mathbb{E}[X] = \mathbb{E}[\mathbb{E}(X|Y)]$. This is because

$$\mathbb{E}[\mathbb{E}(X|Y)] = \mathbb{E}\left[\int x f_{X|Y}(x|y) \mathrm{d}x\right] = \int \left[\int x f_{X|Y}(x|y) \mathrm{d}x\right] f_Y(y) \mathrm{d}y$$

$$= \int x f_{X|Y}(x|Y) f_Y(y) \mathrm{d}x \mathrm{d}y = \int \int x f_{X,Y}(x, y) \mathrm{d}x \mathrm{d}y \qquad (2.308)$$

$$= \int x f_X(x) \mathrm{d}x = \mathbb{E}(X)$$

Next, note that

$$\mathbb{E}[\mathbb{E}(X^2|Y)] = \mathbb{E}\left[\int x^2 f_{X|Y}(x|y)dx\right] = \int \left[\int x^2 f_{X|Y}(x|y)dx\right] f_Y(y)dy$$
$$= \int\int x^2 f_{X|Y}(x|y)f_Y(y)dxdy = \int\int x^2 f_{X,Y}(x,y)dxdy$$
$$= \int x^2 f_X(x)dx = \mathbb{E}(X^2)$$

(2.309)

Thus,

$$\mathbb{V}(X) = \mathbb{E}(X^2) - [\mathbb{E}(X)]^2 = \mathbb{E}[\mathbb{E}(X^2|Y)] - [\mathbb{E}(X)]^2$$
$$= \mathbb{E}\{\mathbb{V}(X|Y) + [\mathbb{E}(X|Y)]^2\} - [\mathbb{E}(X)]^2$$
$$= \mathbb{E}[\mathbb{V}(X|Y)] + \mathbb{E}\{[\mathbb{E}(X|Y)]^2\} - \{\mathbb{E}[\mathbb{E}(X|Y)]\}^2$$
$$= \mathbb{E}[\mathbb{V}(X|Y)] + \mathbb{V}[\mathbb{E}(X|Y)]$$

(2.310)

which completes the proof.   □

Example 2.46

For two random variables X and Y, show that $\mathbb{C}(X, Y) = \mathbb{C}(X, \mathbb{E}(Y|X))$.

Solution

We assume in the following that X and Y are two continuous random variables. The proof for the case of discrete X and Y would be in a similar manner.

Expanding $\mathbb{C}(X, \mathbb{E}(Y|X))$ gives

$$\mathbb{C}(X, \mathbb{E}(Y|X)) = \mathbb{E}[X \cdot \mathbb{E}(Y|X)] - \mathbb{E}(X) \cdot \mathbb{E}[\mathbb{E}(Y|X)]$$
$$= \int \left[x \cdot \int y f_{Y|X}(y|x)dy\right] f_X(x)dx - \mathbb{E}(X) \cdot \mathbb{E}(Y)$$
$$= \int\int xy f_{Y|X}(y|x) f_X(x)dydx - \mathbb{E}(X) \cdot \mathbb{E}(Y)$$
$$= \int\int xy f_{X,Y}(x,y)dydx - \mathbb{E}(X) \cdot \mathbb{E}(Y)$$
$$= \mathbb{E}(XY) - \mathbb{E}(X) \cdot \mathbb{E}(Y) = \mathbb{C}(X, Y)$$

(2.311)

which completes the proof.

Example 2.47

Recall Example 2.31, where the cumulative damage within a reference period of T years, denoted by D_c, was discussed. Estimate the mean value and variance of D_c.

Solution

Method 1. By definition,

$$D_c = \sum_{i=0}^{N} D_i$$

(2.312)

where N is the number of cyclone events within time period $[0, T]$, which is a Poisson random variable with a mean value of λT; $D_{c,i}$ is the cyclone damage costs associated with the ith cyclone event, which is independent of each other and identically distributed with D (i.e., the damage loss conditional on the occurrence of one cyclone event). Also, it has been given that D has a mean value of μ_D and a standard deviation of σ_D. With Eq. (2.306),

$$\mathbb{E}(D_c) = \mathbb{E}[\mathbb{E}(D_c|N)] = \mathbb{E}[N \cdot \mathbb{E}(D_i)] = \mathbb{E}(N\mu_D) = \lambda T \mu_D \tag{2.313}$$

Next, with Eq. (2.307),

$$\begin{aligned} \mathbb{V}(D_c) &= \mathbb{E}[\mathbb{V}(D_c|N)] + \mathbb{V}[\mathbb{E}(D_c|N)] = \mathbb{E}[N\mathbb{V}(D_i)] + \mathbb{V}[N\mathbb{E}(D_i)] \\ &= \sigma_D^2 \mathbb{E}(N) + \mu_D^2 \mathbb{V}(N) = (\mu_D^2 + \sigma_D^2)\lambda T \end{aligned} \tag{2.314}$$

Method 2. Note that in Example 2.31, D_c was assessed by $D_c = \sum_{i=1}^n \widetilde{D}_i$, which is by nature equivalent to Eq. (2.312). With this regard, the mean value and variance of D_c are estimated by considering the summation of each $\widetilde{D}_i$. Since $\widetilde{D}_i = B_i \cdot D$,

$$\mathbb{E}(\widetilde{D}_i) = \mathbb{E}[\mathbb{E}(\widetilde{D}_i|B_i)] = \mathbb{E}[B_i\mu_D] = \mu_D\mathbb{P}(B_i = 1) = \mu_D\lambda T/n \tag{2.315}$$

and

$$\begin{aligned} \mathbb{V}(\widetilde{D}_i) &= \mathbb{E}[\mathbb{V}(\widetilde{D}_i|B_i)] + \mathbb{V}[\mathbb{E}(\widetilde{D}_i|B_i)] = \mathbb{E}(B_i^2\sigma_D^2) + \mathbb{V}(B_i\mu_D) \\ &= \sigma_D^2\mathbb{E}(B_i^2) + \mu_D^2\mathbb{V}(B_i) = (\mu_D^2 + \sigma_D^2)\frac{\lambda T}{n} - \mu_D^2\left(\frac{\lambda T}{n}\right)^2 \end{aligned} \tag{2.316}$$

Thus, by noting that n is large enough, as well as the independence between each $\widetilde{D}_i$, it follows,

$$\mathbb{E}(D_c) = \sum_{i=1}^n \mathbb{E}(\widetilde{D}_i) = n \cdot (\mu_D\lambda T/n) = \mu_D\lambda T \tag{2.317}$$

and

$$\mathbb{V}(D_c) = \sum_{i=1}^n \mathbb{V}(\widetilde{D}_i) = n \cdot \left((\mu_D^2 + \sigma_D^2)\frac{\lambda T}{n} - \mu_D^2\left(\frac{\lambda T}{n}\right)^2\right) = (\mu_D^2 + \sigma_D^2)\lambda T \tag{2.318}$$

Example 2.48

Recall Example 2.21. The building will fail if the wind speed exceeds 50 m/s, and the corresponding damage loss, D, is random with a mean value of μ and a standard deviation of σ. Estimate the mean value and variance of the wind-induced damage loss in terms of present value, D_p, with an annual discount rate of r. Evaluate your result with $\mu = 1$, $\sigma = 0.3$ and $r = 5\%$.

Solution

Let random variable X denote the year in which the wind speed first exceeds 50 m/s, which is a geometric variable with a PMF of $\mathbb{P}(X = k) = (1 - p)^{k-1} p$ for $k = 1, 2, \ldots$ (c.f. Eq. (2.138)). The present value of the damage loss is

$$D_p = \frac{D}{(1 + r)^X} \tag{2.319}$$

With Eq. (2.306),

$$\mathbb{E}(D_p) = \mathbb{E}[\mathbb{E}(D_p|X)] = \mathbb{E}\left[\frac{\mu}{(1 + r)^X}\right] = \mu \sum_{k=1}^{\infty} (1 + r)^{-k} (1 - p)^{k-1} p$$

$$= \frac{\mu}{1 + r} \sum_{k=1}^{\infty} p \left(\frac{1 - p}{1 + r}\right)^{k-1} = \frac{\mu p}{r + p} \tag{2.320}$$

Next, with Eq. (2.307),

$$\mathbb{V}(D_p) = \mathbb{E}[\mathbb{V}(D_p|X)] + \mathbb{V}[\mathbb{E}(D_p|X)] = \mathbb{E}\left(\frac{\sigma^2}{(1 + r)^{2X}}\right) + \mathbb{V}\left(\frac{\mu}{(1 + r)^X}\right)$$

$$= \sigma^2 \mathbb{E}\left(\frac{1}{(1 + r)^{2X}}\right) + \mu^2 \mathbb{V}\left(\frac{1}{(1 + r)^X}\right)$$

$$= (\mu^2 + \sigma^2)\mathbb{E}\left((1 + r)^{-2X}\right) - \mu^2 \mathbb{E}^2\left((1 + r)^{-X}\right)$$

$$= \frac{\mu^2 + \sigma^2}{(1 + r)^2} \sum_{k=1}^{\infty} p \left(\frac{1 - p}{(1 + r)^2}\right)^{k-1} - \left(\frac{\mu p}{r + p}\right)^2 = \frac{p(\mu^2 + \sigma^2)}{(1 + r)^2 - 1 + p} - \left(\frac{\mu p}{r + p}\right)^2 \tag{2.321}$$

Now we estimate the mean value and variance of D_p with some specific values. It is given that the annual probability of wind speed exceeding 50 m/s is $p = 0.1$. Also, $\mu = 1$, $\sigma = 0.3$ and $r = 5\%$. Thus, $\mathbb{E}(D_p) = 0.6667$ and $\mathbb{V}(D_p) = 0.0938$.

2.3.3 Constructing Joint CDF with Copula Function

For two random variables, their joint CDF equals the product of the marginal CDFs, provided that the two variables are independent (c.f. Eq. (2.100)). However, in the presence of variable dependence, one cannot simply construct their joint CDF by multiplying the two CDFs, as the joint CDF is highly related to the dependence structure of the two variables. With this regard, the *copula function* provides a powerful tool for constructing the joint CDF of dependent random variables, provided that the marginal CDFs as well as the variable dependence are known [23].

For two random variables X and Y with marginal CDFs of $F_X(x)$ and $F_Y(y)$ respectively and a joint CDF of $F_{X,Y}(x, y)$, it is easy to see that F_X, F_Y and $F_{X,Y}$ all vary within $[0, 1]$. Provided that $X = x$, $Y = y$, where $x, y \in (-\infty, \infty)$, we can uniquely determine $(F_X(x), F_Y(y)) \in [0, 1] \times [0, 1]$ and $F_{X,Y}(x, y) \in [0, 1]$ respectively. This correspondence motivates us to establish a function that links the CDF pair $(F_X(x), F_Y(y))$ to the joint CDF $F_{X,Y}(x, y)$. Such a function is called a copula function.

The word "copula" is a Latin noun meaning "link, tie, bond", and a copula function is used to "couple" multivariate distribution functions. By definition, a copula function, in its two-dimensional version, is such a function from $[0, 1] \times [0, 1]$ to $[0, 1]$ that satisfies the following two properties:

(1) For $u, v \in [0, 1]$, $C(u, 0) = C(0, v) = 0$ and $C(u, 1) = u$, $C(1, v) = v$;
(2) For $u_1, u_2, v_1, v_2 \in [0, 1]$, if $u_1 \leq u_2, v_1 \leq v_2$, then

$$C(u_1, v_1) + C(u_2, v_2) \geq C(u_1, v_2) + C(u_2, v_1) \tag{2.322}$$

Property (1) is relatively straightforward, by noting that the joint CDF returns 0 if one of the marginal CDFs equals zero, and that the joint CDF becomes a marginal CDF if the other CDF equals 1 (c.f. Eq. (2.84)). Property (2) is to guarantee that $\frac{\partial C^2}{\partial u \partial v} \geq 0$ (recall that for a joint CDF, according to Eq. (2.93), $f_{X,Y}(x, y) = \frac{\partial^2}{\partial x \partial y} F_{X,Y}(x, y) \geq 0$).

Example 2.49

For a copula function $C(u, v)$, and $u_1, u_2, v_1, v_2 \in [0, 1]$, show that

$$|C(u_1, v_1) - C(u_2, v_2)| \leq |u_1 - u_2| + |v_1 - v_2| \tag{2.323}$$

Solution

First, with the triangle inequality (i.e., for three real numbers a, b, c, $|a - b| \leq |a - c| + |c - b|$), one has

$$|C(u_1, v_1) - C(u_2, v_2)| \leq |C(u_2, v_2) - C(u_1, v_2)| + |C(u_1, v_2) - C(u_1, v_1)| \tag{2.324}$$

Now, we consider $|C(u_2, v_2) - C(u_1, v_2)|$. Without loss of generality, we assume that $u_2 \geq u_1$, and construct a function $f(x) = C(u_2, x) - C(u_1, x)$ with respect to $x \in [0, 1]$. It is noticed, according to Eq. (2.322), that $f(x)$ is a non-decreasing function, because

$$C(u_1, t) + C(u_2, t + dt) \geq C(u_1, t + dt) + C(u_2, t)$$
$$\Rightarrow C(u_2, t + dt) - C(u_2, t) \geq C(u_1, t + dt) - C(u_1, t) \Rightarrow f(t + dt) \geq f(t) \tag{2.325}$$

Thus, $f(v_2) \leq f(1) = C(u_2, 1) - C(u_1, 1) = u_2 - u_1$, or equivalently, $|C(u_2, v_2) - C(u_1, v_2)| \leq |u_2 - u_1|$. Similarly, $|C(u_1, v_2) - C(u_1, v_1)| \leq |v_2 - v_1|$ by noting that the function $g(x) = C(x, v_2) - C(x, v_1)$ is non-decreasing if $v_2 \geq v_1$. This, finally, leads to $|C(u_1, v_1) - C(u_2, v_2)| \leq |u_1 - u_2| + |v_1 - v_2|$, which completes the proof.

Regarding the use of copula function to construct a joint CDF, we have the following important theorem.

Theorem 2.4 (Sklar's Theorem ([30])) *Let $F_{X,Y}(x, y)$ be a joint CDF of X and Y with marginal CDFs of $F_X(x)$ and $F_Y(y)$ respectively. Then there exists a copula function $C(u, v)$ such that for $\forall x, y \in (-\infty, \infty)$,*

$$F_{X,Y}(x, y) = C[F_X(x), F_Y(y)] \tag{2.326}$$

If $F_X(x)$ and $F_Y(y)$ are continuous, then $C(u, v)$ is unique.

The Sklar's Theorem guarantees that one can construct the joint CDF of two (dependent) random variables by properly selecting a copula function first. Moreover, with Eq. (2.100), it can be shown that for two random variables X and Y with a copula function of C, X and Y are independent if and only if $C(u, v) = uv$. In such a case, C is said to be an *independence copula*.

Example 2.50

For two random variables X and Y with a joint CDF of

$$F_{X,Y}(x, y) = \frac{1}{1 + \exp(-x) + \exp(-y) + (1 - \theta)\exp(-x - y)} \tag{2.327}$$

where θ is a parameter, show that (1) when $\theta = 1$, the copula function for X and Y is $C(u, v) = \frac{uv}{u+v-uv}$; (2) when $\theta = 0$, X and Y are independent.

Solution

We first obtain the marginal CDFs for X and Y according to Eq. (2.84) as follows.

$$F_X(x) = F_{X,Y}(x, \infty) = \frac{1}{1 + \exp(-x)} \tag{2.328}$$

$$F_Y(y) = F_{X,Y}(\infty, y) = \frac{1}{1 + \exp(-y)} \tag{2.329}$$

(1) When $\theta = 1$,

$$F_{X,Y}(x, y) = C(F_X(x), F_Y(y)) = \frac{1}{1 + \exp(-x) + \exp(-y)}$$

$$= \frac{1}{1 + \left(\frac{1}{F_X(x)} - 1\right) + \left(\frac{1}{F_Y(y)} - 1\right)} \tag{2.330}$$

Thus,

$$C(u, v) = \frac{1}{1 + \left(\frac{1}{u} - 1\right) + \left(\frac{1}{v} - 1\right)} = \frac{uv}{u + v - uv} \tag{2.331}$$

(2) When $\theta = 0$,

$$F_{X,Y}(x, y) = C(F_X(x), F_Y(y)) = \frac{1}{1 + \exp(-x) + \exp(-y) + \exp(-x - y)}$$

$$= \frac{1}{[1 + \exp(-x)] \cdot [1 + \exp(-y)]} = F_X(x) \cdot F_Y(y)$$

(2.332)

Thus, X and Y are independent (correspondingly, $C(u, v) = uv$).

Example 2.51

Fréchet–Hoeffding bounds. Define two copula functions $C_{lb}(u, v) = \max(u + v - 1, 0)$ and $C_{ub}(u, v) = \min(u, v)$. For any copula function $C(u, v)$ with $u, v \in [0, 1]$, show that

$$C_{lb} \le C(u, v) \le C_{ub}$$

(2.333)

Solution

Note that $C(u, v) \le C(u, 1) = u$ and $C(u, v) \le C(1, v) = v$, which yields $C(u, v) \le \min(u, v)$. Similarly, $C(u, v) \ge C(u, 0) = 0$. Moreover, in Eq. (2.322), assigning $u_1 = u, u_2 = 1, v_1 = v, v_2 = 1$ gives $C(u, v) + 1 \ge u + v$ or equivalently, $C(u, v) \ge u + v - 1$. Thus, $C_{lb} \le C(u, v)$.

Remark

With Example 2.49, as well as the Sklar's Theorem, we have the following corollary. Let $F_{X,Y}(x, y)$ be a joint CDF of X and Y with marginal CDFs of $F_X(x)$ and $F_Y(y)$ respectively. For $x_1, x_2, y_1, y_2 \in (-\infty, \infty)$, we have

$$|F_{X,Y}(x_1, y_1) - F_{X,Y}(x_2, y_2)| \le |F_X(x_1) - F_X(x_2)| + |F_Y(y_1) - F_Y(y_2)| \quad (2.334)$$

Also, with Example 2.49, one has

$$0 \le \frac{\partial C(u, v)}{\partial u} \le 1, \quad 0 \le \frac{\partial C(u, v)}{\partial v} \le 1$$

(2.335)

The first part of Eq. (2.335) is proven as follows: first, note that the function $f(x) = C(u + du, x) - C(u, x)$ is non-decreasing with respect to x (c.f. Example 2.49) and thus $f(v) = C(u + du, v) - C(u, v) \ge f(0) = 0 \Rightarrow \frac{\partial C(u,v)}{\partial u} \ge 0$; next, assign $v_1 = v_2 = v$, $u_1 = u, u_2 = u + du$ in Eq. (2.323), which yields $|C(u, v) - C(u + du, v)| \le |u - (u + du)| \Rightarrow \frac{\partial C(u,v)}{\partial u} \le 1$. The proof for the second part of Eq. (2.335) would be in a similar manner.

Furthermore, based on the Fréchet–Hoeffding bounds (c.f. Eq. (2.333)), the Sklar's Theorem guarantees that for $x, y \in (-\infty, \infty)$,

$$\max(F_X(x) + F_Y(y) - 1, 0) \le F_{X,Y}(x, y) \le \min(F_X(x), F_Y(y)) \quad (2.336)$$

Example 2.52

For a sequence of statistically independent and identically distributed random variables $X_1, X_2, \ldots, X_n$ with a common CDF of $F_X(x)$, let $X_{\min}$ and $X_{\max}$ denote the minimum and maximum value of $\{X_i\}$ respectively [29]. Find the copula function for $X_{\min}$ and $X_{\max}$, $C_{mm}(u, v)$, and comment on the dependence between $X_{\min}$ and $X_{\max}$ when $n \to \infty$.

Solution

Before proceeding, we have the following auxiliary statement. For two random variables X and Y with a copula $C(u, v)$, the copula of $-X$ and Y is $C_1(u, v) = v - C(1 - u, v)$. The proof is as follows. Since $F_{-X}(x) = \mathbb{P}(-X \le x) = 1 - \mathbb{P}(X \le -x) = 1 - F_X(-x)$, where $F_{-X}(x)$ is the CDF of $-X$, one has

$$C_1(F_{-X}(x), F_Y(y)) = \mathbb{P}(-X \le x \cap Y \le y) = \mathbb{P}(Y \le y) - \mathbb{P}(X \le -x \cap Y \le y)$$
$$= F_Y(y) - C(F_X(-x), F_Y(y)) = F_Y(y) - C(1 - F_X(-x), F_Y(y)) \quad (2.337)$$

Thus, $C_1(u, v) = v - C(1 - u, v)$.

Let $F_X(x)$, $F_{X_{\min}}(x)$ and $F_{X_{\max}}(x)$ denote the CDFs of X_i, $X_{\min}$ and $X_{\max}$ respectively. Since each X_i is independent of each other, $F_{X_{\max}}(x) = F_X^n(x)$ and $F_{X_{\min}}(x) = 1 - [1 - F_X(x)]^n$. Consider the copula of $-X_{\min}$ and $X_{\max}$, $C_2(u, v)$, first. Let $F_{-X_{\min}}(x)$ be the CDF of $-X_{\min}$. Note that

$$C_2(F_{-X_{\min}}(x), F_{X_{\max}}(y)) = \mathbb{P}(-X_{\min} \le x \cap X_{\max} \le y) = \mathbb{P}(X_{\min} \ge -x \cap X_{\max} \le y)$$
$$= \mathbb{P}\left[\cap_{i=1}^n (-x \le X_i \le y)\right] = \begin{cases} [F_X(y) - F_X(-x)]^n, & y \ge -x \\ 0, & y < -x \end{cases}$$
$$= [\max(F_X(y) - F_X(-x), 0)]^n$$
$$= \left\{\max\left[F_{X_{\max}}^{1/n}(y) - \left(1 - (1 - F_{X_{\min}}(-x))^{1/n}\right), 0\right]\right\}^n \quad (2.338)$$

Let $u = F_{-X_{\min}}(x) = 1 - F_{X_{\min}}(-x)$, and $v = F_{X_{\max}}(y)$. With this, $C_2(u, v) = \left\{\max\left(v^{1/n} - 1 + u^{1/n}, 0\right)\right\}^n$, and $C_{mm}(u, v) = v - C_2(1 - u, v) = v - \left\{\max\left(v^{1/n} - 1 + (1 - u)^{1/n}, 0\right)\right\}^n$.

When $n \to \infty$, $v^{1/n} = \exp\left(\frac{1}{n} \ln v\right) \approx 1 + \frac{1}{n} \ln v$, and $(1 - u)^{1/n} \approx 1 + \frac{1}{n} \ln(1 - u)$. Thus,

$$\lim_{n \to \infty} C_{mm}(u, v) = v - \lim_{n \to \infty}\left\{1 + \frac{1}{n} \ln[v(1 - u)]\right\}^n = v - v(1 - u) = uv \quad (2.339)$$

implying that $X_{\min}$ and $X_{\max}$ trend to be independent when n is large enough.

Example 2.53

For n independent and identically distributed pairs of random variables (X_1, Y_1), (X_2, Y_2) $\ldots (X_n, Y_n)$, if each pair has a common copula function $C(u, v)$, find the copula for $X_{\max} = \max\{X_1, X_2, \ldots, X_n\}$ and $Y_{\max} = \max\{Y_1, Y_2, \ldots, Y_n\}$, $C_{\max}(u, v)$,

Solution

Let $F_{X_{max}}(x)$ and $F_{Y_{max}}(y)$ denote the CDFs of X_{max} and Y_{max} respectively. Due to the independence between each X_i, $F_{X_{max}}(x) = F_X^n(x)$, where $F_X(x)$ is the CDF of X_i. Similarly, $F_{Y_{max}}(y) = F_Y^n(y)$, where $F_Y(y)$ is the CDF of Y_i. The joint CDF of X_{max} and Y_{max}, $F_{X_{max},Y_{max}}(x, y)$, equals $C_{max}(F_{X_{max}}(x), F_{Y_{max}}(y))$. On the other hand,

$$
\begin{aligned}
F_{X_{max},Y_{max}}(x, y) &= \mathbb{P}(X_{max} \le x \cap Y_{max} \le y) = \mathbb{P}\left[\cap_{i=1}^n (X_i \le x \cap Y_i \le y)\right] \\
&= [F_{X,Y}(x, y)]^n = [C(F_X(x), F_Y(y))]^n \\
&= [C(F_{X_{max}}^{1/n}(x), F_{Y_{max}}^{1/n}(y))]^n
\end{aligned}
$$

(2.340)

Thus, $C_{max}(u, v) = C^n(u^{1/n}, v^{1/n})$.

Recall that when X and Y are independent, their copula function equals $C(u, v) = uv$. Motivated by this, for the case of dependent X and Y, constructing their copula C would be easy if there exists a function $\alpha(\bullet)$ such that $\alpha(C) = \alpha(u) \cdot \alpha(v)$. Furthermore, if defining another function $\beta(\bullet) = \ln[\alpha(\bullet)]$, one has $\beta(C) = \beta(u) + \beta(v)$. In such a case, C is said to be an *Archimedean copula function*, which is an important class of copulas with an easy way to construct and some good properties. The formal definition of Archimedean copula is discussed in the following.

For a continuous and monotonically decreasing function $\varphi(x)$ from $[0, 1]$ to $[0, \infty)$ satisfying $\varphi(1) = 0$, we define its pseudo-inverse, $\varphi^{[-1]}(x)$, as follows,

$$
\varphi^{[-1]}(x) = \begin{cases} \varphi^{-1}(x), & 0 \le x \le \varphi(0) \\ 0, & x > \varphi(0) \end{cases} = \max(\varphi^{-1}(x), 0)
$$

(2.341)

With this,

$$
\varphi\left[\varphi^{[-1]}(x)\right] = \begin{cases} x, & 0 \le x \le \varphi(0) \\ \varphi(0), & x > \varphi(0) \end{cases} = \min(x, \varphi(0))
$$

(2.342)

Now we define a function

$$
C(u, v) = \varphi^{[-1]}(\varphi(u) + \varphi(v))
$$

(2.343)

It can be proven that C is a copula function if $\varphi(x)$ is convex [1], as detailed in the following. In such a case, $C(u, v)$ is said to be an Archimedean copula, and $\varphi(x)$ is the generator of C.

Remark

(1) A function $f(x)$, $x \in A$, is convex if for $\forall x_1, x_2 \in A$ and $t \in [0, t]$, $f[tx_1 + (1-t)x_2] \le tf(x_1) + (1-t)f(x_2)$ holds. For example, $f(x) = x^2$ is convex since $tf(x_1) + (1-t)f(x_2) - f[tx_1 + (1-t)x_2] = t(1-t)(x_1 - x_2)^2 \ge 0$.

(2) If $\varphi(x)$ in Eq. (2.341) is convex, then $\varphi^{[-1]}(x)$ is also convex.

Proof We will show that the function defined in Eq. (2.343) is a copula function if $\varphi(x)$ is convex. First, note that for $u \in [0, 1]$, $C(u, 0) = \varphi^{[-1]}(\varphi(u) + \varphi(0)) = 0$ and $C(u, 1) = \varphi^{[-1]}(\varphi(u) + \varphi(1)) = u$. By symmetry, $C(0, v) = 0$ and $C(1, v) = 1$ hold for $v \in [0, 1]$.

Before proceeding, we have the following auxiliary statement. For $u_1 \leq u_2, v \in [0, 1]$, the generator $\varphi(x)$ as in Eq. (2.343) satisfies

$$\varphi^{[-1]}(\varphi(u_2) + \varphi(v)) + u_1 \leq \varphi^{[-1]}(\varphi(u_1) + \varphi(v)) + u_2 \qquad (2.344)$$

The proof is as follows. For simplicity, let $x = \varphi(u_1)$, $y = \varphi(u_2)$, and $z = \varphi(v)$, with which Eq. (2.344) is equivalent to

$$\varphi^{[-1]}(y + z) + \varphi^{[-1]}(x) \leq \varphi^{[-1]}(x + z) + \varphi^{[-1]}(y) \qquad (2.345)$$

Let $t = \frac{x-y}{x-y+z}$, which varies within $[0, 1]$ since $x \geq y$. With this,

$$x = (1 - t)y + t(x + z); \quad y + z = ty + (1 - t)(x + z) \qquad (2.346)$$

Since $\varphi^{[-1]}(x)$ is convex,

$$\varphi^{[-1]}(x) = \varphi^{[-1]}[(1 - t)y + t(x + z)] \leq t\varphi^{[-1]}(x + z) + (1 - t)\varphi^{[-1]}(y) \qquad (2.347)$$

$$\varphi^{[-1]}(y + z) = \varphi^{[-1]}[ty + (1 - t)(x + z)] \leq t\varphi^{[-1]}(y) + (1 - t)\varphi^{[-1]}(x + z) \qquad (2.348)$$

Adding the two inequalities gives the proof to Eq. (2.345), or equivalently, Eq. (2.344).

With Eq. (2.344), the function C in Eq. (2.343) satisfies that for $u_1 \leq u_2, v \in [0, 1]$,

$$C(u_2, v) - C(u_1, v) \leq u_2 - u_1 \qquad (2.349)$$

For $0 \leq v_1 \leq v_2 \leq 1$, the function $g(x) = C(x, v_2)$ with respect to x is continuously non-decreasing for $x \in [0, 1]$. Since $g(0) = C(0, v_2) = 0$, and $g(1) = C(1, v_2) = v_2$, one has $0 \leq g(x) \leq v_2$. Furthermore, as $v_1 \leq v_2$, there exists a real number $t \in [0, 1]$ so that $g(t) = C(t, v_2) = v_1$, or equivalently, according to Eq. (2.343), $\varphi(t) + \varphi(v_2) = \varphi(v_1)$. Thus, for $u_1, u_2, v_1, v_2 \in [0, 1]$, if $u_1 \leq u_2, v_1 \leq v_2$,

$$\begin{aligned} C(u_2, v_1) - C(u_1, v_1) &= \varphi^{[-1]}(\varphi(u_2) + \varphi(v_1)) - \varphi^{[-1]}(\varphi(u_1) + \varphi(v_1)) \\ &= \varphi^{[-1]}(\varphi(u_2) + \varphi(t) + \varphi(v_2)) - \varphi^{[-1]}(\varphi(u_1) + \varphi(t) + \varphi(v_2)) \\ &= C(C(u_2, v_2), t) - C(C(u_1, v_2), t) \leq C(u_2, v_2) - C(u_1, v_2) \end{aligned}$$
$$(2.350)$$

With this, it is seen that Eq. (2.322) holds. Thus, the function defined in Eq. (2.343) is a copula. □

Example 2.54

Consider the following three generators defined within $[0, 1]$: (1) $\varphi(x) = -\ln x$, (2) $\varphi(x) = 1 - x$, (3) $\varphi(x) = \frac{1}{x} - 1$. Find the corresponding copula functions.

Solution

(1) With $\varphi(x) = -\ln x$, one has $\varphi(0) = \infty$ and thus $\varphi^{[-1]}(x) = \varphi^{-1}(x) = \exp(-x)$. Furthermore, according to Eq. (2.343), $C(u, v) = \varphi^{[-1]}(\varphi(u) + \varphi(v)) = \exp(-(-\ln u) - (-\ln v)) = uv$.

(2) Since $\varphi(x) = 1 - x$, according to Eq. (2.341), $\varphi^{[-1]}(x) = \max(0, 1 - x)$, and furthermore, $C(u, v) = \varphi^{[-1]}(1 - u + 1 - v) = \varphi^{[-1]}(2 - u - v) = \max(0, u + v - 1)$. This copula is exactly C_{lb} in Example 2.51.

(3) As $\varphi(x) = \frac{1}{x} - 1$, it is easy to see that $\varphi(0) = \infty$ and thus $\varphi^{[-1]}(x) = \varphi^{-1}(x) = \frac{1}{1+x}$. With Eq. (2.343), $C(u, v) = \varphi^{[-1]}\left(\frac{1}{u} + \frac{1}{v} - 1\right) = \frac{1}{\frac{1}{u}+\frac{1}{v}-1} = \frac{uv}{u+v-uv}$. This copula equals that in Question (1) of Example 2.50.

For an Archimedean copula C with a generator $\varphi(x), x \in [0, 1]$, $\varphi(x^a)$ and $[\varphi(x)]^b$ are also generators of Archimedean copulas, where $a \in (0, 1]$ and $b \in [1, \infty)$. This fact provides an approach to construct parametric families of generators. For example, with $\varphi(x) = -\ln x$ in Example 2.54, a new generator $\varphi_1(x) = [\varphi(x)]^b = (-\ln x)^b$ gives a copula of

$$C_{gh}(u, v) = \exp\left[-\left((-\ln u)^b + (-\ln v)^b\right)^{\frac{1}{b}}\right] \qquad (2.351)$$

which is referred to as *Gumbel–Hougaard copula*.

Remark

Recall that in Example 2.53, the copula for X_{max} and Y_{max} is $C_{max}(u, v) = C^n(u^{1/n}, v^{1/n})$. Now, if C is a Gumbel–Hougaard copula (c.f. Eq. (2.351)), then

$$C_{max}(u, v) = C^n(u^{1/n}, v^{1/n}) = \left\{\exp\left[-\left((-\ln u^{1/n})^b + (-\ln v^{1/n})^b\right)^{\frac{1}{b}}\right]\right\}^n$$

$$= \exp\left[-n \cdot \left(\frac{1}{n^b}\right)^b \left((-\ln u)^b + (-\ln v)^b\right)^{\frac{1}{b}}\right]$$

$$= C(u, v)$$

$$(2.352)$$

This implies that the Gumbel–Hougaard copula is max-stable.

We can extend the copula function to an n-dimensional case. With this regard, the Sklar's Theorem states that for an n-dimensional joint CDF $F_n(x_1, x_2, \ldots, x_n)$ with marginal CDFs of $F_1(x_1)$, $F_2(x_2)$, ..., $F_n(x_n)$, there exists an n-dimensional copula function C_n such that for $\forall x_i \in (-\infty, \infty)$, $i = 1, 2, \ldots, n$, $F_n(x_1, x_2, \ldots, x_n) = C_n(F_1(x_1), F_2(x_2), \ldots, F_n(x_n))$. Furthermore, similar to Eq. (2.343), we can construct an n-dimensional Archimedean copula as follows,

$$C_n(u_1, u_2, \ldots, u_n) = \varphi^{[-1]}(\varphi(u_1) + \varphi(u_2) + \cdots + \varphi(u_n)) \qquad (2.353)$$

It is emphasized that the generator $\varphi(x)$ in Eq. (2.353) should be selected with carefulness. For example, the case of $\varphi(x) = 1 - x$ (c.f. Example 2.54), if substituted to Eq. (2.353), leads to a function C_n that is not a copula for $n \geq 3$. In fact, to ensure that the function C_n in Eq. (2.353) is an n-dimensional copula, $\varphi(x)$ should satisfy the following two requirements [13]. (1) $\varphi(x)$ strictly decreases from [0, 1] to [0, ∞) with $\varphi(0) = \infty$ and $\varphi(1) = 0$; (2) For $k = 0, 1, 2, \ldots$, $(-1)^k \frac{d^k}{dx^k} \varphi(x) \geq 0$ holds. Illustratively, $\varphi(x) = (-\ln x)^b$, which is the generator of the Gumbel–Hougaard copula (c.f. Eq. (2.351)), yields an n-dimensional copula of

$$C_{\text{gh},n}(u_1, u_2, \ldots, u_n) = \exp\left[-\left((-\ln u_1)^b + (-\ln u_2)^b + \cdots + (-\ln u_n)^b\right)^{\frac{1}{b}}\right] \qquad (2.354)$$

However, in Eq. (2.354) (as well as many other n-dimensional copulas), the dependence of the (u_i, u_j) pairs is fully determined by only one parameter (or limited parameters), which may halter the applicability in practical use. Alternatively, the Gaussian copula function has been widely used in the literature, which well describes the linear correlation of n random variables (measured by totally $C_n^2 - n$ correlation coefficients). The n-dimensional Gaussian copula takes the form of

$$C_{\text{Ga},n}(u_1, u_2, \ldots, u_n) = \Phi_\rho[\Phi^{-1}(u_1), \Phi^{-1}(u_2), \ldots, \Phi^{-1}(u_n)] \qquad (2.355)$$

where Φ_ρ is the joint CDF of a multivariate standard Gaussian distribution with an $n \times n$ correlation coefficient matrix ρ. We will further discuss the application of Gaussian copula in Chap. 3.

Note that in Sect. 2.3.1, we have introduced the linear (Pearson) correlation coefficient of two random variables. However, the Pearson correlation coefficient is sensitive to the nonlinear transformation of the variables. That is, for two variables X and Y with a linear correlation of ρ, and a monotonically increasing function f, the linear correlation of $f(X)$ and $f(Y)$ is not necessarily ρ when f is nonlinear. For example, recall the two variables X and Y in Example 2.36, whose linear correlation coefficient is $-\frac{1}{7}$ (c.f. Example 2.38). Now, we consider the linear correlation between X^2 and Y^2. Since

$$\mathbb{E}(X^2) = \int_0^\infty \frac{x^2}{2}(x+1)\exp(-x)dx = 4, \quad \mathbb{E}(Y^2) = \mathbb{E}(X^2) = 4$$

$$\mathbb{E}(X^4) = \int_0^\infty \frac{x^4}{2}(x+1)\exp(-x)dx = 72, \quad \mathbb{E}(Y^4) = \mathbb{E}(X^4) = 72 \qquad (2.356)$$

$$\mathbb{E}(X^2Y^2) = \int_0^\infty \int_0^\infty \frac{x^2y^2}{2}(x+y)\exp(-x-y)dxdy = 12$$

one has $\rho(X^2, Y^2) = \frac{12-4\times4}{72-4^2} = -\frac{1}{14} \neq \rho(X, Y)$.

To overcome the disadvantage of the linear correlation, there are also some other measures for the dependence between random variables. Remarkably, the Kendall's tau (τ_K) and Spearman's rho (ρ_S) have been widely used in the literature, and will be discussed in the following.

Let X and Y be two random variables, and (X_1, Y_1) and (X_2, Y_2) two independent random samples of (X, Y). The Kendall's tau for X and Y, $\tau_K(X, Y)$, is defined as the probability of concordance between (X_1, Y_1) and (X_2, Y_2) minus the probability of discordance between (X_1, Y_1) and (X_2, Y_2), that is,

$$\tau_K(X, Y) = \mathbb{P}[(X_1 - X_2)(Y_1 - Y_2) > 0] - \mathbb{P}[(X_1 - X_2)(Y_1 - Y_2) < 0]$$
$$= 2\mathbb{P}[(X_1 - X_2)(Y_1 - Y_2) > 0] - 1 \qquad (2.357)$$

Clearly, $\tau_K(X, Y) \in [-1, 1]$ since $\mathbb{P}[(X_1 - X_2)(Y_1 - Y_2) > 0] \in [0, 1]$. Also, with a monotonically increasing function f, the Kendall's tau of $f(X)$ and $f(Y)$ equals that of X and Y.

If X and Y have a copula of $C(u, v)$, then it follows,

$$\tau_K(X, Y) = 4\int_0^1 \int_0^1 C(u, v)dC(u, v) - 1 = 4\mathbb{E}(C(U, V)) - 1 \qquad (2.358)$$

where U and V are two uniform random variables within $[0, 1]$ with a joint CDF of C, and $dC(u, v) = \frac{\partial C^2(u,v)}{\partial u \partial v}dudv$. The proof is as follows.

Proof Note that

$$\mathbb{P}[(X_1 - X_2)(Y_1 - Y_2) > 0] = \mathbb{P}(X_1 < X_2 \cap Y_1 < Y_2) + \mathbb{P}(X_1 > X_2 \cap Y_1 > Y_2) \qquad (2.359)$$

where

$$\mathbb{P}(X_1 < X_2 \cap Y_1 < Y_2) = \int\int F_{X,Y}(x_2, y_2)dF_{X,Y}(x_2, y_2) = \int_0^1 \int_0^1 C(u, v)dC(u, v) \qquad (2.360)$$

and $\mathbb{P}(X_1 > X_2 \cap Y_1 > Y_2) = \int_0^1 \int_0^1 C(u, v)dC(u, v)$. Thus, according to Eq. (2.357),

$$\tau_K(X, Y) = 2\mathbb{P}[(X_1 - X_2)(Y_1 - Y_2) > 0] - 1$$

$$= 4 \int_0^1 \int_0^1 C(u, v) \mathrm{d}C(u, v) - 1 = 4\mathbb{E}(C(U, V)) - 1 \qquad (2.361)$$

which completes the proof. $\qquad\qquad\qquad\qquad\qquad\qquad\qquad\qquad\qquad\qquad\qquad\square$

Furthermore, if C is an Archimedean copula with a generator $\varphi(x)$, then

$$\tau_K(X, Y) = 4 \int_0^1 \frac{\varphi(x)}{\varphi'(x)} \mathrm{d}x + 1 \qquad (2.362)$$

The proof of Eq. (2.362) can be found in, e.g., [4, 23].

Example 2.55

For two random variables X and Y with a Gumbel–Hougaard (c.f. Eq. (2.351)) with a parameter $b \geq 1$, find the Kendall's tau for X and Y.

Solution

Note that the Gumbel–Hougaard copula has a generator of $\varphi(x) = (-\ln x)^b$, with which $\varphi'(x) = -\frac{b}{x}(-\ln x)^{b-1}$. According to Eq. (2.362),

$$\tau_K(X, Y) = 4 \int_0^1 \frac{(-\ln x)^b}{-\frac{b}{x}(-\ln x)^{b-1}} \mathrm{d}x + 1 = \frac{4}{b} \int_0^1 x \ln x \mathrm{d}x + 1$$

$$= \frac{4}{b} \cdot \left(-\frac{x^2}{4} + \frac{1}{2}x^2 \ln x \right)\Bigg|_0^1 + 1 = 1 - \frac{1}{b} \qquad (2.363)$$

Now we discuss the Spearman's correlation ρ_S. Let X and Y be two random variables, and (X_1, Y_1), (X_2, Y_2) and (X_3, Y_3) three independent random samples of (X, Y). The Spearman's rho for X and Y, $\rho_S(X, Y)$, is in proportional to the probability of concordance between (X_1, Y_1) and (X_2, Y_3) minus the probability of discordance between (X_1, Y_1) and (X_2, Y_3), that is,

$$\rho_S(X, Y) = 3 \{\mathbb{P}[(X_1 - X_2)(Y_1 - Y_3) > 0] - \mathbb{P}[(X_1 - X_2)(Y_1 - Y_3) < 0]\} \qquad (2.364)$$

If X and Y have a copula of $C(u, v)$, then

$$\rho_S(X, Y) = 12 \int_0^1 \int_0^1 uv \mathrm{d}C(u, v) - 3 = 12\mathbb{E}(UV) - 3$$

$$= 12 \int_0^1 \int_0^1 C(u, v) \mathrm{d}u \mathrm{d}v - 3 \qquad (2.365)$$

where U and V are two uniform random variables within $[0, 1]$ with a joint CDF of C. The proof is as follows.

Proof According to Eq. (2.364),

$$\rho_S(X, Y) = 3\{2\mathbb{P}[(X_1 - X_2)(Y_1 - Y_3) > 0] - 1\}$$
$$= 6\{\mathbb{P}(X_1 < X_2 \cap Y_1 < Y_3) + \mathbb{P}(X_1 > X_2 \cap Y_1 > Y_3)\} - 3$$
$$= 6\left[\int_0^1 \int_0^1 C(u, v)\mathrm{d}u\mathrm{d}v + \int_0^1 \int_0^1 uv\mathrm{d}C(u, v)\right] - 3 \qquad (2.366)$$
$$= 12 \int_0^1 \int_0^1 uv\mathrm{d}C(u, v) - 3 = 12 \int_0^1 \int_0^1 C(u, v)\mathrm{d}u\mathrm{d}v - 3$$

since $\int_0^1 \int_0^1 C(u, v)\mathrm{d}u\mathrm{d}v = \int_0^1 \int_0^1 uv\mathrm{d}C(u, v)$. $\square$

Example 2.56

For two variables X and Y with a copula $C(u, v) = uv + \theta uv(1 - u)(1 - v)$, where $\theta \in [-1, 1]$ is a parameter, find the Spearman's rho of X and Y.

Solution

Since

$$\int_0^1 \int_0^1 C(u, v)\mathrm{d}u\mathrm{d}v = \int_0^1 \int_0^1 [uv + \theta uv(1 - u)(1 - v)]\mathrm{d}u\mathrm{d}v = \frac{1}{4} + \frac{\theta}{36} \qquad (2.367)$$

according to Eq. (2.365),

$$\rho_S(X, Y) = 12 \int_0^1 \int_0^1 C(u, v)\mathrm{d}u\mathrm{d}v - 3 = \frac{\theta}{3} \qquad (2.368)$$

Remark

(1) According to Eq. (2.365), since $\mathbb{E}(U) = \mathbb{E}(V) = \frac{1}{2}$ and $\mathbb{V}(U) = \mathbb{V}(V) = \frac{1}{12}$ (c.f. Sect. 2.2.2), one has

$$\rho_S(X, Y) = 12\mathbb{E}(UV) - 3 = \frac{\mathbb{E}(UV) - \frac{1}{4}}{\frac{1}{12}} = \frac{\mathbb{E}(UV) - \mathbb{E}(U)\mathbb{E}(V)}{\sqrt{\mathbb{V}(U)}\sqrt{\mathbb{V}(V)}} \qquad (2.369)$$

implying that the Spearman's rho of variables X and Y equals the Pearson's linear correlation of $F_X(X)$ and $F_Y(Y)$, where F_X and F_Y are the CDFs of X and Y respectively.

(2) According to Eq. (2.365),

$$\rho_S(X, Y) = 12 \int_0^1 \int_0^1 [C(u, v) - uv]\mathrm{d}u\mathrm{d}v \qquad (2.370)$$

indicating that the Spearman's rho measures the integrated difference between the realistic copula and the independence copula.

Fig. 2.6 ρ_S-τ_K region

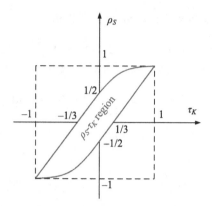

Regarding the difference between the two rank correlations τ_K and ρ_S as described above, we have the following theorem.

Theorem 2.5 *For two continuous random variables X and Y, let τ_K and ρ_S denote their Kendall's tau and Spearman's rho respectively. Then, (1)* $-1 \leq 3\tau_K - 2\rho_S \leq 1$; *(2)* $\frac{1+\rho_S}{2} \geq \left(\frac{1+\tau_K}{2}\right)^2$ *and* $\frac{1-\rho_S}{2} \geq \left(\frac{1-\tau_K}{2}\right)^2$.

The proof of Theorem 2.5 can be found in, e.g., [14]. Moreover, according to Theorem 2.5, the bounds of τ_K and ρ_S define the region for (τ_K, ρ_S), as shown in Fig. 2.6. It is seen that the region is a "narrow band" around the line "$\tau_K = \rho_S$", implying the overall synchronization of the two measures.

Remark

Now we have introduced three measures for the dependence of random variables, namely Person's linear correlation (ρ), Kendall'tau (τ_K) and Spearman's rho (ρ_S). Clearly, two independent variables lead to a zero value of the three indicators. However, the inverse proposition does not hold. That is, $\rho = 0$ or $\tau_K = 0$ or $\rho_S = 0$ cannot guarantee the independence of two random variables. The case of $\rho = 0$ has been discussed before (c.f. Eq. (2.258)). If $\tau_K = 0$, according to Eq. (2.358), we can only conclude that $\mathbb{E}(C(U, V)) = \frac{1}{4}$ but not necessarily $C(u, v) = uv$. Similarly, when $\rho_S = 0$, with Eq. (2.365), one has $\mathbb{E}(UV) = \frac{1}{4}$ only. A counterexample is that the copula $C(u, v) = C_{ub} + C_{lb} = \max(u + v - 1, 0) + \min(u, v)$ (c.f. Example 2.51) yields a zero value of τ_K and ρ_S.

Problems

2.1 Consider the damage state of a post-hazard structure, which be classified into five categories: none (D_0), minor (D_1), moderate (D_2), severe (D_3) and total (D_4). The post-hazard structural performance is deemed as "unsatisfactory" if the damage reaches a moderate or severer state. Let E_1 be the event that the structure suffers from damage, and E_2 the event that the structural performance is unsatisfactory.
(1) What is the sample space? What are the event spaces for E_1 and E_2 respectively?
(2) If the probability of D_i ($i = 0, 1, 2, 3, 4$) is proportional to $(i + 1)^{-1}$, compute the probabilities of E_1 and E_2 respectively.

2.2 Consider two post-hazard structures, namely 1 and 2, whose performances are as described in Problem 2.1. Let p_1 be the probability that the structural performance is unsatisfactory. Furthermore, let $\widetilde{E}_1$ be the event that both structural performances are unsatisfactory, and $\widetilde{E}_2$ the event that either structural performance is unsatisfactory.
(1) What is the sample space for the joint behaviour of the two post-hazard structures? What are the event spaces for $\widetilde{E}_1$ and $\widetilde{E}_2$?
(2) Derive the lower bound for $\mathbb{P}(\widetilde{E}_1)$ in terms of p_1.
(3) Derive the upper bound for $\mathbb{P}(\widetilde{E}_2)$ in terms of p_1.
(4) If the structural performances are independent, and the probability of D_i ($i = 0, 1, 2, 3, 4$) is proportional to $(i + 1)^{-1}$, compute the probabilities of $\widetilde{E}_1$ and $\widetilde{E}_2$ respectively, and verify the results in (2) and (3).

2.3 Consider a structure subjected to repeated loads. In the presence of the effect of each load, the structure may fail in either mode 1 or mode 2. Let p_1 and p_2 denote the probability of failure associated with mode 1 and mode 2 respectively due to the effect of one load. Show that, if the structural fails, the probability of mode 1 failure is $\frac{p_1}{p_1+p_2}$.

2.4 In Problem 2.1, if the probability of D_i ($i = 0, 1, 2, 3, 4$) is p_i, what is the probability of severe damage conditional on the fact that the structural performance is unsatisfactory? What is the conditional probability of E_2 on E_1?

2.5 Recall Example 2.7.
(1) Determine the probability that V occurs only conditional on the failure mode of "beam".
(2) Determine the probability that V occurs only conditional on the failure of the frame (in any mode).
(3) Compare the two probabilities obtained in (1) and (2).

2.6 Consider the resistance (moment-bearing capacity) of a bridge girder, which follows a lognormal distribution with a mean value of 4000 kN·m and a standard deviation of 400 kN·m.
(1) If the bridge girder has survived a load of 3700 kN·m, compute the mean value and standard deviation of the post-load resistance.
(2) If the bridge girder has survived a load with a mean value of 3700 kN·m and a

standard deviation of 370 kN·m, compute the mean value and standard deviation of the post-load resistance.

(3) Compare the results in (1) and (2), and comment on the difference.

2.7 Recall Example 2.8. If the bridge girder have survived two independent proof loads, namely S_1 and S_2, with CDFs of $F_{S_1}(s)$ and $F_{S_2}(s)$ respectively, determine the updated distribution of girder resistance.

2.8 Consider two post-hazard structures, namely 1 and 2, whose performances are as described in Problem 2.1. Let E_i^* be the event that the performance of structure i is satisfactory for $i = 1, 2$. Show that $\mathbb{P}(E_1^*|E_2^*) = \mathbb{P}(E_2^*|E_1^*)$.

2.9 In Problem 2.1, suppose that the probability of D_i is proportional to $(i + 1)^{-1}$, and that the economic loss of the structure is L_i conditional on D_i for $i = 0, 1, 2, 3, 4$. Assume that $L_i = ki^2$ for $i = 0, 1, 2, 3, 4$, where $k > 0$ is a constant. We introduce a random variable X to represent the post-hazard damage state, and a random variable Y to represent the post-hazard economic loss. The variable X takes a value of i corresponding to the damage state D_i.

(1) What are the probability mass function, mean and standard deviation of X?
(2) Compute $\mathbb{P}(1 < X \le 3)$.
(3) What are the mean and standard deviation of Y?

2.10 Consider the post-hazard damage state of a building portfolio with 100 buildings. Each building has an identical failure probability of p, and the behaviour of each building is assumed to be independent of others. The probability that at least 95% of the post-hazard buildings remain safe is required to be greater than 99%. What is the range for p?

2.11 Suppose that random variable X has a mean value of 3 and its PDF is

$$f_X(x) = a \exp\left(-(x - b)^2\right), \quad -\infty < x < \infty \tag{2.371}$$

(1) Compute a and b;
(2) What is the standard deviation of X?
(3) Compute the mean value of $\exp(X)$.

2.12 For two variables $X, Y \in [0, 1]$, if their joint PDF is $f_{X,Y}(x, y) = axy^2 \cdot \mathbb{I}(0 \le x \le 1, 0 \le y \le 1)$, where a is a constant, decide whether X and Y are independent or not.

2.13 Markov's inequality and Chebyshev's inequality

(1) Suppose that the tensile strength of a steel bar is a random variable with a mean value of 300 MPa. Using the Markov's inequality, what is the upper bound for the event that the tensile strength exceeds 600 MPa?

(2) Suppose that the tensile strength of a steel bar is a random variable with a mean value of 300 MPa and a standard deviation of 50 MPa. Using the Chebyshev's inequality, what is the lower bound for the event that the tensile strength is within [250 Mpa, 350 Mpa]?

2.14 Consider the post-hazard damage state of a building portfolio with 100 build-ings. Each building has independent performance and an identical probability of 0.1 for unsatisfactory performance. Let X be the number of buildings with unsatisfactory performance. Compute the probability of $X \leq 40$.

2.15 Reconsider Problem 2.14.
(1) Compute the probability of $X = 40$.
(2) Using a normal distribution to approximate X, recalculate the probability of $X = 40$ [*Hint. The probability can be approximately replaced by* $\mathbb{P}(39.5 \leq X < 40.5)$.].
(3) Compare the results in (1) and (2).

2.16 Suppose that random variable X is binomially distributed with parameters n and p. For $k = 0, 1, \ldots, n$, what is the maximum value for $\mathbb{P}(X = k)$?

2.17 Repeating the experiment in Sect. 2.2.1.1 until the rth outcome of success, the number of experiments, denoted by X, follows a negative binomial distribution, and its PMF is

$$\mathbb{P}(X = k) = C_{k-1}^{r-1} p^r (1 - p)^{k-r}, \quad k \geq r \tag{2.372}$$

(1) What are the mean value and variance of X?
(2) Derive the MGF of X.

2.18 Suppose that a Poisson random variable X has a mean value of λ. For $k = 0, 1, 2, \ldots$, what is the maximum value for $\mathbb{P}(X = k)$?

2.19 For two independent binomial random variables X (with parameters p, n_1) and Y (with parameters p, n_2), show that $X + Y$ is also a binomial random variable.

2.20 For two independent random variables X and Y with mean values of μ_X, μ_Y and standard deviations of σ_X, σ_Y, derive $\mathbb{V}(XY)$ in terms of μ_X, μ_Y, σ_X and σ_Y.

2.21 If random variable X is uniformly distributed within $[0, 1]$,
(1) Show that X^n follows a Beta distribution, where n is a positive integer.
(2) Compute $\mathbb{E}(X^n)$ and $\mathbb{V}(X^n)$.

2.22 For two random variables X and Y that are uniformly distributed within $[0, 1]$, find the PDF of $X + Y$ [*Hint. X + Y follows a triangle distribution.*].

2.23 Let N be a Poisson random variable with a mean value of λ, and $X_1, X_2, \ldots, X_N$ a sequence of statistically independent and identically distributed random variables (independent of N). Let $Y = \sum_{i=1}^{N} X_i$, then Y follows a compound Poisson distri-bution. If the MGF of each X_i is $\psi_X(\tau)$, derive the MGF of Y.

2.24 For a random variable $X \in (0, 1)$ whose PDF is as follows, it is deemed to follow a standard arcsine distribution,

$$f_X(x) = \frac{1}{\pi \sqrt{x(1 - x)}} \tag{2.373}$$

Find the mean value and standard deviation of X respectively [*Hint. This corresponds to a special case of* $\eta = \zeta = \frac{1}{2}$ *in Eq.* (2.223).].

2.25 A random variable X follows a Laplace distribution with parameters μ and $b > 0$ if its PDF is as follows,

$$f_X(x) = \frac{1}{2b} \exp\left(-\frac{|x - \mu|}{b}\right), \quad -\infty < x < \infty \tag{2.374}$$

Show that (1) the mean of X is b; (2) the variance of X is $2b^2$; (3) the MGF of X is $\frac{\exp(\mu\tau)}{1 - b^2\tau^2}$ for $|\tau| < \frac{1}{b}$.

2.26 An Erlang distributed random variable X with shape parameter $k \in \{1, 2, 3, \ldots\}$ and rate parameter $r \geq 0$ has the following PDF,

$$f_X(x) = \frac{r^k x^{k-1} \exp(-rx)}{(k - 1)!}, \quad x \geq 0 \tag{2.375}$$

(1) What are the mean value and standard deviation of X?
(2) What is the MGF of X?
(3) Show that for two Erlang variables X_1 and X_2 with the same rate parameter, $X_1 + X_2$ is also an Erlang random variable [*Remark. The Erlang distribution is a special case of the Gamma distribution with the shape parameter being an integer.*].

2.27 Let $X_1, X_2, \ldots, X_n$ $(n \geq 2)$ be a sequence of statistically independent and identically distributed continuous random variables with a CDF of $F_X(x)$ and a PDF of $f_X(x)$. Let Y be the largest variable among $\{X_i, i = 1, 2, \ldots, n\}$, and W the smallest variable among $\{X_i\}$.
(1) What are the CDF and PDF of Y?
(2) Use the result from (1) to resolve Problem 2.7.
(3) What are the CDF and PDF of W?

2.28 Suppose that $X_1, X_2, \ldots, X_n$ is a sequence of independent geometrically distributed variables. Show that $Y = \min_{i=1}^{n} X_i$ also follows a geometric distribution.

2.29 For a discrete random variable X that is geometrically distributed with parameter p, show that as $p \to 0$, the distribution of pX approaches an exponential distribution.

2.30 Reconsider Problem 2.27. Let Z be the second-largest variable among $\{X_i, i = 1, 2, \ldots, n\}$. What are the CDF and PDF of Z?

2.31 Recall Eq. (2.162), which gives the CDF of the arriving time of the first event, T_1. Now, let T_2 be the arriving time of the second event. What are the PDF, mean and variance of T_2?

2.32 The water seepage in a lining structure can be modeled by the Darcy's law as follows,

$$\chi(t) = \sqrt{2K \frac{p}{\omega_0} t} \tag{2.376}$$

where $\chi(t)$ is the depth of water seepage at time t, K is the hydraulic conductivity, p is the water pressure on the top of the lining structure, and ω_0 is the water density [34]. Treating p and ω_0 as constants, show that if K is lognormally distributed, then $\chi(t)$ also follows a lognormal distribution for $t > 0$.

2.33 Show that if X is an exponentially distributed variable, then $Y = \sqrt{X}$ follows a Rayleigh distribution.

2.34 Let $X_1, X_2, \ldots, X_n$ be a sequence of statistically independent and identically distributed variables. If each X_i follows a Rayleigh distribution with a scale parameter of σ, show that $Y = \sum_{i=1}^{n} X_i^2$ is a Gamma random variable.

2.35 If X is an exponentially distributed variable, show that $-\ln X$ follows a Gumbel distribution.

2.36 The CDF of the generalized extreme value (GEV) distribution is as follows,

$$F(x) = \exp\left(-(1+\xi x)^{-\frac{1}{\xi}}\right), \quad 1+\xi x > 0 \tag{2.377}$$

For the cases of $\xi < 0$, $\xi \to 0$ and $\xi > 0$ respectively, what distribution does Eq. (2.377) reduce to?

2.37 For a Pareto (Type I) random variable X with parameters $\alpha > 0$ and $x_m > 0$, its CDF takes a form of

$$F_X(x) = \begin{cases} 1 - \left(\dfrac{x}{x_m}\right)^{\alpha}, & x \geq x_m \\ 0, & x < x_m \end{cases} \tag{2.378}$$

(1) Derive the mean value and variance of X.

(2) Show that $\ln\left(\dfrac{X}{x_m}\right)$ is an exponential random variable.

2.38 The CDF of the generalized Pareto distribution (GPD) is defined as follows,

$$F(x) = 1 - (1+\xi x)^{-\frac{1}{\xi}}, \quad 1+\xi x > 0 \tag{2.379}$$

Derive the mean value and variance of the GPD in Eq. (2.379) [*Remark. The GPD is often used to describe a variable's tail behaviour, and is an asymptotical distribution of the GEV distribution as in Eq. (2.377).*].

2.39 A random variable X is said to follow a standard Cauchy distribution if its PDF is

$$f_X(x) = \frac{1}{\pi(1+x^2)} \tag{2.380}$$

Show that for two independent standard normal random variables Y_1 and Y_2 (with a mean value of 0 and a variance of 1), $\frac{Y_1}{Y_2}$ follows a standard Cauchy distribution [*Hint. Consider the joint distribution of $U = \frac{Y_1}{Y_2}$ and $V = Y_2$ using the Jacobian matrix.*].

2.40 Reconsider Problem 2.23. Compute $\mathbb{C}(N, Y)$.

2.41 Suppose that random variable $X \in [0, 1]$ has a mixture distribution with $\mathbb{P}(X = 0) = p_0$ and $\mathbb{P}(0 < X \leq 1) = 1 - p_0$. Let μ_X and σ_X be the mean value and standard deviation of X respectively. Compute $\mathbb{E}(X|X > 0)$ and $\mathbb{V}(X|X > 0)$.

2.42 If random variable X is exponentially distributed with a mean value of μ_X, compute $\mathbb{E}(X|X > a)$, where $a > 0$ is a constant.

2.43 Recall the random variable X in Problem 2.25. Compute $\mathbb{E}(X|X > \mu + b)$.

2.44 Consider two independent random variables X and Y. Show that $\mathbb{E}(X|Y = y) = \mathbb{E}(X)$ for $\forall$appropriate y.

2.45 For a normal random variable X with a CDF of $F_X(x)$ and a PDF of $f_X(x)$, show that for two real numbers $a < b$, $\mathbb{E}(X|a \leq X < b) = \mu_X - \sigma_X^2 \frac{f_X(b) - f_X(a)}{F_X(b) - F_X(a)}$, where μ_X and σ_X are the mean value and standard deviation of X respectively.

2.46 For a lognormal random variable X with a mean value of μ_X and a standard deviation of σ_X, compute $\mathbb{E}(X|X > a)$, where $a > 0$ is a constant.

2.47 In Problem 2.1, suppose that the probability of D_i ($i = 0, 1, 2, 3, 4$) is p_i, and that the economic loss of the structure is L_i conditional on D_i for $i = 0, 1, 2, 3, 4$. If each L_i is a random variable with a mean value of μ_i and a standard deviation of σ_i, (a) derive the mean value and standard deviation of the post-hazard economic loss; (b) derive the mean value and standard deviation of the post-hazard economic loss conditional on E_1.

2.48 Define function $C_\theta(u, v) = [\min(u, v)]^\theta \cdot (uv)^{1-\theta}$ for $\theta \in [0, 1]$. Show that $C_\theta(u, v)$ is a copula [*Remark. This is called the Cuadras–Augé family of copulas.*].

2.49 If the joint CDF of random variables X, Y is

$$F_{X,Y}(x, y) = \exp\left(-[\exp(-\theta x) + \exp(-\theta y)]^{\frac{1}{\theta}}\right), \quad \theta \geq 1 \qquad (2.381)$$

show that the copula for X and Y is

$$C_\theta(u, v) = \exp\left(-\left((-\ln u)^\theta + (-\ln v)^\theta\right)^{\frac{1}{\theta}}\right) \qquad (2.382)$$

2.50 Consider the following two generators defined within $[0, 1]$, (1) $\varphi(x) = \ln(1 - \theta \ln x)$ with $0 < \theta \leq 1$; (2) $\varphi(x) = e^{1/x} - e$. Find the corresponding copula functions.

2.51 The generator of the Clayton family copula is $\varphi_\theta(x) = \frac{1}{\theta}(x^{-\theta} - 1)$ for $\theta \geq -1$. (1) Derive the corresponding copula function; (2) What is the Kendall's tau?

2.52 Recall the generator in Problem 2.51. By referring to Eq. (2.353), what is the copula function when extended into n-dimension ($n \geq 2$)?

References

1. Alsina C, Frank MJ, Schweizer B (2006) Associative functions: triangular norms and copulas. World Scientific, Singapore. https://doi.org/10.1142/6036
2. Bender MA, Knutson TR, Tuleya RE, Sirutis JJ, Vecchi GA, Garner ST, Held IM (2010) Modeled impact of anthropogenic warming on the frequency of intense Atlantic hurricanes. Science 327:454–458. https://doi.org/10.1126/science.1180568
3. Billingsley P (1986) Probability and measure, 2nd edn. Wiley, New York
4. Cherubini U, Luciano E, Vecchiato W (2004) Copula methods in finance. Wiley, New York
5. Chung KL (2001) A course in probability theory. Academic, New York
6. Cornell CA (1968) Engineering seismic risk analysis. Bull Seismol Soc Am 58(5):1583–1606
7. Doane DP, Seward LE (2011) Measuring skewness: a forgotten statistic? J Stat Educ 19(2):1–18. https://doi.org/10.1080/10691898.2011.11889611
8. Ellingwood B, Galambos TV, MacGregor JG, Cornell CA (1980) Development of a probability based load criterion for American National Standard A58: building code requirements for minimum design loads in buildings and other structures. US Department of Commerce, National Bureau of Standards
9. Elsner JB, Kossin JP, Jagger TH (2008) The increasing intensity of the strongest tropical cyclones. Nature 455:92–95. https://doi.org/10.1038/nature07234
10. de Haan L, Ferreira A (2007) Extreme value theory: an introduction. Springer Science & Business Media, New York
11. Hall WB (1988) Reliability of service-proven structures. J Struct Eng 114(3):608–624. https://doi.org/10.1061/(ASCE)0733-9445(1988)114:3(608)
12. Heyde CC (1963) On a property of the lognormal distribution. J R Stat Soc: Ser B (Methodol) 25(2):392–393
13. Kimberling CH (1974) A probabilistic interpretation of complete monotonicity. Aequationes Mathematicae 10(2–3):152–164
14. Kruskal WH (1958) Ordinal measures of association. J Am Stat Assoc 53(284):814–861
15. Leone F, Nelson L, Nottingham R (1961) The folded normal distribution. Technometrics 3(4):543–550. https://doi.org/10.2307/1266560
16. Li Q, Wang C (2015) Updating the assessment of resistance and reliability of existing aging bridges with prior service loads. J Struct Eng 141(12):04015072. https://doi.org/10.1061/(ASCE)ST.1943-541X.0001331
17. Li Q, Ye X (2018) Surface deterioration analysis for probabilistic durability design of RC structures in marine environment. Struct Saf 75:13–23. https://doi.org/10.1016/j.strusafe.2018.05.007
18. Li Q, Wang C, Ellingwood BR (2015) Time-dependent reliability of aging structures in the presence of non-stationary loads and degradation. Struct Saf 52:132–141. https://doi.org/10.1016/j.strusafe.2014.10.003
19. Li Q, Wang C, Zhang H (2016) A probabilistic framework for hurricane damage assessment considering non-stationarity and correlation in hurricane actions. Struct Saf 59:108–117. https://doi.org/10.1016/j.strusafe.2016.01.001
20. Li Y, Ellingwood BR (2006) Hurricane damage to residential construction in the US: importance of uncertainty modeling in risk assessment. Eng Struct 28(7):1009–1018. https://doi.org/10.1016/j.engstruct.2005.11.005
21. Melchers RE, Beck AT (2018) Structural reliability analysis and prediction, 3rd edn. Wiley, New York. https://doi.org/10.1002/9781119266105
22. Mudd L, Wang Y, Letchford C, Rosowsky D (2014) Assessing climate change impact on the U.S. East Coast hurricane hazard: temperature, frequency, and track. Nat Hazards Rev 15(3):04014001. https://doi.org/10.1061/(ASCE)NH.1527-6996.0000128
23. Nelsen RB (2007) An introduction to copulas, 2nd edn. Springer Science & Business Media, New York
24. NOAA (National Oceanic and Atmospheric Administration) (2018) Historical hurricane tracks. https://coast.noaa.gov/hurricanes

25. Nowak AS (1999) Calibration of LRFD bridge design code. Technical report, Washington, D.C., Transportation Research Board, National Research Council, National Academy Press
26. Rosowsky DV, Ellingwood BR (2002) Performance-based engineering of wood frame housing: fragility analysis methodology. J Struct Eng 128(1):32–38. https://doi.org/10.1061/(ASCE)0733-9445(2002)128:1(32)
27. Ross SM (2010) Introduction to probability models, 10th edn. Academic, New York
28. Roussas GG (2014) An introduction to measure-theoretic probability. Academic, New York
29. Schmitz V (2004) Revealing the dependence structure between $x_{(1)}$ and $x_{(n)}$. J Stat Plan Inference 123(1):41–47. https://doi.org/10.1016/S0378-3758(03)00143-5
30. Sklar A (1973) Random variables, joint distribution functions, and copulas. Kybernetika 9(6):449–460
31. Stacy EW et al (1962) A generalization of the Gamma distribution. Ann Math Stat 33(3):1187–1192
32. Val DV, Stewart MG (2009) Reliability assessment of ageing reinforced concrete structures - Current situation and future challenges. Struct Eng Int 19(2):211–219. https://doi.org/10.2749/101686609788220114
33. Von Hippel PT (2005) Mean, median, and skew: correcting a textbook rule. J Stat Educ 13(2)
34. Wang C (2020) Reliability-based design of lining structures for underground space against water seepage. Undergr Space. https://doi.org/10.1016/j.undsp.2020.03.004
35. Wang C, Li Q, Zhang H, Ellingwood BR (2017) Modeling the temporal correlation in hurricane frequency for damage assessment of residential structures subjected to climate change. J Struct Eng 143(5):04016224. https://doi.org/10.1061/(ASCE)ST.1943-541X.0001710
36. Wang C, Zhang H, Feng K, Li Q (2017) Assessing hurricane damage costs in the presence of vulnerability model uncertainty. Nat Hazards 85(3):1621–1635. https://doi.org/10.1007/s11069-016-2651-z
37. Wang N (2010) Reliability-based condition assessment of existing highway bridges. PhD thesis, Georgia Institute of Technology

Chapter 3
Monte Carlo Simulation

Abstract This chapter discusses the basic concept and techniques for Monte Carlo simulation. The simulation methods for a single random variable as well as those for a random vector (consisting of multiple variables) are discussed, followed by the simulation of some special stochastic processes, including Poisson process, renewal process, Gamma process and Markov process. Some advanced simulation techniques, such as the importance sampling, Latin hypercube sampling, and subset simulation, are also addressed in this chapter.

3.1 Introduction

The *Monte Carlo simulation* method is a powerful tool for solving problems including random variables [22]. The basic idea to implement a Monte Carlo simulation is to first generate samples of random inputs from their distribution functions and then perform a deterministic calculation on the generated random inputs, based on the mathematical modelling of the problem, to obtained numerical results. The generation of random inputs will be discussed in this chapter; the analysis of the generated random inputs relies on the nature of the problem and will be addressed in the subsequent chapters.

An early version of Monte Carlo simulation is the needle experiment performed by the French mathematician Comte de Buffon (1707–1788). Consider a plane with parallel horizontal lines distanced at d, and a needle with a length of $l < d$ that is randomly positioned on the plane, as illustrated in Fig. 3.1. Let A denote the event that the needle lies across a line. One can show that $\mathbb{P}(A) = \frac{2l}{\pi d}$ by noting that (1) the shortest distance from the needle center to a line, X, is uniformly distributed over $\left[0, \frac{d}{2}\right]$, (2) the angle between the needle and the vertical direction, Θ, also follows a uniform distribution over $\left[0, \frac{\pi}{2}\right]$, and (3) the needle crosses a line when $X \leq \frac{l}{2} \cos \Theta$. That is,

Fig. 3.1 Illustration of the
Buffon's needle problem

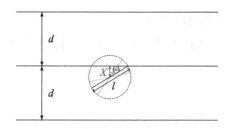

$$\mathbb{P}(A) = \int_0^{\frac{\pi}{2}} \int_0^{\frac{d}{2}} \mathbb{I}\left(x \le \frac{l}{2}\cos\theta\right) \cdot \frac{2}{\pi} \cdot \frac{2}{d} dx d\theta$$

$$= \frac{4}{\pi d} \int_0^{\frac{\pi}{2}} \min\left(\frac{d}{2}, \frac{l}{2}\cos\theta\right) d\theta = \frac{4}{\pi d} \int_0^{\frac{\pi}{2}} \frac{l}{2}\cos\theta d\theta \qquad (3.1)$$

$$= \frac{4}{\pi d} \cdot \frac{l}{2}\sin\theta \Big|_0^{\frac{\pi}{2}} = \frac{2l}{\pi d}$$

Buffon verified this probability by eventually throwing a needle on a plane with parallel lines. This experiment, in fact, reflects the basic idea of implementing Monte Carlo simulation. That is, performing the experiment for n times, if the the needle crosses a line for m times, then the probability of A is approximated by $\mathbb{P}(A) = \frac{m}{n}$. Furthermore, with Eq. (3.1), one can, as marked by Laplace in 1812, estimate the value of π by conducting the Buffon's needle experiment since $\frac{m}{n} = \mathbb{P}(A) = \frac{2l}{\pi d} \Rightarrow \pi = \frac{2nl}{md}$. In 1901, the Italian mathematician Lazzarini performed Buffon's needle experiment using a needle whose length is $\frac{5}{6}$ that of the line separation. He tossed a needle for 3408 times, of which the cross between the needle and a line occurred for 1808 times. With this, he obtained an estimate of π as 3.141592920..., with an accuracy to six significant digits. While this result seems extraordinarily lucky to obtain and has raised some concerns [5], it however offers us such an impression that we can estimate a probability (of an event) or a random quantity via random simulation.

Remark

In 1946, the physicists from Los Alamos were working on the distance likely to be traveled by the neutrons in different materials under the Manhattan Project. They were unable to solve the problem using conventional deterministic mathematical methods. Then Stanislaw Ulam proposed an idea of using random experiments. This idea was subsequently adopted and developed by von Neumann, Metropolis and others to solve many complex problems in making the atomic bomb. Since the work was secret, the "random experiment" method required a code name. Metropolis suggested the name of *Monte Carlo*, which refers to the Monte Carlo Casino in Monaco where Ulam's uncle would borrow money from relatives to gamble [10].

Now, more scientifically, what is the basis for the simulation-based methods?

In fact, we have the strong law of large numbers (detailed subsequently), which guarantees that the average of a set of independent and identically distributed random variables converges to the mean value with probability 1.

Theorem 3.1 (Strong law of large numbers) *For a sequence of statistically independent and identically distributed random variables $X_1, X_2, \ldots X_n$ with a mean value of μ, then*

$$\frac{X_1 + X_2 + \cdots + X_n}{n} \to \mu, \quad n \to \infty \tag{3.2}$$

with probability 1.

With the strong law of large numbers, to estimate the occurrence probability of an event A (e.g., failure of a structure), denoted by $\mathbb{P}(A)$, one can performance the experimental trials for n times, and record the number of trials in which the event occurs, M. As $n \to \infty$, with probability 1,

$$\mathbb{P}(A) = \frac{M}{n} \tag{3.3}$$

Equation (3.3) is explained by introducing a Bernoulli variable B_i for the ith trial, $i = 1, 2, \ldots n$. $B_i = 1$ if the event occurs and 0 otherwise. Clearly, $\mathbb{E}(B_i) = \mathbb{P}(B_i = 1) = \mathbb{P}(A)$. Applying the strong law of large numbers, with probability 1,

$$\lim_{n\to\infty} \frac{B_1 + B_2 + \cdots + B_n}{n} = \frac{M}{n} \to \mathbb{E}(B_i) = \mathbb{P}(A) \tag{3.4}$$

It is noticed that the estimator $\frac{M}{n}$ in Eq. (3.3) is by nature a random variable, since $M = B_1 + B_2 + \cdots + B_n$. With this regard, Eq. (3.3) implies that we are using a sample of $\frac{M}{n}$ to approximate $\mathbb{P}(A)$. Note that due to the independence between each B_i,

$$\mathbb{E}\left(\frac{M}{n}\right) = \frac{\sum_{i=1}^n \mathbb{E}(B_i)}{n} = \mathbb{E}(B_i), \quad \mathbb{V}\left(\frac{M}{n}\right) = \frac{\sum_{i=1}^n \mathbb{V}(B_i)}{n^2} = \frac{\mathbb{V}(B_i)}{n} \tag{3.5}$$

which implies that the variance of $\frac{M}{n}$ approaches to 0 as n is large enough, that is, $\frac{M}{n}$ converges to a deterministic value $\mathbb{E}(B_i)$ with a sufficiently large n.

Furthermore, if the event A involves r random inputs, namely $X_1, X_2, \ldots X_r$, we introduce an indicative function g such that $B_i = g(X_1, X_2, \ldots X_r)$, that is, g returns 1 if A occurs in the ith trial and 0 otherwise. With this, consistent with Eq. (3.4),

$$\mathbb{P}(A) = \mathbb{E}(B_i) = \underbrace{\int \cdots \int}_{r-\text{fold}} g(\mathbf{x}) f_{\mathbf{X}}(\mathbf{x}) d\mathbf{x} \tag{3.6}$$

where $f_{\mathbf{X}}(\mathbf{x})$ is the joint PDF of a random vector $\mathbf{X} = \{X_1, X_2, \ldots X_r\}$.

Equation (3.6) implies that the estimate of an event's probability via Monte Carlo simulation can be transformed into the calculation of a multiple-fold integral. In fact, motivated by Eq. (3.6), another important application of Monte Carlo simulation is to estimate an n-fold integral (e.g., the one in Eq. (3.6) if replacing g with a general function $h = h(\mathbf{X})$), especially when n is large. We will further discuss this point in Sect. 3.5.1.

With advances in computational abilities in recent years, Monte Carlo methods offer a rich variety of approaches to deal with problems incorporating uncertain inputs. Remarkably, parallel computation contributes significantly to reducing the relevant calculation times since the different simulation runs are often independent of each other and thus can be processed in parallel. Nonetheless, simulation-based approaches still have some disadvantages. The first is that the implementation of Monte Carlo simulation, in many cases, needs a massive use of computational resource and thus takes long calculation times, especially in the presence of a large amount of simulation replications. The second is that simulation-based approach provides a linkage, usually a "black-box", between the inputs and the outputs. In terms of the former, there are some advanced simulation techniques that can be used to improve the simulation efficiency, as will be discussed in Sect. 3.5. In terms of the latter, closed-form analytical solutions may alternatively offer insights that may otherwise be difficult to achieve through Monte Carlo simulation alone, and may support and even accelerate the development of more complex problems including random variables. This argument will be demonstrated in later chapters.

3.2   Simulation of a Single Random Variable

3.2.1   Inverse Transformation Method

For a continuous random variable X with a CDF of $F_X(x)$, a powerful tool to simulate a realization of X is the *inverse transformation method* [17]. The basic idea is to first generate a realization of a uniform variable within $[0, 1]$, denoted by u, and then find x by letting $x = F_X^{-1}(u)$. This method is applicable provided that the expression of the CDF (as well as the inverse of the CDF) is computable. Figure 3.2 illustrates the inverse transformation method, which is based on the fact that $F_X(X)$ is a uniform random variable within $[0, 1]$. In fact, letting $Y = F_X(X)$, it follows,

$$\mathbb{P}(Y \le u) = \mathbb{P}(F_X(X) \le u) = \mathbb{P}(X \le F_X^{-1}(u)) = F_X[F_X^{-1}(u)] = u, \quad u \in [0, 1] \tag{3.7}$$

which is evident of the uniform distribution of $F_X(X)$ (c.f. Sect. 2.2.2).

Fig. 3.2 Illustration of the inverse transformation method

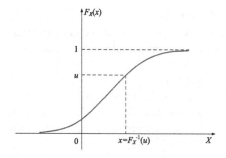

Simulating a uniform random variable over $[0, 1]$ is the basis for sampling other distribution types. An intuitive approach is to prepare ten slips of paper, marked $0, 1, 2, \ldots, 9$, put them in a box, and then equally randomly select one of them with replacement for n times. The selected numbers are treated as a sequence of digits after the decimal point of a desired sample x rounded off to the nearest 10^{-n}. For instance, with $n = 5$, if the selected numbers are orderly $4, 3, 7, 4, 9$, then the sample is 0.43749 rounded off to the nearest 0.00001. Motivated by this, the RAND Corporation published a random number tables known as *A Million Random Digits with 100,000 Normal Deviates* in 1955. However, this is not the way that digital computers generate random variables. Instead, the pseudo random variables are generated rather than the truly random variables. With this regard, the most widely used approach to simulate a sequence of random variables $\{R_1, R_2, \ldots\}$ is to assign

$$R_{k+1} = aR_k + b \quad \text{mod } m, \quad k \geq 0 \tag{3.8}$$

where a and b are two properly chosen constants, and R_0 is the initial value (the "seed"). Clearly, each R_i may take a value out of $\{0, 1, , 2, \ldots m - 1\}$ and thus R_i/m is a pseudo uniform variable over $[0, 1]$ if m is large enough.

Example 3.1

For an exponential random variable X whose PDF is $f_X(x) = \lambda \exp(-\lambda x)$ (c.f. Eq. (2.163)), develop an algorithm to generate a sample for X, x.

Solution

The CDF of X is $F_X(x) = 1 - \exp(-\lambda x)$, and correspondingly, the inverse is $F_X^{-1}(x) = -\frac{1}{\lambda} \ln(1 - x)$. Thus, the procedure for sampling X is as follows.

(1) Generate a uniform variable within $[0, 1]$, u.
(2) Let $x = -\frac{1}{\lambda} \ln u$.

The second step is based on the fact that for a uniform variable U over $[0, 1]$, $1 - U$ has the same distribution as U.

Example 3.2

If random variable X follows a Gamma distribution with a shape parameter a being a positive integer and a scale parameter $b > 0$, propose a method to sample a realization of X, x.

Solution

Note that X can be treated as the sum of a independent and identically exponentially distributed variables $X_1, X_2, \ldots X_a$ with a mean value of b (c.f. Sect. 1.2.7). With this, one can first simulate a exponential samples $x_1, x_2, \ldots, x_a$ (c.f. Example 3.1), and then let $x = \sum_{i=1}^{a} x_i$.

Example 3.3

For a random variable X that follows an Extreme Type I distribution with a mean value of m_X and a standard deviation of s_X, develop an algorithm to generate a sample for X, x.

Solution

Note that the CDF of X is $F_X(x) = \exp\left[-\exp\left(-\frac{x-\mu}{\sigma}\right)\right]$, with which the inverse of $F_X(x)$ is $F_X^{-1}(x) = \mu - \sigma \ln[-\ln(u)]$. Thus, the procedure for sampling X is as follows, using the inverse transformation method.

(1) Find the two parameters μ and σ in Eq. (2.231) according to $\sigma = \frac{\sqrt{6}}{\pi} \cdot s_X$ and $\mu = m_X - \gamma\sigma$, where γ is the Euler-Mascheroni constant (≈ 0.5772).

(2) Generate a uniform variable within $[0, 1]$, u.

(3) Let $x = \mu - \sigma \ln[-\ln(u)]$.

Example 3.4

By referring to Fig. 2.5, develop an algorithm to simulate a realization of a normal random variable X with a mean value of μ_X and a standard deviation of σ_X.

Solution

The procedure is as follows.

(1) Generate two independent uniform variables within $[0, 1]$, u_1 and u_2.

(2) Let $r = \sqrt{-2 \ln u_1} \cdot \sigma_X$ and $\theta = 2\pi u_2$.

(3) Set $x = r \cdot \cos\theta + \mu_X$.

For the discrete case, if X is a discrete random variable with n possible values $\{x_1, x_2, \ldots x_n\}$ and a PMF of $p_X(x_i)$, $i = 1, 2, \ldots n$, since $F_X(X) = \sum_{\forall x_i \leq x} p_X(x_i)$ (c.f. Eq. (2.45)), we can extend the "inverse transformation method" to sample a realization of X, x, as follows.

(1) Sample a realization of uniform distribution over $[0, 1]$, u;

(2) Let $x = x_i$ if $\sum_{k=1}^{i-1} p_X(x_k) \leq u < \sum_{k=1}^{i} p_X(x_k)$, where $i = 1, 2, \ldots n$ and $\sum_{k=1}^{0} p_X(x_k) = 0$.

Example 3.5

For a geometric variable X with a mean value of $\frac{1}{p}$ (c.f. Sect. 2.2.1.3), develop an algorithm to simulate a realization of X.

Solution

Note that X has a PMF of $p_X(x_k) = \mathbb{P}(X = k) = (1 - p)^{k-1}p, k = 1, 2, \ldots$, with which

$$\sum_{k=1}^{i} p_X(x_k) = \sum_{k=1}^{i} (1 - p)^{k-1} p = 1 - (1 - p)^i \qquad (3.9)$$

Next, we need to find such an i that $\sum_{k=1}^{i-1} p_X(x_k) \leq U < \sum_{k=1}^{i} p_X(x_k)$, or equivalently, $1 - (1 - p)^{i-1} \leq U < 1 - (1 - p)^i$. With some simple algebra, one has

$$(1 - p)^i < U' \leq (1 - p)^{i-1} \qquad (3.10)$$

where $U' = 1 - U$ has the same distribution as U. Thus, $i - 1 \leq \frac{\ln U'}{\ln(1-p)} < i$, and further, $i = \left\lfloor \frac{\ln U'}{\ln(1-p)} \right\rfloor + 1$, where $\lfloor x \rfloor$ means the maximum integer that does not exceed x. With the aforementioned analysis, the procedure to sample a realization of X is as follows.

(1) Sample a realization of uniform distribution over $[0, 1]$, u.
(2) Let $x = \left\lfloor \frac{\ln u}{\ln(1-p)} \right\rfloor + 1$.

3.2.2 Composition Method

The *composition method* is used to simulate random variables based on the conditional probability (c.f. Sect. 2.3.2) of a random variable.

For a continuous variable X with a PDF of $f_X(x)$, one can also express $f_X(x)$ as follows (c.f. Eq. (2.296)),

$$f_X(x) = \int f_{X|Y}(x|y) f_Y(y) \mathrm{d}y \qquad (3.11)$$

where $f_Y(y)$ is the PDF of another variable Y, and $f_{X|Y}(x|y)$ is the conditional PDF of X on $Y = y$. With this, the procedure to sample a realization of X, x, is as follows,

(1) Simulate a sample for Y, y.
(2) Simulate a sample for X with a PDF of $f_{X|Y}(x|y)$.

Example 3.6

If the PDF of a non-negative random variable X, $f_X(x)$, is

$$f_X(x) = 8 \int_2^\infty \frac{\exp(-xy)}{y^2} dy, \quad x \geq 0 \tag{3.12}$$

Establish a procedure to sample X using the composition method.
Solution
We first rewrite the expression of $f_X(x)$ as follows,

$$f_X(x) = \int_2^\infty \underbrace{\frac{8}{y^3}}_{f_Y(y)} \cdot \underbrace{[y\exp(-xy)]}_{f_{X|Y}(x|y)} dy, \quad x \geq 0 \tag{3.13}$$

With this, we can first generate a sample for Y whose PDF is $f_Y(y) = \frac{8}{y^3}$, $y \geq 2$, and then simulate a sample for X whose conditional PDF is $f_{X|Y}(x|y) = y\exp(-xy)$. Regarding the sampling of Y, it is noticed that the CDF of Y is $F_Y(y) = 1 - \frac{4}{y^2}$, $y \geq 2$, with which the "inverse transform method" can be used. Next, conditional on $Y = y$, X follows an exponential distribution.
With the aforementioned information, the procedure to sample X is as follows.

(1) Simulate two independent samples for uniform variable over $[0, 1]$, u_1 and u_2.
(2) Let $y = \frac{2}{\sqrt{u_1}}$.
(3) Set $x = -\frac{1}{y} \ln u_2$.

We can further extend Eq. (3.11) to the discrete case. Suppose that $f_X(x)$ is expressed as

$$f_X(x) = \sum_{i=1}^n p_Y(y_i) f_{X|Y}(x|y_i) \tag{3.14}$$

where $p_Y(y_i)$ is the PMF of Y, which is a discrete variable with n possible values $y_1, y_2, \ldots y_n$. With this, one can sample X with the following steps: (1) Simulate a sample for Y, y; (2) Simulate a sample for X with a PDF of $f_{X|Y}(x|y_i)$.

Example 3.7

For a continuous random variable X that follows a *Lindley distribution*, its PDF takes the form of

$$f_X(x) = \frac{\theta^2}{\theta + 1}(x + 1)\exp(-\theta x), \quad x \geq 0, \theta > 0 \tag{3.15}$$

where θ is a parameter. The CDF of X is

$$F_X(x) = 1 - \frac{\theta + 1 + \theta x}{\theta + 1}\exp(-\theta x), \quad x \geq 0, \theta > 0 \tag{3.16}$$

Clearly, it is not straightforward to sample X using the inverse transform method due to the difficulty in solving $F_X(x) = u$. On the other hand, note that the following relationship holds [7],

$$f_X(x) = p f_1(x) + (1 - p) f_2(x) \tag{3.17}$$

where $p = \frac{\theta}{\theta+1}$, $f_1(x) = \theta \exp(-\theta x)$, and $f_2(x) = \theta^2 x \exp(-\theta x)$. Equation (3.17) indicates that the PDF of X is a mixture of exponential distribution (f_1) and Gamma distribution (f_2). With this, using the composition method, develop an algorithm to simulate a sample for X.

Solution

The simulation procedure is as follows.

(1) Simulate a uniform variable over $[0, 1]$, u.
(2) Generate an exponentially distributed sample x_1 with a mean value of $\frac{1}{\theta}$ (c.f. Example 3.1).
(3) Generate a Gamma-distributed sample x_2 with a shape parameter of 2 and a scale parameter of $\frac{1}{\theta}$ (c.f. Example 3.2).
(4) If $u \le p = \frac{\theta}{\theta+1}$, set $x = x_1$; otherwise, set $x = x_2$.

3.2.3 Acceptance-Rejection Method

For the case where the explicit expression of the CDF of X is not available, one may alternatively use the *acceptance-rejection method* [17] to generate a sample for X, which is based on the sampling of another variable Y. We will first consider the continuous case and then extend the method to the case of discrete random variables.

Mathematically, if the PDFs of X and Y are $f_X(x)$ and $f_Y(y)$, respectively, and there exists a constant ξ such that $f_X(x) \le \xi f_Y(x)$ holds for $\forall x$, then the procedure of sampling X is as follows,

(1) Simulate a realization for Y with a PDF of $f_Y(y)$, y, and a uniform variable within $[0, 1]$, u.
(2) Assign $x = y$ if $u \le \frac{f_X(y)}{\xi f_Y(y)}$; otherwise, return to step (1).

The acceptance-rejection method is guaranteed by the fact that

$$\mathbb{P}(X_0 = x) = \mathbb{P}\left(Y = x \,\middle|\, u \le \frac{f_X(x)}{\xi f_Y(x)}\right) = \frac{\mathbb{P}\left[Y = x \cap u \le \frac{f_X(x)}{\xi f_Y(x)}\right]}{\mathbb{P}\left[u \le \frac{f_X(x)}{\xi f_Y(x)}\right]}$$

$$= \frac{f_Y(x) dx \cdot \frac{f_X(x)}{\xi f_Y(x)}}{\int f_Y(x) \cdot \frac{f_X(x)}{\xi f_Y(x)} dx} = f_X(x) dx \tag{3.18}$$

where X_0 is the simulated random variable. In practice, we can choose such an $f_Y(y)$ that it is easy to generate a sample for Y.

Example 3.8

If the PDF of random variable X is $f_X(x) = \frac{3}{2\pi(x^6+1)}$, $x \in (-\infty, \infty)$, establish a procedure to generate a realization of X, x.

Solution

Consider a variable Y with a PDF of $f_Y(y) = \frac{1}{\pi} \cdot \frac{1}{1+y^2}$, $y \in (-\infty, \infty)$. The CDF of Y is

$$F_Y(y) = \int_{-\infty}^{y} f_Y(\tau)d\tau = \frac{1}{\pi}\left(\arctan y + \frac{\pi}{2}\right) \tag{3.19}$$

For an arbitrary value z, the following relationship holds,

$$\frac{f_X(z)}{f_Y(z)} = \frac{3}{2} \cdot \frac{z^2+1}{z^6+1} \le 2 \tag{3.20}$$

Thus, the procedure of simulating a sample for X is as follows.

(1) Simulate two independent uniform variables within $[0, 1]$, u_1 and u_2.
(2) Let $y = \tan\left(u_1\pi - \frac{\pi}{2}\right)$.
(3) If $u_2 \le \frac{3}{4} \cdot \frac{y^2+1}{y^6+1}$, then set $x = y$; otherwise, return to step (1).

Example 3.9

Recall that the PDF of a standard half-normal variable X is $f_X(x) = \frac{\sqrt{2}}{\sqrt{\pi}}\exp\left(-\frac{x^2}{2}\right)$ (c.f. Eq. (2.182)). It is noticed that for an exponentially distributed variable Y whose PDF is $f_Y(y) = \exp(-y)$, $y \ge 0$, one has

$$\frac{f_X(x)}{f_Y(x)} = \sqrt{\frac{2}{\pi}}\exp\left(\frac{1}{2}\right) \cdot \exp\left[-\frac{1}{2}(x-1)^2\right] \le \sqrt{\frac{2e}{\pi}} \tag{3.21}$$

With this, propose a procedure to sample a realization of X using the acceptance-rejection method.

Solution

The procedure is as follows.

(1) Simulate two independent uniform variables within $[0, 1]$, u_1 and u_2.
(2) Let $y = -\ln u_1$.
(3) If $u_2 \le \exp\left[-\frac{1}{2}(y-1)^2\right]$, then set $x = y$; otherwise, return to step (1).

Remark

A simpler version of the acceptance-rejection method was originally developed by von Neumann, which is stated in the following. For a variable X whose PDF $f(x)$ is defined within $[a, b]$ and has a maximum value of max $f(x)$, one can generate a sample for X as follows,

(1) Sample a realization of uniform variable over $[a, b]$, denoted by y.
(2) Generate another sample u of uniform variable within $[0, 1]$.
(3) Set $x = y$ if $u \leq \frac{f(y)}{\max f(y)}$, and return to step (1) otherwise.

However, this version is not applicable if the range of X is not limited.

Example 3.10

Consider a Beta distributed variable X whose PDF is $f_X(x) = \frac{35}{4}x(1-x)^{\frac{3}{2}}$. Use the acceptance-rejection method to generate a sample for X.

Solution

We herein use the acceptance-rejection method in von Neumann's version. First, we find the maxima of $f_X(x)$ by letting

$$\frac{df_X(x)}{dx} = \frac{35}{4}(1-x)^{\frac{1}{2}}\left(1 - x - \frac{3}{2}x\right) = 0$$

$$\Rightarrow x = \frac{2}{5} \Rightarrow \max f_X(x) = \frac{21}{50}\sqrt{15} \approx 1.6267$$

(3.22)

Next, the procedure to sample X is as follows. (1) Generate two samples of uniform variable over $[0, 1]$, denoted by u_1 and u_2. (2) Set $x = u_2$ if $u_1 \leq 5.379u_2(1 - u_2)^{\frac{3}{2}}$; otherwise, return to step (1).

For the discrete case, the acceptance-rejection method works in an analogous manner. That is, if the PMFs of discrete random variables X and Y are $p_X(x_i)$ and $p_Y(y_j)$, respectively, and there exists a constant ξ such that $p_X(x_i) \leq \xi p_Y(x_i)$ holds for $\forall i$, then the procedure of sampling X is as follows,

(1) Simulate a realization for Y with a PMF of $p_Y(y_j)$, y, and a uniform variable within $[0, 1]$, u.
(2) Assign $x = y$ if $u \leq \frac{p_X(y)}{\xi p_Y(y)}$; otherwise, return to step (1).

Example 3.11

Consider a discrete random variable X which takes a value out of $\{2, 3, 4, 5, 6, 7\}$ and has a PMF of $p_X(k) = \frac{k}{27}, k = 2, 3, \ldots 7$. Establish a method to generate a sample of X, x.

Solution

We introduce another discrete variable Y which has a PMF of $p_Y(k) = \frac{1}{6}, k = 2, 3, \ldots 7$. With this, it is easy to see that

$$\frac{p_X(y)}{p_Y(y)} \le \frac{\max p_X(y)}{p_Y(y)} = \frac{\frac{7}{27}}{\frac{1}{6}} = \frac{14}{9} \tag{3.23}$$

Thus, the procedure to sample X is as follows. (1) Simulate two samples of uniform variable within [0, 1], u_1 and u_2. (2) Let $y = [6u_1] + 2$. (3) Set $x = y$ if $u_2 \le \frac{y}{7}$; otherwise, return to step (1).

Example 3.12

For a Poisson random variable X with a mean value of λ, give a procedure to generate a sample of X.

Solution

First, note that the PMF of X is $p_X(k) = \frac{\lambda^k}{k!} \exp(-\lambda)$, $k = 0, 1, 2, \ldots$ (c.f. Eq. (2.152)). Now, we introduce another discrete variable Y such that $Y + 1$ is a geometric variable with a mean value of $\frac{1}{p}$ (in such a case, the domain of Y is the same as X). With this, $p_Y(k) = \mathbb{P}(Y + 1 = k + 1) = (1 - p)^k p$, $k = 0, 1, 2, \ldots$.

We let

$$h(k) = \frac{p_X(k)}{p_Y(k)} = \frac{1}{pk!} \cdot \left(\frac{\lambda}{1-p}\right)^k \exp(-\lambda) \tag{3.24}$$

and seek for the maxima of $h(k)$. Note that

$$\frac{h(k+1)}{h(k)} = \frac{\frac{1}{p(k+1)!} \cdot \left(\frac{\lambda}{1-p}\right)^{(k+1)} \exp(-\lambda)}{\frac{1}{pk!} \cdot \left(\frac{\lambda}{1-p}\right)^k \exp(-\lambda)} = \frac{\lambda}{(k+1)(1-p)} \tag{3.25}$$

Thus, the maximum $h(k)$ is achieved when $k = \widetilde{k} = \left\lfloor \frac{\lambda}{1-p} \right\rfloor$, i.e.,

$$\frac{p_X(k)}{p_Y(k)} = h(k) \le h(\widetilde{k}) = \frac{1}{p\widetilde{k}!} \cdot \left(\frac{\lambda}{1-p}\right)^{\widetilde{k}} \exp(-\lambda) \tag{3.26}$$

or equivalently,

$$\frac{p_X(k)}{h(\widetilde{k}) p_Y(k)} = \frac{\widetilde{k}!}{k!} \left(\frac{\lambda}{1-p}\right)^{k-\widetilde{k}} \tag{3.27}$$

Also, note that the generation of a geometric variable has been discussed in Example 3.5. Thus, the procedure to sample X is as follows.

(1) Sample two samples of uniform distribution over [0, 1], u_1 and u_2.

(2) Let $y = \left\lfloor \frac{\ln u_1}{\ln(1-p)} \right\rfloor$, and $\widetilde{k} = \left\lfloor \frac{\lambda}{1-p} \right\rfloor$.

(3) If $u_2 \le \frac{\widetilde{k}!}{y!} \left(\frac{\lambda}{1-p}\right)^{y-\widetilde{k}}$, set $x = y$; otherwise, return to step (1).

3.3 Simulation of Correlated Random Variables

In the time or space domain, the random events associated with different time points or locations are often correlated, and their statistical characteristics can be represented by a sequence of random variables or a *random vector*. We will first consider the continuous case and then move on to the discrete case.

For a correlated random vector $\mathbf{X} = \{X_1, X_2, \ldots X_n\}$, if the CDF of X_i is $F_{X_i}(x)$ for $i = 1, 2, \ldots n$, the *Rosenblatt transformation method* can be used to map $\mathbf{X}$ to an independent uniform random vector $\mathbf{U} = \{U_1, U_2, \ldots U_n\}$ over $[0, 1]$ according to

$$
\begin{aligned}
u_1 &= F_{X_1}(x_1); \\
u_i &= F_{X_i|\mathbf{X}_{i-1}}(x_i|\mathbf{x}_{i-1}), \quad i = 2, 3, \ldots n
\end{aligned}
\tag{3.28}
$$

where $F_{X_i|\mathbf{X}_{i-1}}(\cdot)$ is the conditional CDF of X_i on $\mathbf{X}_{i-1} = \mathbf{x}_{i-1} = \{x_1, x_2, \ldots x_{i-1}\}$. With Eq. (3.28), the algorithm to generate samples $\{x_1, x_2, \ldots x_n\}$ for $\mathbf{X}$ is as follows:

(1) Generate n independent uniform random variables over $[0, 1]$, $u_1, u_2, \ldots u_n$.
(2) Set $x_1 = F_{X_1}^{-1}(u_1)$ and $x_i = F_{X_1|\mathbf{X}_{i-1}}^{-1}(u_i)$ for $i = 2, 3, \ldots n$ respectively.

Example 3.13

Consider two continuous random variables X and Y with marginal CDFs of $F_X(x)$ and $F_Y(y)$ respectively and a joint CDF of $C(F_X(x), F_Y(y))$, where C is a copula function, $C(u, v) = \frac{uv}{u+v-uv}$ (c.f. Example 2.50). Develop an algorithm to simulate a sample pair for (X, Y), (x, y).

Solution

Let $U = F_X(X)$ and $V = F_Y(Y)$. The basic idea is to first simulate a sample pair for (U, V), (u, v), and then let $x = F_X^{-1}(u)$ and $y = F_Y^{-1}(v)$. With this regard, we can first sample U and then simulate V conditional on U. The conditional CDF of V on $U = u$, $F_{V|U}(v|u)$, is determined by

$$
\begin{aligned}
F_{V|U}(v|u) &= \mathbb{P}(V \leq v|U = u) = \lim_{du \to 0} \mathbb{P}(V \leq v|u \leq U \leq u + du) \\
&= \lim_{du \to 0} \frac{\mathbb{P}(V \leq v \cap u \leq U \leq u + du)}{P(u \leq U \leq u + du)} = \lim_{du \to 0} \frac{C(u + du, v) - C(u, v)}{du} \\
&= \frac{\partial C(u, v)}{\partial u}
\end{aligned}
\tag{3.29}
$$

Now, since $C(u, v) = \frac{uv}{u+v-uv}$, one has

$$
F_{V|U}(v|u) = \frac{\partial}{\partial u}\left(\frac{uv}{u+v-uv}\right) = \left(\frac{v}{u+v-uv}\right)^2
\tag{3.30}
$$

and correspondingly,

$$
F_{V|U}^{-1}(x) = \frac{u\sqrt{x}}{1 + (u-1)\sqrt{x}}
\tag{3.31}
$$

Thus, the procedure to sample (X, Y) is as follows.

(1) Generate two independent samples of uniform distribution over $[0, 1]$, u and u_1.

(2) Let $v = F_{V|U}^{-1}(u_1) = \frac{u\sqrt{u_1}}{1+(u-1)\sqrt{u_1}}$.

(3) Set $x = F_X^{-1}(u)$ and $y = F_Y^{-1}(v)$, with which (x, y) is the desired sample pair.

However, the Rosenblatt transformation is often underpowered due to the $n!$ possible ways of conditioning X_i in Eq. (3.28), which may result in considerable difference in the difficulty of sampling $\mathbf{X}$. Also, for most cases, the Rosenblatt transformation suffers from inaccessibility of conditional CDF $F_{X_i|\mathbf{X}_{i-1}}(\cdot)$ with limited statistical data.

As a worthwhile alternative of sampling $\mathbf{X}$, the *Nataf transformation method* only requires the marginal CDFs and correlation matrix of $\mathbf{X}$, and thus is more straightforward and applicable. The Nataf transformation assumes a Gaussian Copula function for the joint distribution of $\mathbf{X}$.

To begin with, we consider $\mathbf{X} = \{X_1, X_2, \ldots X_n\}$ as a standard normal random vector (i.e., each element in $\mathbf{X}$ has a normal distribution with a mean value of 0 and standard deviation of 1). Recall that in Sect. 2.2.4, it has been shown that the sum of two normal random variables still follows a normal distribution. Motivated by this, we can first generate n independent standard normal variables $U_1, U_2, \ldots U_n$, and then assign each X_i as the linear combination of $\{U_1, U_2, \ldots U_n\}$. In such a case, the obtained X_i is also normally distributed and is correlated with each other due to the common terms (U_i's) involved. Mathematically, if $\mathbf{X} = \mathbf{BU}$, or, in an expanded form,

$$\begin{bmatrix} X_1 \\ X_2 \\ \vdots \\ X_n \end{bmatrix} = \begin{bmatrix} b_{11} & b_{12} & \ldots & b_{1n} \\ b_{21} & b_{22} & \ldots & b_{2n} \\ \vdots & \vdots & \ddots & \vdots \\ b_{n1} & b_{n2} & \ldots & b_{nn} \end{bmatrix} \cdot \begin{bmatrix} U_1 \\ U_2 \\ \vdots \\ U_n \end{bmatrix} \tag{3.32}$$

then the variance of X_i is

$$\mathbb{V}(X_i) = \mathbb{V}(b_{i1}U_1 + b_{i2}U_2 + \cdots + b_{in}U_n) = b_{i1}^2 + b_{i2}^2 + \cdots + b_{in}^2 = 1 \tag{3.33}$$

Also, the covariance between X_i and X_j $(i \neq j)$ is obtained as

$$\begin{aligned} \mathbb{C}(X_i, X_j) &= \mathbb{C}(b_{i1}U_1 + b_{i2}U_2 + \cdots + b_{in}U_n, b_{j1}U_1 + b_{j2}U_2 + \cdots + b_{jn}U_n) \\ &= b_{i1}b_{j1}\mathbb{C}(U_1, U_1) + b_{i2}b_{j2}\mathbb{C}(U_2, U_2) + \cdots + b_{in}b_{jn}\mathbb{C}(U_n, U_n) \\ &= b_{i1}b_{j1} + b_{i2}b_{j2} + \cdots + b_{in}b_{jn} \end{aligned} \tag{3.34}$$

and accordingly, the linear correlation coefficient between X_i and X_j is

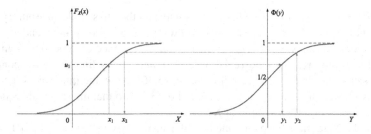

Fig. 3.3 Mapping an arbitrary continuous variable X to a standard normal variable Y

$$\rho(X_i, X_j) = \frac{\mathbb{C}(X_i, X_j)}{\sqrt{\mathbb{V}(X_i)\mathbb{V}(X_j)}} = \mathbb{C}(X_i, X_j) = b_{i1}b_{j1} + b_{i2}b_{j2} + \cdots + b_{in}b_{jn}$$

(3.35)

As such, we can properly choose b_{ij} so that both Eqs. (3.33) and (3.35) are satisfied. On the other hand, note that Eqs. (3.33) and (3.35) totally give $n + C_n^2 = \frac{1}{2}n(n+1)$ constraints for b_{ij} and thus the coefficient matrix **B** is theoretically not uniquely determined.

Furthermore, if **X** is not a standard normal vector but follows an arbitrary distribution, we can map each X_i to a standard normal variable Y_i by letting

$$F_{X_i}(x_i) = \Phi(Y_i)$$

(3.36)

in which F_{X_i} is the CDF of X_i. Equation (3.36) implies that X_i and the corresponding Y_i are connected via the same values of CDF evaluated at X_i and Y_i respectively, as illustrated in Fig. 3.3.

With Eq. (3.36), supposing that the linear correlation coefficient between Y_i and Y_j is ρ'_{ij}, we consider the correlation between X_i and X_j, ρ_{ij}, which is obtained as follows,

$$
\begin{aligned}
\rho_{ij} = \rho(X_i, X_j) &= \rho\left(\frac{X_i - \mathbb{E}(X_i)}{\sqrt{\mathbb{V}(X_i)}}, \frac{X_j - \mathbb{E}(X_j)}{\sqrt{\mathbb{V}(X_j)}}\right) = \mathbb{E}\left(\frac{X_i - \mathbb{E}(X_i)}{\sqrt{\mathbb{V}(X_i)}} \cdot \frac{X_j - \mathbb{E}(X_j)}{\sqrt{\mathbb{V}(X_j)}}\right) \\
&= \mathbb{E}\left(\frac{F_{X_i}^{-1}(\Phi(Y_i)) - \mathbb{E}(X_i)}{\sqrt{\mathbb{V}(X_i)}} \cdot \frac{F_{X_j}^{-1}(\Phi(Y_j)) - \mathbb{E}(X_j)}{\sqrt{\mathbb{V}(X_j)}}\right) \\
&= \int_{-\infty}^{\infty}\int_{-\infty}^{\infty} \frac{F_{X_i}^{-1}(\Phi(y_i)) - \mathbb{E}(X_i)}{\sqrt{\mathbb{V}(X_i)}} \cdot \frac{F_{X_j}^{-1}(\Phi(y_j)) - \mathbb{E}(X_j)}{\sqrt{\mathbb{V}(X_j)}} f_{Y_i,Y_j}(y_i, y_j)\mathrm{d}y_i\mathrm{d}y_j
\end{aligned}
$$

(3.37)

where $f_{Y_i,Y_j}(y_i, y_j)$ is the joint PDF of Y_i and Y_j,

$$f_{Y_i,Y_j}(y_i, y_j) = \frac{1}{2\pi\sqrt{1 - \rho_{ij}'^2}} \exp\left[-\frac{y_i^2 - 2\rho'_{ij}y_iy_j + y_j^2}{2(1 - \rho_{ij}'^2)}\right]$$

(3.38)

With the aforementioned analysis, a promising method to sample an arbitrary random vector $\mathbf{X} = \{X_1, X_2, \ldots X_n\}$ is as follows. First, generate n independent standard normal variables $u_1, u_2, \ldots u_n$; then obtain a sequence of dependent standard normal variables $y_1, y_2, \ldots y_n$ (c.f. Eq. (3.32)), and finally find the corresponding vector sample $x_1, x_2, \ldots x_n$ according to Eq. (3.36). Now, the question is, can we properly choose the coefficient matrix $\mathbf{B}$ in Eq. (3.32) so that the sampled sequence $x_1, x_2, \ldots x_n$ is exactly a sample of $\mathbf{X}$?

With the Nataf transformation method, the answer would be "yes". In fact, if the correlation matrix of $\mathbf{Y}$ is $\boldsymbol{\rho}' = [\rho'_{ij}]_{n \times n}$, we can perform the Cholesky decomposition of $\boldsymbol{\rho}'$ to find an $n \times n$ lower triangular matrix $\mathbf{A}$, i.e., $\mathbf{AA}^\mathsf{T} = \boldsymbol{\rho}'$, where the superscript T denotes the transpose of the matrix, and let $\mathbf{B} = \mathbf{A}$. With this, Eq. (3.32) becomes

$$\mathbf{Y} = \mathbf{AU} \qquad (3.39)$$

It can be shown that the vector $\mathbf{Y}$ in Eq. (3.39) is a standard normal vector having a correlation matrix of $\boldsymbol{\rho}'$ (see Example 3.14). Moreover, since $\mathbf{A}$ is a lower triangle matrix, totally $n + \frac{n(n-1)}{2}$ elements are involved, which can be uniquely determined by the constraints in Eqs. (3.33) and (3.35). Also, note that the elements in $\boldsymbol{\rho}'$, ρ'_{ij}, can be determined through Eq. (3.37). Specially, for the case where the COV of X_i is less than 0.5, ρ'_{ij} can also be approximately calculated by $R_{ij} \cdot \rho_{ij}$, where R_{ij} is a polynomial function of ρ_{ij} and the COVs of X_i and X_j [6, 12].

In summary, the procedure of simulating a sample for $\mathbf{X}$, $\{x_1, x_2, \ldots x_n\}$, is as follows using the Nataf transformation method.

(1) Determine the correlation matrix of $\mathbf{Y}$, $\boldsymbol{\rho}'$, according to Eq. (3.37).
(2) Find the lower triangle matrix $\mathbf{A}$ through Cholesky factorization of $\boldsymbol{\rho}'$.
(3) Generate n independent standard normal samples $u_1, u_2, \ldots u_n$.
(4) Set $x_i = F_{X_i}^{-1}\left[\Phi\left(\sum_{j=1}^{i} a_{ij} u_j \right) \right]$ for $i = 1, 2, \ldots n$.

Example 3.14

Show that the vector $\mathbf{Y}$ in Eq. (3.39) is a standard normal vector having a correlation matrix of $\boldsymbol{\rho}'$.

Solution

First, by noting that $\mathbf{AA}^\mathsf{T} = \boldsymbol{\rho}'$, it follows,

$$a_{ii} = \sqrt{\rho'_{ii} - \sum_{k=1}^{i-1} a_{ik}^2}, \quad i = 1, 2, \ldots n \qquad (3.40)$$

and

$$a_{ij} = \frac{1}{a_{ii}} \left(\rho'_{ij} - \sum_{k=1}^{i-1} a_{ik} a_{jk} \right), \quad 1 \le j < i \le n \qquad (3.41)$$

With Eq. (3.39), $Y_i = \sum_{j=1}^{i} a_{ij} U_j$, with which $\mathbb{E}(Y_i) = 0$. Furthermore, according to Eq. (3.40), $\sum_{k=1}^{i} a_{ik}^2 = \rho_{ii}' = 1$, and thus

$$\mathbb{V}(Y_i) = \sum_{j=1}^{i} a_{ij}^2 \mathbb{V}(U_j) = \sum_{j=1}^{i} a_{ij}^2 = 1 \tag{3.42}$$

Thus, each Y_i follows a standard normal distribution. Next, we consider the linear correlation between Y_i and Y_j ($i \leq j$). According to Eq. (3.41), $\sum_{k=1}^{i} a_{ik} a_{jk} = \rho_{ij}'$. Thus,

$$\rho(Y_i, Y_j) = \mathbb{C}(Y_i, Y_j) = \mathbb{C}\left(\sum_{k=1}^{i} a_{ik} U_k, \sum_{l=1}^{j} a_{jl} U_l\right) = \sum_{k=1}^{i} \sum_{l=1}^{j} \mathbb{C}(a_{ik} U_k, a_{jl} U_l)$$

$$= \sum_{k=1}^{i} \mathbb{C}(a_{ik} U_k, a_{jk} U_k) = \sum_{k=1}^{i} a_{ik} a_{jk} = \rho_{ij}'$$

$$\tag{3.43}$$

which indicates that the correlation coefficient of Y_i and Y_j is exactly ρ_{ij}' as desired.

Example 3.15

In Eq. (3.37), if both X_i and X_j are identically lognormally distributed with a COV of v, show that

$$\rho_{ij}' = \frac{\ln(1 + \rho_{ij} v^2)}{\ln(1 + v^2)} \tag{3.44}$$

Solution

Suppose that both X_i and X_j have a CDF of

$$F_{X_i}(x) = F_{X_j}(x) = \Phi\left(\frac{\ln x - \lambda}{\varepsilon}\right) \tag{3.45}$$

where ε equals $\sqrt{\ln(v^2 + 1)}$ and λ is the mean value of $\ln X_i$ (c.f. Sect. 2.2.5). Thus,

$$F_{X_i}^{-1}(x) = F_{X_j}^{-1}(x) = \exp\left(\lambda + \varepsilon \Phi^{-1}(x)\right) \tag{3.46}$$

According to Eq. (3.37),

$$\rho_{ij} = \int\int \frac{F_{X_i}^{-1}(\Phi(y_i)) - \mathbb{E}(X_i)}{\sqrt{\mathbb{V}(X_i)}} \cdot \frac{F_{X_j}^{-1}(\Phi(y_j)) - \mathbb{E}(X_j)}{\sqrt{\mathbb{V}(X_j)}} f_{Y_i, Y_j}(y_i, y_j) dy_i dy_j$$

$$= \int\int \frac{\exp(\lambda + \varepsilon y_i) - \exp\left(\lambda + \frac{1}{2}\varepsilon^2\right)}{v \exp\left(\lambda + \frac{1}{2}\varepsilon^2\right)} \cdot \frac{\exp(\lambda + \varepsilon y_j) - \exp\left(\lambda + \frac{1}{2}\varepsilon^2\right)}{v \exp\left(\lambda + \frac{1}{2}\varepsilon^2\right)} f_{Y_i, Y_j}(y_i, y_j) dy_i dy_j$$

$$= \frac{1}{v^2} \int\int \left[\exp\left(\varepsilon y_i - \frac{1}{2}\varepsilon^2\right) - 1\right] \cdot \left[\exp\left(\varepsilon y_j - \frac{1}{2}\varepsilon^2\right) - 1\right] f_{Y_i, Y_j}(y_i, y_j) dy_i dy_j$$

$$\tag{3.47}$$

Since

$$\left[\exp\left(\varepsilon y_i - \frac{1}{2}\varepsilon^2\right) - 1\right] \cdot \left[\exp\left(\varepsilon y_j - \frac{1}{2}\varepsilon^2\right) - 1\right] \tag{3.48}$$

$$= \underbrace{\exp\left(\varepsilon y_i + \varepsilon y_j - \varepsilon^2\right)}_{\text{part 1}} - \underbrace{\exp\left(\varepsilon y_i - \frac{1}{2}\varepsilon^2\right)}_{\text{part 2}} - \underbrace{\exp\left(\varepsilon y_j - \frac{1}{2}\varepsilon^2\right)}_{\text{part 3}} + 1$$

$$\int\int (\text{part 1}) \cdot f_{Y_i,Y_j}(y_i, y_j) dy_i dy_j = \int\int \frac{\exp\left(\varepsilon y_i + \varepsilon y_j - \varepsilon^2\right)}{2\pi\sqrt{1-\rho_{ij}'^2}} \exp\left[-\frac{y_i^2 - 2\rho_{ij}' y_i y_j + y_j^2}{2(1-\rho_{ij}'^2)}\right] dy_i dy_j$$

$$= \exp(\varepsilon^2 \rho_{ij}') \tag{3.49}$$

$$\int\int (\text{part 2}) \cdot f_{Y_i,Y_j}(y_i, y_j) dy_i dy_j = \int\int \frac{\exp\left(\varepsilon y_i - \frac{1}{2}\varepsilon^2\right)}{2\pi\sqrt{1-\rho_{ij}'^2}} \exp\left[-\frac{y_i^2 - 2\rho_{ij}' y_i y_j + y_j^2}{2(1-\rho_{ij}'^2)}\right] dy_i dy_j$$

$$= 1 = \int\int (\text{part 3}) \cdot f_{Y_i,Y_j}(y_i, y_j) dy_i dy_j \tag{3.50}$$

one has,

$$\rho_{ij} = \frac{1}{v^2}\left[\exp(\varepsilon^2 \rho_{ij}') - 1\right] \Rightarrow \rho_{ij}' = \frac{\ln(1 + \rho_{ij} v^2)}{\ln(1 + v^2)} \tag{3.51}$$

Example 3.16

Consider three identically exponentially distributed variables X_1, X_2 and X_3 with a mean value of 2. Suppose that the correlation coefficients between (X_1, X_2), (X_2, X_3) and (X_1, X_3) are 0.9, 0.7 and 0.5 respectively. In order to sample a realization of $\mathbf{X} = \{X_1, X_2, X_3\}$ with the Nataf transformation method, one can first generate three independent standard normal samples u_1, u_2, u_3. Now, given that $u_1 = 0.5$, $u_2 = 1.8$, $u_3 = -2.2$, find the corresponding sample of $\mathbf{X}$.

Solution

The CDF of each X_i is $F_X(x) = 1 - \exp\left(-\frac{1}{2}x\right)$ for $x \geq 0$, and the inverse of F_X is $F_X^{-1}(x) = -2\ln(1-x)$. Moreover, the correlation matrix for $\mathbf{X}$ is

$$\boldsymbol{\rho} = \begin{bmatrix} 1.0 & 0.9 & 0.5 \\ 0.9 & 1.0 & 0.7 \\ 0.5 & 0.7 & 1.0 \end{bmatrix} \tag{3.52}$$

Now we map $\{X_1, X_2, X_3\}$ to standard normal variables $\{Y_1, Y_2, Y_3\}$ with $F_X(X_i) = \Phi(Y_i)$. With this, it follows [20]

$$R_{ij} = \frac{\rho'_{ij}}{\rho_{ij}} \approx -0.0553\rho_{ij}^3 + 0.1520\rho_{ij}^2 - 0.3252\rho_{ij} + 1.2285 \tag{3.53}$$

where ρ'_{ij} is the correlation coefficient between Y_i and Y_j. As such, the correlation matrix for $\mathbf{Y}$ is obtained as

$$\boldsymbol{\rho}' = \begin{bmatrix} 1.0 & 0.9168 & 0.5485 \\ 0.9168 & 1.0 & 0.7395 \\ 0.5485 & 0.7395 & 1.0 \end{bmatrix} \tag{3.54}$$

Performing the Cholesky decomposition gives

$$\mathbf{A} = \begin{bmatrix} 1.0 & 0 & 0 \\ 0.9168 & 0.3993 & 0 \\ 0.5485 & 0.5926 & 0.5899 \end{bmatrix} \tag{3.55}$$

Thus,

$$x_1 = F_X^{-1}[\Phi(a_{11} \cdot u_1)] = -2\ln[1 - \Phi(1 \times 0.5)] = 2.3518$$
$$x_2 = F_X^{-1}[\Phi(a_{21} \cdot u_1 + a_{22} \cdot u_2)] = -2\ln[1 - \Phi(0.9168 \times 0.5 + 0.3993 \times 1.8)] = 4.2477$$
$$x_3 = F_X^{-1}[\Phi(a_{31} \cdot u_1 + a_{32} \cdot u_2 + a_{33} \cdot u_3)] = -2\ln[1 - \Phi(0.5485 \times 0.5 + 0.5926 \times 1.8 - 2.2 \times 0.5899)] = 1.4563$$
$$\tag{3.56}$$

Example 3.17

For two standard normal variables Y_i and Y_j with a linear correlation coefficient of ρ'_{ij}, show that their joint PDF is as in Eq. (3.38).

Solution

We first introduce a standard variable $\widetilde{Y}$ independent of Y_i such that

$$Y_j = \rho'_{ij} Y_i + \sqrt{1 - \rho'^2_{ij}} \cdot \widetilde{Y} \tag{3.57}$$

It is easy to verify that

$$\mathbb{E}(Y_j) = \rho'_{ij}\mathbb{E}(Y_i) + \sqrt{1 - \rho'^2_{ij}} \cdot \mathbb{E}(\widetilde{Y}) = 0$$
$$\mathbb{V}(Y_j) = \rho'^2_{ij}\mathbb{V}(Y_i) + \left(\sqrt{1 - \rho'^2_{ij}}\right)^2 \cdot \mathbb{V}(\widetilde{Y}) = 1 \tag{3.58}$$
$$\rho(Y_i, Y_j) = \mathbb{C}(Y_i, Y_j) = \mathbb{C}(Y_i, \rho'_{ij}Y_i) = \rho'_{ij}$$

Note that Y_i and $\widetilde{Y}$ are independent, with which their joint PDF is simply

$$f_{Y_i, \widetilde{Y}} = f_{Y_i}(y_i) \cdot f_{\widetilde{Y}}(\widetilde{y}) = \frac{1}{2\pi} \exp\left(-\frac{y_i^2}{2} - \frac{\widetilde{y}^2}{2}\right) \tag{3.59}$$

The joint PDF of Y_i and Y_j is obtained by first considering the determinant of the Jacobian matrix from $(Y_i, \tilde{Y})$ to (Y_i, Y_j) as follows (c.f. Sect. 2.3),

$$J = \begin{vmatrix} \frac{\partial y_i}{\partial y_i} & \frac{\partial y_i}{\partial \tilde{y}} \\ \frac{\partial y_j}{\partial y_i} & \frac{\partial y_j}{\partial \tilde{y}} \end{vmatrix} = \begin{vmatrix} 1 & 0 \\ \rho'_{ij} & \sqrt{1 - \rho'^2_{ij}} \end{vmatrix} = \sqrt{1 - \rho'^2_{ij}} \tag{3.60}$$

Thus,

$$f_{Y_i,Y_j}(y_i, y_j) = \frac{1}{|J|} f_{Y_i,\tilde{Y}}\left(y_i, \frac{y_j - \rho'_{ij} y_i}{\sqrt{1 - \rho'^2_{ij}}}\right) = \frac{1}{\sqrt{1 - \rho'^2_{ij}}} \cdot \frac{1}{2\pi} \exp\left[-\frac{y_i^2}{2} - \frac{1}{2}\left(\frac{y_j - \rho'_{ij} y_i}{\sqrt{1 - \rho'^2_{ij}}}\right)^2\right]$$

$$= \frac{1}{2\pi\sqrt{1 - \rho'^2_{ij}}} \exp\left[-\frac{y_i^2 - 2\rho'_{ij} y_i y_j + y_j^2}{2(1 - \rho'^2_{ij})}\right]$$

$$\tag{3.61}$$

which gives the same result as in Eq. (3.38).

Now we discuss the simulation of correlated discrete random variables. Similar to the observation that the sum of two normal variables is also normally distributed, the sum of two Poisson random variables still follows a Poisson distribution (c.f. Example 2.23). Let X_1 and X_2 be two correlated Poisson variables with mean values of v_1 and v_2 respectively and a linear correlation coefficient of ρ. Now, to model X_1 and X_2, we introduce three independent Poisson variables Y_1, Y_2 and Y_3 with mean values of γ_1, γ_2 and γ_3 respectively such that

$$X_1 = Y_1 + Y_2, \quad X_2 = Y_1 + Y_3 \tag{3.62}$$

With this, one can uniquely determine γ_1, γ_2 and γ_3 by considering

$$\mathbb{E}(X_1) = \mathbb{E}(Y_1) + \mathbb{E}(Y_2) \Rightarrow \gamma_1 + \gamma_2 = v_1$$
$$\mathbb{E}(X_2) = \mathbb{E}(Y_1) + \mathbb{E}(Y_3) \Rightarrow \gamma_1 + \gamma_3 = v_2$$
$$\rho(X_1, X_2) = \frac{\mathbb{E}(X_1 X_2) - v_1 v_2}{\sqrt{v_1 v_2}} = \frac{\gamma_1^2 + \gamma_1 + \gamma_1 \gamma_2 + \gamma_1 \gamma_3 + \gamma_2 \gamma_3 - v_1 v_2}{\sqrt{v_1 v_2}} = \rho$$

$$\tag{3.63}$$

which yield

$$\gamma_1 = \rho\sqrt{v_1 v_2}, \quad \gamma_2 = v_1 - \rho\sqrt{v_1 v_2}, \quad \gamma_3 = v_2 - \rho\sqrt{v_1 v_2} \tag{3.64}$$

With Eq. (3.64), to ensure $\gamma_i > 0$ $(i = 1, 2, 3)$, ρ should satisfy that $0 < \rho < \min\left(\sqrt{\frac{v_1}{v_2}}, \sqrt{\frac{v_2}{v_1}}\right)$. This suggests that the correlation between X_1 and X_2 is strictly limited to $\left[0, \min\left(\sqrt{\frac{v_1}{v_2}}, \sqrt{\frac{v_2}{v_1}}\right)\right]$ if modelling X_1 and X_2 with Eq. (3.62).

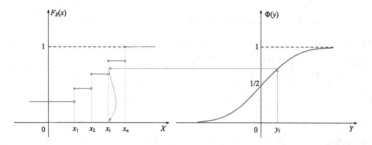

Fig. 3.4 Mapping a standard normal variable Y to a discrete variable X

In general, for the discrete case, we can also use a method as in Eq. (3.36) to establish a link between a random vector and a standard normal vector [19]. For a discrete variable X with n possible values $x_1, x_2, \ldots x_n$ and a PMF of $p_X(x_i)$, $i = 1, 2, \ldots n$, we introduce a standard normal variable Y such that

$$X = F_X^{-1}[\Phi(Y)] \tag{3.65}$$

where $F_X^{-1}(u) = \inf\{x : F_X(x) \geq u\}$ and $F_X(x)$ is the CDF of X. Figure 3.4 illustrates the mapping from Y to X. It is noticed that X is a function of Y but the inverse proposition is not true, i.e., Y is not a function of X.

Now, for two standard normal variables Y_i and Y_j with a linear correlation coefficient of ρ'_{ij}, let X_i and X_j denote the two corresponding discrete variables with a correlation coefficient of ρ_{ij}. Similar to Eq. (3.37),

$$\rho_{ij} = \int_{-\infty}^{\infty} \int_{-\infty}^{\infty} \frac{F_{X_i}^{-1}(\Phi(y_i)) - \mathbb{E}(X_i)}{\sqrt{\mathbb{V}(X_i)}} \cdot \frac{F_{X_j}^{-1}(\Phi(y_j)) - \mathbb{E}(X_j)}{\sqrt{\mathbb{V}(X_j)}} f_{Y_i,Y_j}(y_i, y_j) \mathrm{d}y_i \mathrm{d}y_j \tag{3.66}$$

where $f_{Y_i,Y_j}(y_i, y_j)$ is the joint PDF of Y_i and Y_j, as in Eq. (3.38). Unfortunately, the numerical calculation of Eq. (3.66) is generally very difficult due to the discreteness of the inverse function $F_X^{-1}(x)$ incorporated in the double-fold integral [4]. This fact implies that the Nataf transformation method would be more efficient for the case of continuous random variables. Some later researches are available in the literature to improve the efficiency of generating correlated discrete vectors, e.g., [18, 21].

Example 3.18

For two correlated and identically distributed Poisson variables X_1 and X_2 with a mean value of 2 and a linear correlation correlation of 0.5, develop a method to generate a sample pair for X_1 and X_2, x_1 and x_2.

Solution

According to Eq. (3.64), since $v_1 = v_2 = 2$, and $\rho = 0.5$,

$$\gamma_1 = \rho\sqrt{v_1 v_2} = 1, \quad \gamma_2 = v_1 - \rho\sqrt{v_1 v_2} = 1, \quad \gamma_3 = v_2 - \rho\sqrt{v_1 v_2} = 1 \tag{3.67}$$

Thus, the procedure to simulate a sample pair for (X_1, X_2) is as follows.

(1) Generate three independent samples for Poisson distribution with a mean value of 1, denoted by y_1, y_2 and y_3 respectively.
(2) Set $x_1 = y_1 + y_2$ and $x_2 = y_1 + y_3$.

Example 3.19

Similar to Eq. (3.62), we consider three correlated Poisson variables X_1, X_2 and X_3 with mean values of v_1, v_2 and v_3 respectively. The correlation coefficient of X_i and X_j is ρ_{ij} for $i, j = 1, 2, 3$ (note that $\rho_{ij} = \rho_{ji}$). To model the dependence between X_1, X_2 and X_3, we introduce six independent Poisson variables Y_1 through Y_6 such that

$$X_1 = Y_2 + Y_3 + Y_4, \quad X_2 = Y_1 + Y_3 + Y_5, \quad X_3 = Y_1 + Y_2 + Y_6 \qquad (3.68)$$

(1) Determine the mean values of Y_1 through Y_6; (2) Find the range of ρ_{ij}.
Solution
(1) Let γ_i denote the mean value of Y_i, $i = 1, 2, \ldots 6$. One can uniquely determine γ_i by considering

$$\mathbb{E}(X_1) = \mathbb{E}(Y_2) + \mathbb{E}(Y_3) + \mathbb{E}(Y_4) \Rightarrow \gamma_2 + \gamma_3 + \gamma_4 = v_1$$
$$\mathbb{E}(X_2) = \mathbb{E}(Y_1) + \mathbb{E}(Y_3) + \mathbb{E}(Y_5) \Rightarrow \gamma_1 + \gamma_3 + \gamma_5 = v_2 \qquad (3.69)$$
$$\mathbb{E}(X_3) = \mathbb{E}(Y_1) + \mathbb{E}(Y_2) + \mathbb{E}(Y_6) \Rightarrow \gamma_1 + \gamma_2 + \gamma_6 = v_3$$

and

$$\rho(X_1, X_2) = \frac{\mathbb{E}(X_1 X_2) - v_1 v_2}{\sqrt{v_1 v_2}} = \frac{\gamma_3}{\sqrt{v_1 v_2}} = \rho_{12}$$

$$\rho(X_1, X_3) = \frac{\mathbb{E}(X_1 X_3) - v_1 v_3}{\sqrt{v_1 v_3}} = \frac{\gamma_2}{\sqrt{v_1 v_3}} = \rho_{13} \qquad (3.70)$$

$$\rho(X_2, X_3) = \frac{\mathbb{E}(X_2 X_3) - v_2 v_3}{\sqrt{v_2 v_3}} = \frac{\gamma_1}{\sqrt{v_2 v_3}} = \rho_{23}$$

which yield

$$\gamma_1 = \rho_{23}\sqrt{v_2 v_3}, \quad \gamma_2 = \rho_{13}\sqrt{v_1 v_3}, \quad \gamma_3 = \rho_{12}\sqrt{v_1 v_2},$$
$$\gamma_4 = v_1 - \gamma_2 - \gamma_3, \quad \gamma_5 = v_2 - \gamma_1 - \gamma_3, \quad \gamma_6 = v_3 - \gamma_1 - \gamma_2 \qquad (3.71)$$

(2) To ensure that $\gamma_i > 0$ $(i = 1, 2, \ldots 6)$, ρ_{ij} should satisfy that $0 < \rho_{12}, \rho_{23}, \rho_{13}$ and

$$\begin{cases} \rho_{13}\sqrt{v_1 v_3} + \rho_{12}\sqrt{v_1 v_2} < v_1 \\ \rho_{23}\sqrt{v_2 v_3} + \rho_{12}\sqrt{v_1 v_2} < v_2 \\ \rho_{23}\sqrt{v_2 v_3} + \rho_{13}\sqrt{v_1 v_3} < v_3 \end{cases} \qquad (3.72)$$

Specifically, if $v_1 = v_2 = v_3$ (i.e., each X_i is identically distributed), one has

$$0 < \rho_{13} + \rho_{12} < 1, \quad 0 < \rho_{23} + \rho_{12} < 1, \quad 0 < \rho_{23} + \rho_{13} < 1 \qquad (3.73)$$

3.4   Simulation of Some Special Stochastic Processes

A collection of random variables on the time scale is called a *stochastic process*, denoted by $\{X(t), t \in A\}$, where $X(t)$ evaluated at any time $t \in A$ is a random variable and A is the index set of the process. For the case where A is a set of finite (countable) number of possible values, $X(t)$ is called a *discrete-time process* or simply *discrete process*. If A is a continuum on the time scale, then $X(t)$ is said to be a *continuous-time process* or *continuous process*. For instance, the sequence of monthly largest wind speed within one year, $\{X_1, X_2, \ldots X_{12}\}$, is a discrete process. The sequence of wind speed for an arbitrary time point of the year, written as $X(t), t \in [0, 1 \text{ year}]$, is a continuous process.

In this section, some special stochastic processes will be discussed. The characteristics of a general stochastic process will be further addressed in Chap. 5.

3.4.1   Poisson Process

In Sect. 2.2.3, we discussed the Poisson distribution, which is derived based on a Poisson process. For a reference period of $[0, T]$, we divide it into n identical sections, with which the duration of each interval, $\Delta t = \frac{T}{n}$, is small enough when n is large enough. For the case of a stationary Poisson process with an occurrence rate of λ, at most one event may occur during each interval and the occurrence probability is $p = \lambda \Delta t$. With this, a straightforward approach to simulate the occurrence times of a Poisson process is as follows. We call it as *Method 1*.

(1) Divide the considered time interval $[0, T]$ into n identical sections, where n is a sufficiently large integer.
(2) For each small interval, generate a Bernoulli sample with a mean value (or probability of success) of $p = \lambda \Delta t$, and record the number of the interval orderly if the sample is 1.
(3) With the sequence of recorded interval numbers $\{k_1, k_2, \ldots\}$, the desired sample sequence of occurrence times is $\left\{ \frac{k_1}{n}T, \frac{k_2}{n}T, \ldots \right\}$.

With a Poisson process, if it is known that totally m events occur within $[0, T]$, then the occurrence time points, $\{T_1, T_2, \ldots T_m\}$, are independent of each other and have a joint PDF of [13]

$$f_{T_1, T_2, \ldots T_n}(t_1, t_2, \ldots t_n) = \left(\frac{1}{T}\right)^n \tag{3.74}$$

That is, each T_i is independent of each other and has a uniform distribution over $[0, T]$. With this, another procedure for sampling the Poisson occurrence times within $[0, T]$ is as follows, referred to as *Method 2*.

(1) Simulate a realization of Poisson random variable with a mean value of λT, denoted by m.
(2) Simulate m independent samples of a uniform distribution over $[0, T]$, denoted by $t_1, t_2, \ldots t_m$.
(3) Ranked in an ascending order, the sequence $\{t_1, t_2, \ldots t_m\}$ is the simulated occurrence times of the Poisson process.

Furthermore, it is noticed that the time interval of two adjacent successive Poisson events follows an exponential distribution. That is, for a sequence of independent and identically exponentially distributed variables $T_1, T_2, \ldots T_n$, the number of Poisson events within $[0, T]$, X, has a PMF of

$$\mathbb{P}(X = k) = \max\left\{k : \sum_{i=1}^{k} T_i \leq T\right\}, \quad k = 0, 1, 2, \ldots \tag{3.75}$$

where $\sum_{i=1}^{0} T_i = 0$. Equation (3.75) is the basis for the following approach to generate a sample sequence of a Poisson process over $[0, T]$, called *Method 3*.

(1) Simulate a sample of random variable $\widetilde{T}$ which follows an exponential distribution with a mean value of $\frac{1}{\lambda}$, t_1, and set $s_1 = t_1$.
(2) If $s_1 < T$, generate another sample of $\widetilde{T}$, t_2, and let $s_2 = s_1 + t_2$.
(3) If $s_2 < T$, generate another sample of $\widetilde{T}$, t_3, and let $s_3 = s_2 + t_3$.
(4) Similar to steps (2) and (3), we can generate samples of $\widetilde{T}$, $t_1, t_2, \ldots$ and calculate $s_1, s_2, \ldots$, until such an m is found that s_m does not exceed T but $s_{m+1} > T$.
(5) The sequence $\{s_1, s_2, \ldots s_m\}$ is the sampled occurrence times of the Poisson process.

Now we consider the case of a non-stationary Poisson process. Suppose that the occurrence rate is $\lambda(t)$, $t \in [0, T]$, which is a function of time t. It can be shown that the each occurrence time, T_i, is independent mutually and has a PDF of

$$f_{T_i}(t) = \frac{\lambda(t)}{\int_0^T \lambda(\tau) d\tau} \tag{3.76}$$

The aforementioned three approaches, namely *Method 1*, *Method 2* and *Method 3*, shall be modified slightly to sample a non-stationary Poisson process.

Regarding *Method 1*, for the ith time interval ($i = 1, 2, \ldots n$), the Bernoulli variable has a mean value of $p_i = \lambda\left(\frac{i}{n}T\right) \cdot \Delta t$ instead of a constant p.

For *Method 2*, the steps are revised as follows.

(1) Simulate a realization of Poisson random variable with a mean value of $\int_0^T \lambda(t) dt$, denoted by m.
(2) Simulate m independent samples with a PDF of $f(t) = \frac{\lambda(t)}{\int_0^T \lambda(\tau) d\tau}, t \in [0, T]$, denoted by $t_1, t_2, \ldots t_m$.
(3) Ranked in an ascending order, the sequence $\{t_1, t_2, \ldots t_m\}$ is the occurrence times of the Poisson process.

Finally, for *Method 3*, note that with a time-varying occurrence rate $\lambda(t)$, the CDF of the inter-arrival times is also time-dependent. Suppose that an event occurs at time t' and let $T_{t'}$ denote the interval between t' and the occurring time of the subsequent event. The CDF of $T_{t'}$ is [20]

$$F_{T_{t'}}(t) = 1 - \exp\left[-\int_0^t \lambda(t' + \tau) d\tau\right], \quad t \geq 0 \tag{3.77}$$

Equation (3.77) can be explained by dividing the time after t' into identical small sections with a duration of $\Delta t \to 0$. With this, $\lfloor \frac{t}{\Delta t} \rfloor \approx \frac{t}{\Delta t}$ as Δt is small enough. Thus,

$$F_{T_{t'}}(t) = 1 - \mathbb{P}(T_{t'} > t) = 1 - \prod_{i=1}^{\lfloor \frac{t}{\Delta t} \rfloor} \left[1 - \lambda\left(t' + i \Delta t\right) \Delta t\right]$$

$$= 1 - \exp\left[\sum_{i=1}^{\lfloor \frac{t}{\Delta t} \rfloor} \ln\left[1 - \lambda\left(t' + i \Delta t\right) \Delta t\right]\right] = 1 - \exp\left[-\sum_{i=1}^{\lfloor \frac{t}{\Delta t} \rfloor} \lambda\left(t' + i \Delta t\right) \Delta t\right]$$

$$= 1 - \exp\left[-\int_0^t \lambda(t' + \tau) d\tau\right], \quad t \geq 0$$

$$\tag{3.78}$$

Now, with Eq. (3.77), the procedure for sampling a non-stationary Poisson process, based on *Method 3*, is as follows.

(1) Simulate a sample of random variable T_1 which has a CDF of $F_{T_1}(t) = 1 - \exp\left[-\int_0^t \lambda(\tau) d\tau\right]$, t_1, and set $s_1 = t_1$.
(2) If $s_1 < T$, generate another sample of T_2 with a CDF of $F_{T_2}(t) = 1 - \exp\left[-\int_0^t \lambda(t_1 + \tau) d\tau\right]$, t_2, and let $s_2 = s_1 + t_2$.
(3) If $s_2 < T$, generate another sample of T_3 with a CDF of $F_{T_3}(t) = 1 - \exp\left[-\int_0^t \lambda(t_2 + \tau) d\tau\right]$, t_3, and let $s_3 = s_2 + t_3$.
(4) Similar to steps (2) and (3), we can generate samples of T_i $(i = 1, 2, \ldots)$ with CDFs of $F_{T_i}(t) = 1 - \exp\left[-\int_0^t \lambda(t_{i-1} + \tau) d\tau\right]$, $t_1, t_2, \ldots$ and calculate $s_1, s_2, \ldots$, until such an m is found that s_m does not exceed T but $s_{m+1} > T$.
(5) The sequence $\{s_1, s_2, \ldots s_m\}$ is the sampled occurrence times of the non-stationary Poisson process.

Example 3.20

With a reference period of $[0, T]$, we employ the non-stationary Poisson process to model the occurrence of a sequence of repeated events. The occurrence rate is $\lambda(t) = a + bt$, where a and b are two time-invariant parameters, and $t \in [0, T]$. Establish a method to simulate a sequence of occurrence times of the events.

Solution

We use *Method 2* to sample a non-stationary Poisson process. The CDF of each occurrence time, T_i, is obtained, according to Eq. (3.76), as

$$F_{T_i}(t) = \frac{\int_0^t \lambda(\tau)d\tau}{\int_0^T \lambda(\tau)d\tau} = \frac{at + \frac{1}{2}bt^2}{aT + \frac{1}{2}bT^2} \tag{3.79}$$

To generate a sample of t_i, one can use the inverse transform method, by noting that the inverse function of $F_{T_i}(t)$ is

$$F_{T_i}^{-1}(x) = \frac{\sqrt{a^2 + 2b\left(aT + \frac{1}{2}bT^2\right)x} - a}{b} \tag{3.80}$$

Thus, the procedure to simulate a sequence of occurrence times is as follows.

(1) Simulate a realization of Poisson random variable with a mean value of $aT + \frac{1}{2}bT^2$, denoted by m.
(2) Simulate m independent samples of uniform distribution over $[0, 1]$, $u_1, u_2, \ldots u_m$.
(3) Let $t_i = \frac{\sqrt{a^2 + 2b\left(aT + \frac{1}{2}bT^2\right)u_i} - a}{b}$ for $i = 1, 2, \ldots m$.
(4) Ranked in an ascending order, the sequence $\{t_1, t_2, \ldots t_m\}$ is the occurrence times of the Poisson process.

Example 3.21

For two non-stationary Poisson processes $X(t)$ and $Y(t)$, $t \in [0, T]$, with occurrence rates of $\lambda_X(t)$ and $\lambda_Y(t)$ respectively, if $\lambda_X(t) \leq \lambda_Y(t)$ holds for $\forall t \in [0, T]$, one can simulate $X(t)$ as follows: (1) Simulate a sequence of occurrence times of process $Y(t)$, $\{t_1, t_2, \ldots t_m\}$; (2) Sample m independent samples of uniform distribution over $[0, 1]$, $u_1, u_2, \ldots u_m$. (3) For each $i = 1, 2, \ldots m$, if $u_i > \frac{\lambda_X(t_i)}{\lambda_Y(t_i)}$, delete t_i from the sequence obtained in step (1). (4) The remaining items of the sequence, in an ascending order, is the sampled occurrence times of $X(t)$.

Show the validity of the aforementioned algorithm.

Solution

Using *Method 1* to sample a sequence of occurrence times for $Y(t)$, we first divide $[0, T]$ into n identical intervals and simulate a Bernoulli variable with a mean value of $p_i = \lambda_Y\left(\frac{i}{n}T\right) \cdot \Delta t$ for the ith interval to determine whether an event occurs.

Now, for each interval, an additional requirement of $u_i \leq \frac{\lambda_X(t_i)}{\lambda_Y(t_i)}$ is applied to the Bernoulli variable, with which the mean value (or probability of success) of the modified Bernoulli distribution is $\left(\lambda_Y\left(\frac{i}{n}T\right) \cdot \Delta t\right) \cdot \mathbb{P}\left(U \leq \frac{\lambda_X\left(\frac{i}{n}T\right)}{\lambda_Y\left(\frac{i}{n}T\right)}\right) = \lambda_X\left(\frac{i}{n}T\right) \cdot \Delta t$. Thus, the "acceptance-rejection-method-like" algorithm gives a sample sequence of the occurrence times of $X(t)$.

3.4.2 Renewal Process

A *renewal process* is by nature a generalization of a stationary Poisson process. Recall that in Sects. 2.2.3 and 3.4.1, we mentioned that the time interval of two adjacent successive events independently follows an exponential distribution for the case of a stationary Poisson process. Now, if releasing the condition of "exponentially distributed", i.e., considering a counting process whose time interval of two adjacent successive events is mutually independent and identically distributed (following an arbitrary continuous distribution), we call this generalized process as a renewal process. The occurrence of an event leads to a *renewal* taking place.

Mathematically, let $\{T_1, T_2, \ldots\}$ be a sequence of independent and identically distributed non-negative variables. Define

$$S_0 = 0, \quad S_k = \sum_{i=1}^{k} T_i, \quad k = 1, 2, \ldots \tag{3.81}$$

If treating T_k as the time between the $(k-1)$th and the kth events of a renewal process, $k = 1, 2 \ldots$, then S_k denotes the time of the kth renewal. For a reference period of $[0, T]$, the number of events, X, is determined by

$$X = \max\{k : S_k \leq T\}, \quad k = 0, 1, 2, \ldots \tag{3.82}$$

Regarding the convergence of X as well as its statistics (mean value and variance), we have the following theorem.

Theorem 3.2 (Elementary renewal theorem (e.g., [14])) *For a renewal process over* $[0, T]$ *with an inter-arrival time of* T_i, *as* $T \to \infty$,

$$\frac{X}{T} \to \frac{1}{\mathbb{E}(T_i)}, \quad \frac{\mathbb{E}(X)}{T} \to \frac{1}{\mathbb{E}(T_i)}, \quad \frac{\mathbb{V}(X)}{T} \to \frac{\mathbb{V}(T_i)}{\mathbb{E}^3(T_i)} \tag{3.83}$$

with probability 1.

Now, we discuss the calculation of the mean value of X, $\mathbb{E}(X)$. Note that $\mathbb{E}(X)$ itself is a function of T; with this, we can also write $\mathbb{E}(X)$ as $m(T)$, which is known as the *renewal function*. It can be proven that $m(T)$ uniquely determines a renewal process and vice versa.

Recall that $m(T) = \lambda T$ with a stationary Poisson process (c.f. Sect. 3.4.1), implying that the Poisson process yields a linear renewal function. For the general case of a renewal process, according to Eq. (3.82),

$$m(T) = \mathbb{E}(X) = \sum_{k=0}^{\infty} k\mathbb{P}(X = k) = \sum_{k=0}^{\infty} k[\mathbb{P}(X \geq k) - \mathbb{P}(X \geq k+1)]$$

$$= \sum_{k=0}^{\infty} k[\mathbb{P}(S_k \leq T) - \mathbb{P}(S_{k+1} \leq T)] = \sum_{k=0}^{\infty} k[F_{S_k}(T) - F_{S_{k+1}}(T)] = \sum_{k=1}^{\infty} F_{S_k}(T)$$

$$(3.84)$$

where $F_{S_k}(\bullet)$ is the CDF of S_k.

Another approach to deriving $m(T)$ is to consider the conditional mean on the occurrence time of the first renewal. Let $F(t)$ denote the CDF of the inter-arrival time and $f(t)$ the corresponding PDF. Using the law of total probability, it follows,

$$m(T) = \mathbb{E}(X) = \int_0^{\infty} \mathbb{E}(X|T_1 = t)f(t)\mathrm{d}t = \int_0^T \mathbb{E}(X|T_1 = t)f(t)\mathrm{d}t + \int_T^{\infty} \mathbb{E}(X|T_1 = t)f(t)\mathrm{d}t$$

$$(3.85)$$

where T_1 is the time of the first renewal. Clearly, $\int_T^{\infty} \mathbb{E}(X|T_1 = t)f(t)\mathrm{d}t = 0$. Thus,

$$m(T) = \mathbb{E}(X) = \int_0^T \mathbb{E}(X|T_1 = t)f(t)\mathrm{d}t \tag{3.86}$$

Furthermore, by noting that

$$\mathbb{E}(X|T_1 = t) = 1 + \mathbb{E}(X'|T_1 = t) = 1 + m(T - t) \tag{3.87}$$

where X' is the number of events (renewals) within $[t, T]$, one has

$$m(T) = \mathbb{E}(X) = \int_0^T [1 + m(T - t)]f(t)\mathrm{d}t = F(T) + \int_0^T m(T - t)f(t)\mathrm{d}t$$

$$(3.88)$$

As such, Eq. (3.88), called the *renewal equation*, provides an alternative tool for deriving the renewal function.

Example 3.22

For a renewal process with a uniformly distributed inter-arrival time over $[0, t_0]$, calculate $\mathbb{E}(X)$, where X is the number of renewals over a reference period of $[0, T]$ $(T \leq t_0)$.

Solution

We use Eq. (3.88) to derive $\mathbb{E}(X) = m(T)$.

For $t \in [0, T]$, $F(t) = \frac{t}{t_0}$ and correspondingly, $f(t) = \frac{1}{t_0}$. With this, according to Eq. (3.88),

$$m(T) = F(T) + \int_0^T m(T - t) f(t) \mathrm{d}t = \frac{T}{t_0} + \frac{1}{t_0} \int_0^T m(T - t) \mathrm{d}t \xlongequal{x=T-t} \frac{T}{t_0} + \frac{1}{t_0} \int_0^T m(x) \mathrm{d}x \tag{3.89}$$

Taking the differential form of both sides with respect to T gives

$$t_0 m'(T) = 1 + m(T) \tag{3.90}$$

or equivalently,

$$\frac{1}{1+m} \mathrm{d}m = \frac{1}{t_0} \mathrm{d}T \tag{3.91}$$

Thus,

$$\int \frac{1}{1+m} \mathrm{d}m = \int \frac{1}{t_0} \mathrm{d}T \Rightarrow m(T) = \exp\left(\frac{T}{t_0} + c\right) - 1 \tag{3.92}$$

where c is a constant.

Since $m(0) = 0$, c is obtained as 0. As a result,

$$m(T) = \exp\left(\frac{T}{t_0}\right) - 1 \tag{3.93}$$

Example 3.23

For a sequence of independent uniform variables over $[0, 1]$, $U_1, U_2, U_3, \ldots$, define a variable $N = \min\left\{k : \sum_{i=1}^k U_i > 1\right\}$. Calculate $\mathbb{E}(N)$.

Solution

Method 1. It is easy to see that $N - 1 = N' = \max\left\{k : \sum_{i=1}^k U_i \leq 1\right\}$. According to Example 3.22, $\mathbb{E}(N') = \exp(1) - 1$ (i.e., letting $t_0 = 1$ and $T = 1$ in Eq. (3.93)). Thus, $\mathbb{E}(N) = e$.

Method 2. Let $S_k = \sum_{i=1}^k U_i$ for $k = 1, 2, \ldots$. Suppose that the PDF and CDF of S_k are f_{S_k} and F_{S_k} respectively. Clearly, $f_{S_1}(x) = 1$ for $0 \leq x \leq 1$. We let $F_{S_0}(x) = 1$ for consistency. Furthermore, according to Eq. (2.303), $f_{S_{k+1}}(x) = \int_0^x f_{S_k}(\tau) \mathrm{d}\tau$ holds for $0 \leq x \leq 1$. Using the method of induction, we have

$$f_{S_k}(x) = \frac{x^{k-1}}{(k-1)!} \Rightarrow F_{S_k}(x) = \frac{x^k}{k!}, \quad 0 \leq x \leq 1 \tag{3.94}$$

for $k = 1, 2, \ldots$.

Since $N - 1 = N' = \max\left\{k : \sum_{i=1}^k U_i \leq 1\right\}$, according to Eq. (3.84),

$$\mathbb{E}(N') = \sum_{k=1}^{\infty} F_{S_k}(1) = \sum_{k=0}^{\infty} F_{S_k}(1) - F_{S_0}(1) = \sum_{k=0}^{\infty} \frac{1}{k!} - 1 = \exp(1) - 1 \quad (3.95)$$

Thus, $\mathbb{E}(N) = \exp(1) - 1 + 1 = e$.

In order to sample a renewal process over $[0, T]$, one can use the following procedure, which is similar to *Method 3* in Sect. 3.4.1 by noting that the Poisson process is a specific renewal process.

(1) Simulate a sample of the inter-arrival time, t_1, and set $s_1 = t_1$.
(2) If $s_1 < T$, generate another sample of the inter-arrival time, t_2, and let $s_2 = s_1 + t_2$.
(3) If $s_2 < T$, generate another sample of the inter-arrival time, t_3, and let $s_3 = s_2 + t_3$.
(4) Similar to steps (2) and (3), we can generate samples $t_1, t_2, \ldots$ and calculate $s_1, s_2, \ldots$, until such an m appears that $s_m \leq T < s_{m+1}$.
(5) The sequence $\{s_1, s_2, \ldots s_m\}$ is the sampled occurrence times of the renewal process.

3.4.3 Gamma Process

A stationary *Gamma process*, $X(t), t \geq 0$ with parameters $a > 0$ and $b > 0$, is a continuous stochastic process with stationary, independent and Gamma distributed increments. The process $X(t)$ satisfies that [8]

(1) $\mathbb{P}[X(0) = 0] = 1$.
(2) $\Delta X(t) = X(t + \Delta t) - X(t)$ also follows a Gamma distribution with a shape parameter of $a \Delta t$ and a scale parameter of b, for any $t \geq 0$ and $\Delta t > 0$.
(3) For $n \geq 1$ and time points $0 \leq t_0 < t_1 < \ldots < t_n$, the variables $X(t_0), X(t_1) - X(t_0), \ldots X(t_n) - X(t_{n-1})$ are independent of each other.

Due to the non-negative increments, the Gamma process is a non-decreasing process. Thus, it can be used to describe a monotonic stochastic process in practical engineering (e.g., the difference between the initial resistance and the time-variant degraded resistance of an aging structure without repair/maintenance measures [9]).

With the aforementioned properties of a Gamma process, we have the following corollaries.

(1) For a stationary Gamma process $X(t)$ with a shape parameter of a and a scale parameter of b, $\mathbb{E}(X(t)) = abt$ and $\mathbb{V}(X(t)) = ab^2 t$.

Proof We divide the time interval $[0, t]$ into n identical sections, where n is large enough. Let $A_i = X\left(\frac{i}{n}t\right) - X\left(\frac{i-1}{n}t\right)$ for $i = 1, 2, \ldots n$. Clearly, each A_i is independent of each other and has a mean value of $\frac{abt}{n}$ and a variance of $\frac{ab^2 t}{n}$. Thus,

$\mathbb{E}(X(t)) = \mathbb{E}\left(\sum_{i=1}^{n} A_i\right) = \sum_{i=1}^{n} \mathbb{E}(A_i) = abt$ and $\mathbb{V}(X(t)) = \mathbb{V}\left(\sum_{i=1}^{n} A_i\right) = \sum_{i=1}^{n} \mathbb{V}(A_i) = ab^2 t$.

(2) If $X(t)$ is a Gamma process with a shape parameter of a and a scale parameter of b, then for a constant c, $cX(t)$ is also a Gamma process with a shape parameter of a and a scale parameter of cb.

Now, we generalize the stationary Gamma process to a non-stationary one, by considering that $\Delta X(t) = X(t + \Delta t) - X(t)$ follows a Gamma distribution with a shape parameter of $a(t)\Delta t$ and a scale parameter of b, for any $t \geq 0$ and positive $\Delta t \to 0$. With this, the mean value and variance of $X(t)$ are $b \int_0^t a(\tau)d\tau$ and $b^2 \int_0^t a(\tau)d\tau$ respectively. This can be derived by, as before, dividing the time interval $[0, t]$ into n identical sections, where n is large enough, and letting $A_i = X\left(\frac{i}{n}t\right) - X\left(\frac{i-1}{n}t\right)$ for $i = 1, 2, \ldots n$. Due to the independency between each A_i,

$$\mathbb{E}(X(t)) = \sum_{i=1}^{n} \mathbb{E}(A_i) = \lim_{\Delta t \to 0} \sum_{i=1}^{n} a(t)b\Delta t = b \int_0^t a(\tau)d\tau$$

$$\mathbb{V}(X(t)) = \sum_{i=1}^{n} \mathbb{V}(A_i) = \lim_{\Delta t \to 0} \sum_{i=1}^{n} a(t)b^2 \Delta t = b^2 \int_0^t a(\tau)d\tau = b\mathbb{E}(X(t))$$

$$(3.96)$$

In Eq. (3.96), when $a(t)$ takes some specific forms, the time-dependent mean value and variance of the non-stationary Gamma process can be obtained accordingly, as presented in Table 3.1, where η, κ and θ are three time-invariant parameters.

It can be seen, from Eq. (3.96), that once $\mathbb{E}(X(t))$ and $\mathbb{V}(X(t))$ are given, $a(t)$ and b are uniquely determined accordingly. Furthermore, at time t, the COV of $X(t)$ equals

$$\text{COV}(X(t)) = \frac{\sqrt{\int_0^t a(\tau)d\tau}}{\int_0^t a(\tau)d\tau} = \frac{1}{\sqrt{\int_0^t a(\tau)d\tau}}$$

$$(3.97)$$

In order to sample a (non-stationary) Gamma process $X(t)$ over a reference period of $[0, t]$, we firs divide the interval $[0, t]$ into n identical sections, namely $[t_0 = 0, t_1], [t_1, t_2], \ldots [t_{n-1}, t_n = t]$, where n is large enough. The *random walk approximation* method provides us a practical tool for simulating $X(t)$ by first exactly generating the realizations of $X(t)$ at the grid points $t_1, t_2, \ldots t_n$ and then constructing piece-wise connections between these samples to approximate the trajectory of $X(t)$. The algorithm is as follows,

Table 3.1 The mean and variance of $X(t)$ with some specific functions of $a(t)$

Case No.	$a(t)$	$\mathbb{E}(X(t))$	$\mathbb{V}(X(t))$
(1) Power law	$\eta t^\kappa + \theta$	$b\left(\frac{\eta}{1+\kappa}t^{\kappa+1} + \theta t\right)$	$b^2\left(\frac{\eta}{1+\kappa}t^{\kappa+1} + \theta t\right)$
(2) Exponential law	$\eta \exp(\kappa t) + \theta$	$b\left[\frac{\eta}{\kappa}(\exp(\kappa t) - 1) + \theta t\right]$	$b^2\left[\frac{\eta}{\kappa}(\exp(\kappa t) - 1) + \theta t\right]$
(3) Logarithmic	$\eta \ln(t + \kappa) + \theta$	$b\{\eta[(t+\kappa)\ln(t+\kappa) - t - \kappa\ln\kappa] + \theta t\}$	$b^2\{\eta[(t+\kappa)\ln(t+\kappa) - t - \kappa\ln\kappa] + \theta t\}$

(1) Simulate a Gamma variable A_i for the ith interval with a with a shape parameter of $a\left(\frac{it}{n}\right) \cdot \frac{t}{n}$ and a scale parameter of b.

(2) Set $x(t_0) = 0$ and $x(t_i) = \sum_{j=1}^{i} A_j$ for $i = 1, 2, \ldots n$.

(3) For $i=1, 2, \ldots n$, let $X(t)=\frac{n}{t}\left[X\left(\frac{it}{n}\right) - X\left(\frac{(i-1)t}{n}\right)\right]\left(t - \frac{(i-1)t}{n}\right) + X\left(\frac{(i-1)t}{n}\right)$

if $t \in \left(\frac{(i-1)t}{n}, \frac{it}{n}\right)$.

It can be found from the random walk approximation algorithm that each $X(t_i)$ $(i = 1, 2, \ldots n)$ is sampled by using the sum of many tiny increments, where some round-off errors may be introduced, especially when n is remarkably large. To avoid this disadvantage, the *Gamma bridge sampling* method can be used alternatively to sample the realizations of $X(t)$ at the grid points $t_1, t_2, \ldots t_n$.

To begin with, consider $X(t)$ at time points $t_1 < t_2 < t_3$, $X(t_1)$, $X(t_2)$ and $X(t_3)$. Our target herein is to simulate $X(t_2)$ provided that $X(t_1) = x(t_1)$ and $X(t_3) = x(t_3)$. Let $U = X(t_2) - X(t_1)$ and $V = X(t_3) - X(t_2)$. Clearly, both U and V are Gamma variables, have a common scale parameter, and are independent of each other. Furthermore, one has

$$X(t_2) - X(t_1) = U = \frac{U}{U + V}(U + V) = Y \cdot (U + V) \qquad (3.98)$$

where $Y = \frac{U}{U+V}$. Recall that in Example 2.41, we showed that for two independent Gamma variables X_1 and X_2 with a common scale parameter, $X_1 + X_2$ is independent of $\frac{X_1}{X_1+X_2}$. With this, the terms $U + V$ and Y in Eq. (3.98) are also independent mutually. Also, Y follows a Beta distribution with two shape parameters of η_1 and η_2, where η_1 and η_2 are the shape parameters of U and V respectively.

Given that $X(t_1) = x(t_1)$ and $X(t_3) = x(t_3)$, we can first simulate a sample of Y, y, and then, according to Eq. (3.98), set $x(t_2) = x(t_1) + y \cdot [x(t_3) - x(t_1)]$. This is the basis for the Gamma bridge sampling method.

Now we consider the sampling of $X(t)$ at the grid time points $t_1, t_2, \ldots t_{2^m} = t$. The steps, using the Gamma bridge sampling method, are as follows.

(1) Generate $x(t_0)$ and $x(t_{2^m})$.

(2) For $i = 1, 2, \ldots m$, simulate $x_{(2j+1)\times 2^{n-i}}$ from $x_{j\times 2^{n-i+1}}$ and $x_{(j+1)\times 2^{n-i+1}}$ (that is, from $x_{(2j+1)\times 2^{n-i}}$ and $x_{(2j+1)\times 2^{n-i}}$) for each $0 \le j \le 2^{i-1} - 1$.

To better visualize the Gamma bridge sampling algorithm, we consider a special case where the time interval of interest is divided into $2^5 = 32$ sections. The simulation order for $X(t)$ at each time point $t_1, t_2, \ldots t_{32}$, $x(t_1), x(t_2), \ldots x(t_{32})$, is illustrated in Fig. 3.5.

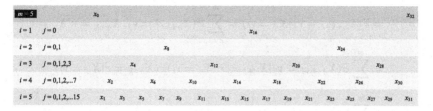

Fig. 3.5 Simulation order of $x(t_1), x(t_2), \ldots x(t_{32})$ using the Gamma bridge sampling method

3.4.4 Markov Process

For a discrete stochastic process $\{X_1, X_2, \ldots X_n\}$, suppose that each X_i may take a value out of $\{v_1, v_2, \ldots v_m\}$. If X_n ($n = 2, 3, \ldots$) depends on X_{n-1} only but not the previous states $X_{n-1}, X_{n-2}, \ldots X_1$, i.e.,

$$\mathbb{P}(X_{n+1} = v_j | X_n = v_i, X_{n-1} = v_{i_{n-1}}, \ldots X_1 = v_{i_1}) = \mathbb{P}(X_{n+1} = v_j | X_n = v_i) = p_{ij}$$
(3.99)

where $j, i, i_{n-1}, \ldots i_1 \in \{1, 2, \ldots m\}$, then the stochastic process $\{X_1, X_2, \ldots\}$ is called a *Markov chain* or a *Markov process*. The process is homogeneous if p_{ij} is independent of n.

Conditional on $X_n = v_i$, the PMF of X_{n+1} is $p_{X_{n+1}|X_n}(v_j | v_i) = \mathbb{P}(X_{n+1} = v_j | X_n = v_i) = p_{ij}$. Obviously, $\sum_{j=1}^{m} p_{ij} = 1$.

We further write the transition probabilities p_{ij} in a matrix form as follows,

$$\mathbf{P} = \begin{bmatrix} p_{11} & p_{12} & \cdots & p_{1m} \\ p_{21} & p_{22} & \cdots & p_{2m} \\ \vdots & \vdots & \ddots & \vdots \\ p_{m1} & p_{m2} & \cdots & p_{mm} \end{bmatrix}$$
(3.100)

which is called the *transition matrix*.

Similar to the one-step transition probability p_{ij} in Eq. (3.99), the k-step transition probability $p_{ij}^{(k)}$ measures the probability of $X_{n+k} = v_j$ conditional on $X_n = v_i$. Mathematically,

$$p_{ij}^{(k)} = \mathbb{P}(X_{n+k} = v_j | X_n = v_i)$$
(3.101)

Using the *Chapman-Kolmogorov equation*, one can calculate $p_{ij}^{(k)}$ as follows for all positive integers k, l,

$$p_{ij}^{(k+l)} = \sum_{r=1}^{m} p_{ir}^{(k)} p_{rj}^{(l)}$$
(3.102)

Equation (3.102) is explained by the fact that

$$p_{ij}^{(k+l)} = \mathbb{P}(X_{n+k+l} = v_j | X_n = v_i) = \sum_{r=1}^{m} \mathbb{P}(X_{n+k+l} = v_j, X_{n+k} = r | X_n = v_i)$$

$$= \sum_{r=1}^{m} \mathbb{P}(X_{n+k+l} = v_j | X_{n+k} = r, X_n = v_i) \mathbb{P}(X_{n+k} = r | X_n = v_i)$$

$$= \sum_{r=1}^{m} \mathbb{P}(X_{n+k+l} = v_j | X_{n+k} = r) \mathbb{P}(X_{n+k} = r | X_n = v_i)$$

$$= \sum_{r=1}^{m} p_{ir}^{(k)} p_{rj}^{(l)}$$

$$(3.103)$$

Similar to Eq. (3.100), we can also write the k-step transition probabilities $p_{ij}^{(k)}$ in a matrix form,

$$\mathbf{P}^{(k)} = \begin{bmatrix} p_{11}^{(k)} & p_{12}^{(k)} & \cdots & p_{1m}^{(k)} \\ p_{21}^{(k)} & p_{22}^{(k)} & \cdots & p_{2m}^{(k)} \\ \vdots & \vdots & \ddots & \vdots \\ p_{m1}^{(k)} & p_{m2}^{(k)} & \cdots & p_{mm}^{(k)} \end{bmatrix} \tag{3.104}$$

With the Chapman-Kolmogorov equation in Eq. (3.102), by noting that $\mathbf{P}^{(1)} = \mathbf{P}$, one has

$$\mathbf{P}^{(k+l)} = \mathbf{P}^{(k)} \cdot \mathbf{P}^{(l)} = \mathbf{P}^k \cdot \mathbf{P}^l = \mathbf{P}^{k+l} \tag{3.105}$$

Example 3.24

Consider the annual extreme winds actioned on a structure. We classify the wind load into two levels (1 and 2) according to the wind speed (level 1 if the wind speed does not exceed 50m/s and level 2 otherwise). Suppose that the wind level in the $(k + 1)$th year depends on that in the kth year only, $k = 1, 2, \ldots$. Given a level 1 wind in the kth year, the probability of a level 1 wind in the $(k + 1)$th year is q_1; if the wind load is level 2 in the kth year, then the probability of a level 1 wind in the $(k + 1)$th year is q_2, where $0 < q_1, q_2 < 1$. We define a Markov process $\{X_1, X_2, \ldots\}$ to represent the wind level for each year, $X_k = 1$ for a level 1 wind in the kth year and $X_k = 2$ for level 2, $k = 1, 2, \ldots$. It is known that $X_1 = 1$.

(1) Find the transition matrix.

(2) Compute the probability of a level 2 wind in the 5th year using $q_1 = 0.7$ and $q_2 = 0.2$.

(3) Calculate the probability of a level 1 wind load in the nth year as n is large enough.

Solution

(1) Note that $p_{11} = \mathbb{P}(X_{k+1} = 1 | X_k = 1) = q_1$, $p_{12} = 1 - p_{11} = 1 - q_1$, $p_{21} = \mathbb{P}(X_{k+1} = 1 | X_k = 2) = q_2$, $p_{22} = 1 - p_{21} = 1 - q_2$. With this, the transition matrix is as follows,

$$\mathbf{P} = \begin{bmatrix} q_1 & 1 - q_1 \\ q_2 & 1 - q_2 \end{bmatrix} \tag{3.106}$$

(2) $\mathbb{P}(X_5 = 2|X_1 = 1) = p_{12}^{(4)}$, and $p_{12}^{(4)}$ can be obtained via the transition matrix $\mathbf{P}^{(4)}$. According to Eq. (3.105), since $q_1 = 0.7$ and $q_2 = 0.2$,

$$\mathbf{P}^{(4)} = \mathbf{P}^4 = \begin{bmatrix} 0.7 & 0.3 \\ 0.2 & 0.8 \end{bmatrix}^4 = \begin{bmatrix} 0.4375 & 0.5625 \\ 0.3750 & 0.6250 \end{bmatrix} \tag{3.107}$$

which gives $p_{12}^{(4)} = 0.5625$. Thus, the probability of a level 2 wind in the 5th year is 56.25%.

(3) Let $\{\theta_1^{(k)}, \theta_2^{(k)}\}$ denote the PMF of X_k. That is, $\mathbb{P}(X_k = 1) = \theta_1^{(k)}$, and $\mathbb{P}(X_k = 2) = \theta_2^{(k)} = 1 - \theta_1^{(k)}$. With this,

$$\begin{aligned} \theta_1^{(n)} &= \mathbb{P}(X_n = 1) = \theta_1^{(n-1)} p_{11} + \theta_2^{(n-1)} p_{21} \\ \theta_2^{(n)} &= \mathbb{P}(X_n = 2) = \theta_1^{(n-1)} p_{12} + \theta_2^{(n-1)} p_{22} \end{aligned} \tag{3.108}$$

or, equivalently in a matrix form,

$$\begin{bmatrix} \theta_1^{(n)} \\ \theta_2^{(n)} \end{bmatrix} = \mathbf{P}^\mathsf{T} \begin{bmatrix} \theta_1^{(n-1)} \\ \theta_2^{(n-1)} \end{bmatrix} \tag{3.109}$$

As $n \to \infty$, we assume that $\{\theta_1^{(n)}, \theta_2^{(n)}\}$ converges to $\{\theta_1, \theta_2\}$, with which

$$\begin{bmatrix} \theta_1 \\ \theta_2 \end{bmatrix} = \mathbf{P}^\mathsf{T} \begin{bmatrix} \theta_1 \\ \theta_2 \end{bmatrix} \Rightarrow (\mathbf{P}^\mathsf{T} - \mathbf{I}) \begin{bmatrix} \theta_1 \\ \theta_2 \end{bmatrix} = \mathbf{0} \tag{3.110}$$

where $\mathbf{I} = \begin{bmatrix} 1 & 0 \\ 0 & 1 \end{bmatrix}$. Thus,

$$\theta_1 = \frac{q_2}{1 + q_2 - q_1}, \quad \theta_2 = \frac{1 - q_1}{1 + q_2 - q_1} \tag{3.111}$$

That is, the probability of a level 1 wind load in the nth year is $\frac{q_2}{1+q_2-q_1}$ as n is large enough. Specifically, if $q_1 = 0.7$ and $q_2 = 0.2$, then $\theta_1 = 0.4$, $\theta_2 = 0.6$, giving a probability of level 1 wind of 40%.

Example 3.25

Let in Example 3.24 $q_1 = 0.7$ and $q_2 = 0.2$.

(1) Given that $\mathbf{P}^6 = \begin{bmatrix} 0.4094 & 0.5906 \\ 0.3937 & 0.6063 \end{bmatrix}$, calculate the linear correlation between X_7 and X_{13}.

(2) Using a simulation-based approach, find numerically and plot the relationship between $\mathbb{E}(X_k)$ and k for k being positive integers up to 20.

(3) Repeat (1) and (2) with $X_1 = 2$, and comment on the result.

Solution

(1) We first find the PMFs for X_7 and X_{13}. Note that $\mathbb{P}(X_7 = 1|X_1 = 1) = p_{11}^{(6)} = 0.4094$, and $\mathbb{P}(X_7 = 2|X_1 = 1) = p_{12}^{(6)} = 0.5906$. Moreover, as

$$\mathbf{P}^{12} = (\mathbf{P}^6)^2 = \begin{bmatrix} 0.4094 & 0.5906 \\ 0.3937 & 0.6063 \end{bmatrix}^2 = \begin{bmatrix} 0.4001 & 0.5999 \\ 0.3999 & 0.6001 \end{bmatrix} \tag{3.112}$$

one has $\mathbb{P}(X_{13} = 1|X_1 = 1) = p_{11}^{(12)} = 0.4001$, and $\mathbb{P}(X_{13} = 2|X_1 = 1) = p_{12}^{(12)} = 0.5999$. It can be seen that X_7 and X_{13} are approximately identically distributed. The correlation coefficient between X_7 and X_{13} is estimated by

$$\rho(X_7, X_{13}) = \rho(X_7 - 1, X_{13} - 1) = \frac{\mathbb{E}[(X_7 - 1)(X_{13} - 1)] - \mathbb{E}(X_7 - 1)\mathbb{E}(X_{13} - 1)}{\sqrt{\mathbb{V}(X_7 - 1)\mathbb{V}(X_{13} - 1)}} \tag{3.113}$$

where

$$\begin{aligned}
& \mathbb{E}(X_7 - 1) = 0.5906, \quad \mathbb{V}(X_7 - 1) = 0.4094 \times 0.5906 = 0.2418 \\
& \mathbb{E}(X_{13} - 1) = 0.5999, \quad \mathbb{V}(X_{13} - 1) = 0.4001 \times 0.5999 = 0.2397 \\
& \mathbb{E}[(X_7 - 1)(X_{13} - 1)] = \mathbb{P}(X_{13} = 2, X_7 = 2) \\
& = \mathbb{P}(X_{13} = 2|X_7 = 2)\mathbb{P}(X_7 = 2) = 0.6063 \times 0.5906 = 0.3581
\end{aligned} \tag{3.114}$$

Thus, $\rho(X_7, X_{13}) = \frac{0.3581 - 0.5906 \times 0.5999}{\sqrt{0.2418 \times 0.2397}} = 0.0157$.

(2) We use simulation-based method to sample a large amount of samples for X_k and use the average of these samples to approximate $\mathbb{E}(X_k)$. The iterative simulation procedure is as follows,

(a) Simulate a sample for X_2, x_2. Note that $X_2 - 1$ is a Bernoulli variable with a mean value of 0.3.

(b) For $i = 3, 4, \ldots k$, simulate a sample for X_i, x_i, based on the realization of x_{i-1}. Note that if $x_{i-1} = 1$, then $X_i - 1$ is a Bernoulli variable with a mean value of 0.3; if $x_{i-1} = 2$, then $X_i - 1$ is a Bernoulli variable with a mean value of 0.8.

Figure 3.6a plots the relationship between $\mathbb{E}(X_k)$ and k for k being positive integers up to 20, using 1,000,000 simulation runs for each k. It can be seen that $\mathbb{E}(X_k)$ gradually converges to 1.6 with an increasing k, which is consistent with the previous observation that the limiting probability of a level 1 wind is 0.4 (so that $\lim_{k \to \infty} \mathbb{E}(X_k) = 1.6$).

(3) Now, we repeat (1) and (2) with the initial value $X_1 = 2$.

First, for (1), note that $\mathbb{P}(X_7 = 1|X_1 = 2) = p_{21}^{(6)} = 0.3937$, and $\mathbb{P}(X_7 = 2|X_1 = 2) = p_{22}^{(6)} = 0.6063$. Moreover, as $\mathbf{P}^{12} = \begin{bmatrix} 0.4001 & 0.5999 \\ 0.3999 & 0.6001 \end{bmatrix}$, one has $\mathbb{P}(X_{13} = 1|X_1 = 2) = p_{21}^{(12)} = 0.3999$, and $\mathbb{P}(X_{13} = 2|X_1 = 2) = p_{12}^{(22)} = 0.6001$. Furthermore,

$$\mathbb{E}(X_7 - 1) = 0.6063, \quad \mathbb{V}(X_7 - 1) = 0.3937 \times 0.6063 = 0.2387$$
$$\mathbb{E}(X_{13} - 1) = 0.6001, \quad \mathbb{V}(X_{13} - 1) = 0.3999 \times 0.6001 = 0.2400$$
$$\mathbb{E}[(X_7 - 1)(X_{13} - 1)] = \mathbb{P}(X_{13} = 2, X_7 = 2) \qquad (3.115)$$
$$= \mathbb{P}(X_{13} = 2|X_7 = 2)\mathbb{P}(X_7 = 2) = 0.6063 \times 0.6063 = 0.3676$$

Thus, $\rho(X_7, X_{13}) = \frac{0.3676 - 0.6063 \times 0.6001}{\sqrt{0.2387 \times 0.2400}} = 0.0157$. Comapared with (1), it can be seen
that the initial value X_1 has negligible impact on the correlation between X_7 and X_{13}.
For (2), step (1) of the aforementioned simulation procedure for sequence $x_1, x_2, \ldots$ is
slightly modified as follows: "Simulate a sample for X_2, x_2. Note that $X_2 - 1$ is a Bernoulli
variable with a mean value of 0.8". Figure 3.6b presents the dependence of $\mathbb{E}(X_k)$ on k,
which also converges to 1.6 with k, regardless of the initial value of X_1.

It is interesting to see, from Examples 3.24 and 3.25, that the PMF of X_n stabilizes
as $n \to \infty$ and is independent of the initial realization of X_1. With this observation,
if one wants to generate a sample for random variable Y, y, with a PMF of $\mathbb{P}(Y =
1) = \frac{q_2}{1+q_2-q_1}, \mathbb{P}(Y = 2) = \frac{1-q_1}{1+q_2-q_1}$, it is fairly straightforward to simulate a sample
sequence of $X_1, X_2, \ldots X_n, x_1, x_2, \ldots x_n$ and set $y = x_n$ when n is large enough.
This algorithm is known as the *Markov Chain Monte Carlo* (MCMC) simulation
method.

Generally, to sample a realization of X with a PMF of $p_X(v_i)$, the MCMC simulation method generates a sample sequence of a Markov chain $\{X_1, X_2, \ldots X_n\}$ with
a stationary (limiting) PMF of $p_X(v_i)$, and treats the sample of X_n as that of X as n
is large enough. This is because X_n and X are approximately identically distributed
with a sufficiently large n. With this regard, the *Metropolis-Hastings algorithm* has
been widely used in the literature. The main steps are as follows [1, 16].

(1) Generate an initial realization of X_1, x_1, and set $i = 1$.
(2) Simulate a sample for X^*, x^*, with a conditional PMF of $q(x^*|x_i)$, and a sam-
    ple for uniform distribution over $[0, 1]$, u. Here, $q(x^*|x_i)$ is called a "proposal
    distribution".

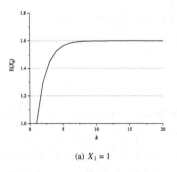
(a) $X_1 = 1$

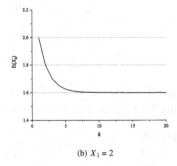
(b) $X_1 = 2$

Fig. 3.6 Dependence of $\mathbb{E}(X_k)$ on k

(3) If $u < \min\left[1, \frac{p_X(x^*)}{p_X(x_i)} \cdot \frac{q(x_i|x^*)}{q(x^*|x_i)}\right]$, let $x_{i+1} = x^*$; otherwise, let $x_{i+1} = x_i$.

(4) Increase i by 1.

(5) If $i < n$, where n is a predefined threshold, return to step (2).

With this algorithm, we can obtain a sample sequence $x_1, x_2, \ldots x_n$. As n is large enough, x_n is deemed as a sample of X.

To explain the Metropolis-Hastings algorithm, we first consider the transition matrix of $\{X_1, X_2, \ldots\}$, $\mathbf{P} = [p_{ij}]$. When $i \neq j$,

$$
\begin{aligned}
p_{ij} &= \mathbb{P}(X_{k+1} = v_j | X_k = v_i) = \mathbb{P}(X_{k+1} = x^*, x^* = v_j | X_k = v_i) \\
&= q(v_j|v_i) \cdot \min\left[1, \frac{p_X(v_j)}{p_X(v_i)} \cdot \frac{q(v_i|v_j)}{q(v_j|v_i)}\right]
\end{aligned}
\tag{3.116}
$$

Let $p'(v_i)$, $i = 1, 2, \ldots$, be the stationary PMF of X_n as $n \to \infty$. Since

$$
\begin{aligned}
\mathbb{P}(X_k = v_i, X_{k+1} = v_j) &= \mathbb{P}(X_k = v_i | X_{k+1} = v_j)\mathbb{P}(X_{k+1} = v_j) = p_{ji} p'(v_j) \\
&= \mathbb{P}(X_{k+1} = v_j | X_k = v_i)\mathbb{P}(X_k = v_i) = p_{ij} p'(v_i)
\end{aligned}
\tag{3.117}
$$

it follows,

$$
q(v_i|v_j) \cdot \min\left[1, \frac{p_X(v_i)}{p_X(v_j)} \cdot \frac{q(v_j|v_i)}{q(v_i|v_j)}\right] p'(v_j) = q(v_j|v_i) \cdot \min\left[1, \frac{p_X(v_j)}{p_X(v_i)} \cdot \frac{q(x_i|v_j)}{q(v_j|v_i)}\right] p'(v_i)
\tag{3.118}
$$

If $\frac{p_X(v_i)}{p_X(x_j)} \cdot \frac{q(x_j|v_i)}{q(v_i|x_j)} > 1$, Eq. (3.118) becomes

$$
\begin{aligned}
q(v_i|v_j)p'(v_j) &= q(v_j|v_i) \cdot \frac{p_X(v_j)}{p_X(v_i)} \cdot \frac{q(v_i|v_j)}{q(v_j|v_i)} p'(v_i) \\
\Rightarrow \frac{p'(v_i)}{p'(v_j)} &= \frac{p(v_i)}{p(v_j)}
\end{aligned}
\tag{3.119}
$$

If, on the contrary, $\frac{p_X(v_i)}{p_X(x_j)} \cdot \frac{q(x_j|v_i)}{q(v_i|x_j)} \leq 1$ in Eq. (3.118), then

$$
\begin{aligned}
q(v_i|v_j) \cdot \frac{p_X(v_i)}{p_X(v_j)} \cdot \frac{q(v_j|v_i)}{q(v_i|v_j)} p'(v_j) &= q(v_j|v_i)p'(v_i) \\
\Rightarrow \frac{p'(v_i)}{p'(v_j)} &= \frac{p(v_i)}{p(v_j)}
\end{aligned}
\tag{3.120}
$$

By noting $\sum_i p'(v_i) = \sum_i p_X(v_i) = 1$, it is seen that $p'(v_i) = p_X(v_i)$. That is, the stationary PMF of the Metropolis-Hastings algorithm is exactly the desired $p_X(v_i)$.

Example 3.26

Suppose that a Poisson variable X has a mean value of 5. To sample X, we employ the MCMC simulation method to generate a sample sequence of a Markov chain $\{X_1, X_2, \ldots X_n\}$ with a stationary PMF of that of X, and approximate X with X_n. Using

Fig. 3.7 Dependence of $\mathbb{E}(X_n)$ on n with a limiting Poisson distribution

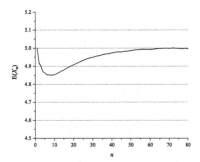

a proposal distribution of $q(x^*|x_i) = \begin{cases} \dfrac{1}{2}\mathbb{I}(|x^* - x_i| = 1), & x_i \geq 1 \\ \dfrac{1}{2}\mathbb{I}(x^* \leq 1), & x_i = 0 \end{cases}$, and $X_1 = 5$, find numerically and plot the relationship between $\mathbb{E}(X_n)$ and n.

Solution

The MCMC simulation method is employed to generate a sample sequence of $\{X_1, X_2, \ldots X_n\}$, with which X is approximated by X_n. Figure 3.7 shows the dependence of $\mathbb{E}(X_n)$ on n for positive integers n up to 80. It can be seen that $\mathbb{E}(X_n)$, starting from 5 because the initial value $X_1 = 5$, decreases at the early stage slightly and then converges to 5 as expected with an increasing n. For better visualization, Fig. 3.8 presents the comparison between the PMF of X_n and that of X for the cases of $n = 5, 10, 30$ and 60 respectively. Consistent with Fig. 3.7, when n becomes larger, the distribution of X_n stabilizes at a Poisson distribution with a mean value of 5 as desired.

Remark

The Metropolis-Hastings algorithm also applies for the case of a continuous random variable X. Illustratively, suppose that X has a Gamma distribution with a mean value of 3 and a standard deviation of 1. Introducing a proposal distribution of $q(x^*|x_i) = \frac{1}{20\sqrt{\pi}}\exp\left(-\frac{(x^*-x_i)^2}{400}\right)$ (that is, X^* follows a normal distribution with a mean value of x_i and a standard deviation of $10\sqrt{2}$), we generate a Markov chain $\{X_1, X_2, \ldots\}$ and approximate X with X_n. The PDF of simulated X_{50} is plotted in Fig. 3.9 for illustration purpose, from which it can be seen the MCMC simulation method works well for sampling a continuous random variable.

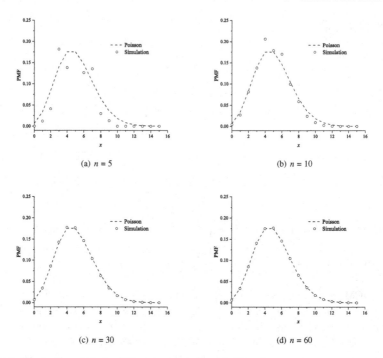

(a) $n = 5$

(b) $n = 10$

(c) $n = 30$

(d) $n = 60$

Fig. 3.8 Distribution of X_n for different values of n with $X_1 = 5$

Fig. 3.9 Theoretical PDF of X and that of simulated X_{50}

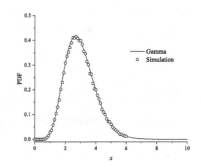

3.5 Simulation Techniques

3.5.1 Naïve Monte Carlo Simulation

In Sect. 3.1, we mentioned that the Monte Carlo simulation can be used to estimate the mean value of a function $h(\mathbf{X}) = h(X_1, X_2, \ldots X_r)$, which may be relatively difficult to solve if calculating the integral directly (e.g., when r is large). The mathematical expression of $\mathbb{E}(h)$, similar to Eq (3.6), is expressed as follows,

$$\mathbb{E}(h(\mathbf{X})) = \underbrace{\int \cdots \int}_{r-\text{fold}} h(\mathbf{x}) f_{\mathbf{X}}(\mathbf{x}) d\mathbf{x} \tag{3.121}$$

where $f_{\mathbf{X}}(\mathbf{x})$ is the joint PDF of $\mathbf{X}$. Note that the estimate of an event's probability using Monte Carlo simulation can be converted into a specific case of Eq. (3.121) (c.f. Eq (3.6)).

According to Eq. (3.121), the simulation-based method to estimate $\mathbb{E}(h(\mathbf{X}))$ is as follows. For the ith simulation run ($i = 1, 2, \ldots$), generate a realization of $\mathbf{X}$, $\mathbf{x}_i$, and calculate $h_i = h(\mathbf{x}_i)$. Repeating for n times, $\mathbb{E}(h(\mathbf{X}))$ is approximated as

$$\mathbb{E}(h(\mathbf{X})) = \widetilde{P} = \frac{1}{n} \sum_{i=1}^{n} h_i \tag{3.122}$$

This algorithm is called *naïve Monte Carlo simulation* or *brute-force Monte Carlo simulation*.

Example 3.27

Consider the serviceability of an RC structure that is subjected to marine environment. As introduced in Example 2.26, we use the Fick's second law to model the chloride difffusion in the concrete cover. Assume that the diffusion coefficient D, the surface chloride concentration C_s, and the concrete cover depth X are described by independent random variables. The structural performance is deemed as "satisfactory" if the chloride concentration at the steel bar surface (with a distance of X from the concrete surface) does not exceed a predefined threshold c_{cr}. Establish a procedure to compute the probability of satisfactory structural performance at time t_0.

Solution

Recall that in Example 2.26, we mentioned that the chloride concentration at distance x from the concrete surface at time t is explicitly expressed as $C(x, t) = C_s \cdot \left[2 - 2\Phi\left(\frac{x}{\sqrt{2Dt}}\right) \right]$. With this, we can use the Monte Carlo simulation method to estimate the structural performance by comparing the sample of $C(X, t_0)$ with c_{cr}. The procedure is as follows for the ith simulation run.

(1) Simulate samples for D, C_s and X independently, denoted by d, c_s and x.

(2) Set $c(x, t_0) = c_s \cdot \left[2 - 2\Phi\left(\frac{x}{\sqrt{2dt_0}}\right) \right]$.

(3) If $c(x, t_0) < c_{\text{cr}}$, set $h_i = 1$; otherwise, let $h_i = 0$.

With n replications, a sequence of $h_1, h_2, \ldots h_n$ is obtained. Further, the probability of satisfactory structural performance at time t_0 is estimated as $\frac{1}{n} \sum_{i=1}^{n} h_i$ as n is large enough.

Example 3.28

Reconsider the portal frame in Example 2.7. Develop a simulation procedure to estimate the probability that H occurs only conditional on the failure of the frame.

Solution

First, set counters $c_1 = 0$ and $c_2 = 0$. The simulation procedure for the ith $(i = 1, 2, \ldots)$ run is as follows.

(1) Simulate independently five samples for uniform distribution over $[0, 1]$, u_1 through u_5.
(2) Let $z_1, z_2, z_{12} \in \{0, 1\}$ represent the occurrence of structural failure associated with modes 1, 2 and "combination" (e.g., $z_1 = 1$ means the "beam" mode failure). Set $z_1 = z_2 = z_{12} = 0$.
(3) Let $b_v, b_h \in \{0, 1\}$ represent the occurrence of V and H respectively. Set $b_v = b_h = 0$.
(4) Let $b_v = 1$ if $u_1 < q_v$, and let $b_h = 1$ if $u_2 < q_h$.
(5) If $b_v = 1$ and $b_h = 0$ (i.e., only V occurs), let $z_1 = 1$ if $u_3 < p_{v1}$.
(6) If $b_v = 0$ and $b_h = 1$ (i.e., only H occurs), let $z_2 = 1$ if $u_4 < p_{h2}$.
(7) If $b_v = 1$ and $b_h = 1$ (i.e., joint action of V and H), let $z_1 = 1$ if $u_5 \leq p_{vh1}$, or let $z_2 = 1$ if $p_{vh1} < u_5 \leq p_{vh1} + p_{vh2}$, or let $z_{12} = 1$ if $p_{vh1} + p_{vh2} < u_5 \leq p_{vh1} + p_{vh2} + p_{vh3}$.
(8) Increase c_2 by 1 if $z_1 + z_2 + z_{12} > 0$ (i.e., failure occurs), and increase c_1 by 1 if $b_v = 0$, $b_h = 1$ and $z_1 + z_2 + z_{12} > 0$.

With a sufficiently large amount of replications, the desired probability is approximated by $\frac{c_1}{c_2}$.

Example 3.29

Estimate the following integral using Monte Carlo simulation method,

$$f = \int_{-\infty}^{-\frac{1}{2}} \int_{-\infty}^{5x_1} \exp\left[-\frac{2}{3}\left(x_1^2 - x_1 x_2 + x_2^2\right)\right] dx_2 dx_1 \tag{3.123}$$

Solution

Solving the two-fold integral in Eq. (3.123) directly gives a result of 5.26659×10^{-4}. Now we use the Monte Carlo simulation method to estimate Eq. (3.123). First, we rewrite Eq. (3.123) as follows,

$$f = 2\pi\sqrt{1-\rho^2} \int_{-\infty}^{-\frac{1}{2}} \int_{-\infty}^{5x_1} \frac{1}{2\pi\sqrt{1-\rho^2}} \exp\left[-\frac{x_1^2 - 2\rho x_1 x_2 + x_2^2}{2(1-\rho^2)}\right] dx_2 dx_1$$

$$= 2\pi\sqrt{1-\rho^2} \int_{-\infty}^{\infty} \int_{-\infty}^{\infty} \mathbb{I}(x_2 < 5x_1) \cdot \mathbb{I}\left(x_1 < -\frac{1}{2}\right) \cdot \frac{1}{2\pi\sqrt{1-\rho^2}} \exp\left[-\frac{x_1^2 - 2\rho x_1 x_2 + x_2^2}{2(1-\rho^2)}\right] dx_2 dx_1$$

$$\tag{3.124}$$

Table 3.2 Results of computing Eq. (3.38) via Monte Carlos simulation ($\times 10^{-4}$)

Number of replications	Trial 1	Trial 2	Trial 3	Trial 4	Trial 5	Trial 6	Average
$n = 10^4$	5.4414	5.4414	0	5.4414	0	10.8828	4.5345
$n = 10^5$	7.6180	4.3531	5.9855	2.7207	5.4414	8.7062	5.8042
$n = 10^6$	5.2782	4.6796	4.5708	5.6046	5.7135	4.6252	5.0786
$n = 10^7$	5.1584	5.3761	5.1965	5.1639	5.0551	5.5774	5.2546
$n = 10^8$	5.2754	5.2809	5.3162	5.1829	5.2765	5.3162	5.2747

where $\rho = \frac{1}{2}$. Note that $\frac{1}{2\pi\sqrt{1-\rho^2}} \exp\left[-\frac{x_1^2 - 2\rho x_1 x_2 + x_2^2}{2(1-\rho^2)}\right]$ is exactly the joint PDF of two correlated standard normal variables with a linear correlation coefficient of $\frac{1}{2}$ (c.f. Eq. (3.38)). With this, the simulation procedure is as follows. For the ith run,

(1) Simulate two independent standard normal variables x and x_1.
(2) Set $x_2 = \frac{1}{2}x_1 + \frac{\sqrt{3}}{2}x$ (c.f. Example 3.17).
(4) If $x_2 < 5x_1 < -\frac{5}{2}$, set $h_i = 1$; otherwise, let $h_i = 0$.

With n simulation runs, one can obtain a sequence of records $h_1, h_2, \ldots h_n$. With this, f in Eq. (3.38) is estimated as

$$f = \sqrt{3}\pi \cdot \frac{1}{n}\sum_{i=1}^{n} h_i \qquad (3.125)$$

Table 3.2 presents the simulation results according to Eq. (3.125) with different values of n. For each case, the results associated with six trials, as well as their average, are given for illustration purpose. It can be seen that the simulation accuracy is improved with a larger number of simulation replications.

It is noticed that the estimator $\widetilde{P}$ in Eq. (3.122) is also a random variable by nature (see, e.g., Table 3.2 for reference). The mean value of $\widetilde{P}$ is obtained as

$$\mathbb{E}(\widetilde{P}) = \mathbb{E}\left(\frac{1}{n}\sum_{i=1}^{n} h_i\right) = \frac{1}{n} \cdot n \cdot \mathbb{E}(h_i) = \mathbb{E}(h(\mathbf{X})) \qquad (3.126)$$

which implies that the term $\widetilde{P}$ in Eq. (3.122) is an unbiased estimator of $\mathbb{E}(h(\mathbf{X}))$.
Next,

$$\mathbb{V}(\widetilde{P}) = \mathbb{V}\left(\frac{1}{n}\sum_{i=1}^{n} h_i\right) = \frac{1}{n^2} \cdot n \cdot \mathbb{V}(h_i) = \frac{\mathbb{V}(h(\mathbf{X}))}{n} = \frac{\mathbb{E}\{[h(\mathbf{X}) - \mathbb{E}(h(\mathbf{X}))]^2\}}{n}$$

$$(3.127)$$

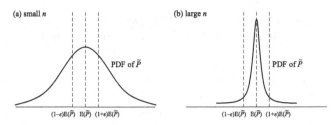

Fig. 3.10 Illustration of the PDF of $\widetilde{P}$ with different n

Since $\widetilde{P}$ estimates $\mathbb{E}(h(\mathbf{X}))$ unbiasedly, as $n \to \infty$, we have

$$\mathbb{V}(\widetilde{P}) \approx \frac{\frac{1}{n}\sum_{i=1}^{n}\left[h_i - \mathbb{E}(\widetilde{P})\right]^2}{n} = \frac{\frac{1}{n}\sum_{i=1}^{n}h_i^2 - \mathbb{E}(\widetilde{P})}{n} \tag{3.128}$$

Equation (3.128) implies that the variance of $\widetilde{P}$ decreases with an increasing n, as schematically shown in Fig. 3.10. According to Eq. (3.122), a sample of $\widetilde{P}$ is in fact used to approximate $\mathbb{E}(h(\mathbf{X}))$ in Monte Carlo simulation. To measure the simulation accuracy, we consider $\Delta = \left|\widetilde{P} - \mathbb{E}(h(\mathbf{X}))\right|$, or equivalently, $\Delta = \left|\widetilde{P} - \mathbb{E}(\widetilde{P})\right|$, which is a random variable representing the difference between the simulated $\mathbb{E}(h(\mathbf{X}))$ and the realistic value. If the requirement is that $\frac{\Delta}{\mathbb{E}(h(\mathbf{X}))} \leq \varepsilon$ holds with a confidence level of θ ($0 < \theta < 1$), where ε is a predefined threshold (e.g., 10%), one has

$$\mathbb{P}\left(\frac{\Delta}{\mathbb{E}(h(\mathbf{X}))} \leq \varepsilon\right) = \theta \tag{3.129}$$

Equation (3.129) implies that the probability that the difference between the simulated $\mathbb{E}(h(\mathbf{X}))$ and the realistic value does not exceed ε is θ. This equation can be further used to determine the minimum value of n (i.e., the minimum number of replications that are needed to achieve a desired simulation accuracy).

An equivalent form of Eq. (3.129) is that

$$\mathbb{P}\left[(1 - \varepsilon)\mathbb{E}(h_i) \leq \widetilde{P} \leq (1 + \varepsilon)\mathbb{E}(h_i)\right] = \theta \tag{3.130}$$

With the central limit theorem (c.f. Sect. 2.2.4), $\widetilde{P}$ follows a normal distribution. With this, Eq. (3.130) becomes

$$\Phi\left(\frac{\varepsilon\mathbb{E}(h_i)}{\sqrt{\frac{\mathbb{V}(h_i)}{n_{\min}}}}\right) - \Phi\left(-\frac{\varepsilon\mathbb{E}(h_i)}{\sqrt{\frac{\mathbb{V}(h_i)}{n_{\min}}}}\right) = \theta \Rightarrow n_{\min} = \left[\frac{\Phi^{-1}\left(\frac{1+\theta}{2}\right)\sqrt{\mathbb{V}(h_i)}}{\varepsilon\mathbb{E}(h_i)}\right]^2 \tag{3.131}$$

Specifically, if Eq. (3.122) is used to estimate the occurrence probability of an event with $\mathbb{E}(h_i) = p$, then $\mathbb{V}(h_i) = p(1 - p)$, and

$$n_{\min} = \frac{1-p}{p} \left[\frac{\Phi^{-1}\left(\frac{1+\theta}{2}\right)}{\varepsilon} \right]^2 \tag{3.132}$$

Illustratively, if $p = 10^{-3}$, $\varepsilon = 20\%$, and $\theta = 95\%$, then $n_{\min} = 9.594 \times 10^4$. That is, to guarantee that the simulation error is less than 20% with a confidence level of 95%, one should perform the simulation for at least 9.594×10^4 times.

Ideally, if $\mathbb{V}(\widetilde{P})$ in Eq. (3.127) equals 0, then only one simulation run would give an exact estimate of $\mathbb{E}(h(\mathbf{X}))$. From a view of practical application, one can improve the simulation accuracy by reducing $\mathbb{V}(\widetilde{P})$. A straightforward approach is to increase the number of simulation replications, as illustrated in Example 3.29 as well as in Fig. 3.10, since $\mathbb{V}(\widetilde{P})$ decreases proportionally to $\frac{1}{n}$. This, however, needs a large amount of computational resources. Another idea is to reduce $\frac{1}{n}\sum_{i=1}^{n} h_i^2 - \mathbb{E}(\widetilde{P})$ via some advanced simulation techniques, e.g., the use of importance sampling method, as will be described subsequently.

3.5.2 Importance Sampling

As an introductory background, we recall Example 3.29, where the naïve Monte Carlo simulation was employed to compute Eq. (3.123), as shown in Fig. 3.11a. It can be seen that most samples of (x_1, x_2) contribute to the simulation scarcely with the sampled h_i being zero. To circumvent this situation, we can move the center of the joint distribution of (X_1, X_2) from the origin to a new position located close to the region of $h(\mathbf{X}) = 1$ (c.f. Fig. 3.11b), with which the probability of $h(\mathbf{X}) = 1$ is enhanced so that the sample average is relatively stabilized. This operation can also be explained by Eq. (3.132). When p is small, $n_{\min} \propto \frac{1}{p}$. Thus, increasing p will subsequently lead to a reduced $n_{\min}$. This method, generating samples from an alternative joint PDF rather than the original one, is called the *importance sampling* method.

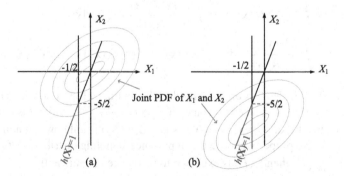

Fig. 3.11 Estimating Eq. (3.123) via simulation: role of $f_{\mathbf{X}}(\mathbf{x})$

Formally, we reconsider the expectation integral in Eq. (3.121), which is rewritten as follows,

$$\mathbb{E}(h(\mathbf{X})) = \int \cdots \int \frac{h(\mathbf{x}) f_{\mathbf{X}}(\mathbf{x})}{f'_{\mathbf{X}}(\mathbf{x})} f'_{\mathbf{X}}(\mathbf{x}) d\mathbf{x} = \int \cdots \int h'(\mathbf{x}) f'_{\mathbf{X}}(\mathbf{x}) d\mathbf{x} = \mathbb{E}(h'(\mathbf{X}))$$

(3.133)

where $h'(\mathbf{x}) = \frac{h(\mathbf{x}) f_{\mathbf{X}}(\mathbf{x})}{f'_{\mathbf{X}}(\mathbf{x})}$, and $f'_{\mathbf{X}}(\mathbf{x})$ is a legitimate joint PDF that is different from $f_{\mathbf{X}}(\mathbf{x})$, known as *importance sampling distribution*.

As before, according to Eq. (3.133), we can use simulation-based method to estimate $\mathbb{E}(h(\mathbf{X}))$, or equivalently, $\mathbb{E}(h'(\mathbf{X}))$. For the ith simulation run ($i = 1, 2, \ldots$), generate a sample of $\mathbf{X}$, $\mathbf{x}_i$ according to $f'_{\mathbf{X}}(\mathbf{x})$, and calculate $h'_i = h'(\mathbf{x}_i)$. Repeating for n times, $\mathbb{E}(h'(\mathbf{X}))$ is approximated as

$$\mathbb{E}(h'(\mathbf{X})) = \widetilde{P}' = \frac{1}{n} \sum_{i=1}^{n} h'_i$$

(3.134)

With Eq. (3.134), the mean value of $\widetilde{P}'$ is

$$\mathbb{E}(\widetilde{P}') = \mathbb{E}\left(\frac{1}{n} \sum_{i=1}^{n} h'_i\right) = \frac{1}{n} \cdot n \cdot \mathbb{E}\left(h'_i\right) = \mathbb{E}(h'(\mathbf{X})) = \mathbb{E}(h(\mathbf{X}))$$

(3.135)

which indicates that $\widetilde{P}'$ is an unbiased estimator of $\mathbb{E}(h(\mathbf{X}))$. Next, the variance of $\widetilde{P}'$ is estimated by

$$\mathbb{V}(\widetilde{P}) = \mathbb{V}\left(\frac{1}{n} \sum_{i=1}^{n} h'_i\right) = \frac{1}{n^2} \cdot n \cdot \mathbb{V}\left(h'_i\right) = \frac{\mathbb{V}(h'(\mathbf{X}))}{n} = \frac{\mathbb{E}[h'^2(\mathbf{X})] - \mathbb{E}^2(h'(\mathbf{X}))}{n}$$

(3.136)

With Eq. (3.136), if $f'_{\mathbf{X}}(\mathbf{x}) = \frac{f_{\mathbf{X}}(\mathbf{x}) h(\mathbf{x})}{\mathbb{E}(h'(\mathbf{x}))}$, then

$$\mathbb{E}[h'^2(\mathbf{X})] = \int h'^2(\mathbf{x}) f'_{\mathbf{X}}(\mathbf{x}) d\mathbf{x} = \int \frac{h^2(\mathbf{x}) f_{\mathbf{X}}^2(\mathbf{x})}{f'^2_{\mathbf{X}}(\mathbf{x})} \cdot f'_{\mathbf{X}}(\mathbf{x}) d\mathbf{x}$$
$$= \int \frac{h^2(\mathbf{x}) f_{\mathbf{X}}^2(\mathbf{x})}{f'_{\mathbf{X}}(\mathbf{x})} d\mathbf{x} = \mathbb{E}(h'(\mathbf{x})) \int h(\mathbf{x}) f_{\mathbf{X}}(\mathbf{x}) d\mathbf{x} = \mathbb{E}^2(h'(\mathbf{X}))$$

(3.137)

and thus, $\mathbb{V}(\widetilde{P}) = 0$. This observation implies that in Eq. (3.133), one will achieve an estimate of $\mathbb{E}(h(\mathbf{X}))$ with a zero-variance if $f'_{\mathbf{X}}(\mathbf{x}) = \frac{f_{\mathbf{X}}(\mathbf{x}) h(\mathbf{x})}{\mathbb{E}(h'(\mathbf{x}))}$. While it is not operational in practice to set $f'_{\mathbf{X}}(\mathbf{x})$ in this way since $\mathbb{E}(h'(\mathbf{X}))$ is unknown, it nonetheless suggests that if we properly choose the importance sampling distribution $f'_{\mathbf{X}}(\mathbf{x})$ that is close to $\frac{f_{\mathbf{X}}(\mathbf{x}) h(\mathbf{x})}{\mathbb{E}(h'(\mathbf{x}))}$, then $\mathbb{V}(\widetilde{P})$ is expected to be reduced significantly.

Table 3.3 Results of computing Eq. (3.38) via importance sampling method ($\times 10^{-4}$)

Number of replications	Trial 1	Trial 2	Trial 3	Trial 4	Trial 5	Trial 6	Average
$n = 10^4$	4.9658	5.3024	5.3614	4.7089	5.3973	4.5880	5.0540
$n = 10^5$	5.2449	5.1798	5.3861	5.0551	5.5921	5.1504	5.2681
$n = 10^6$	5.1946	5.3006	5.2385	5.2685	5.2258	5.2009	5.2381
$n = 10^7$	5.3008	5.2525	5.2563	5.2734	5.2673	5.2562	5.2678
$n = 10^8$	5.2652	5.2634	5.2728	5.2708	5.2660	5.2677	5.2676

Example 3.30

Recall Example 3.29. If we move the joint PDF of (X_1, X_2) to a new position such that $\mathbb{E}(X_1) = -\frac{1}{2}$ and $\mathbb{E}(X_2) = -3$, comment on the improved simulation efficiency employing the importance sampling method.

Solution

We use the importance sampling method to recompute Eq. (3.123). First, we rewrite Eq. (3.123) as follows,

$$
f = 2\pi\sqrt{1-\rho^2} \int_{-\infty}^{\infty} \int_{-\infty}^{\infty} \frac{\mathbb{I}(x_2 < 5x_1) \cdot \mathbb{I}\left(x_1 < -\frac{1}{2}\right) \cdot \frac{1}{2\pi\sqrt{1-\rho^2}} \exp\left[-\frac{x_1^2 - 2\rho x_1 x_2 + x_2^2}{2(1-\rho^2)}\right]}{\frac{1}{2\pi\sqrt{1-\rho^2}} \exp\left[-\frac{\left(x_1+\frac{1}{2}\right)^2 - 2\rho\left(x_1+\frac{1}{2}\right)(x_2+3) + (x_2+3)^2}{2(1-\rho^2)}\right]}
$$

$$
\cdot \frac{1}{2\pi\sqrt{1-\rho^2}} \exp\left[-\frac{\left(x_1+\frac{1}{2}\right)^2 - 2\rho\left(x_1+\frac{1}{2}\right)(x_2+3) + (x_2+3)^2}{2(1-\rho^2)}\right] dx_2 dx_1
$$

(3.138)

where $\rho = \frac{1}{2}$.

With this, the simulation procedure for the ith simulation run is formulated in the following.

(1) Simulate two independent standard normal variables z_1 and z_2.

(2) Set $x_1 = z_1 - \frac{1}{2}$ and $x_2 = \frac{1}{2}z_1 + \frac{\sqrt{3}}{2}z_2 - 3$.

(3) If $x_2 < 5x_1 < -\frac{5}{2}$, set $h_i' = \exp\left[-\frac{2}{3}\left(2x_1 - \frac{11}{2}x_2 - \frac{31}{4}\right)\right]$; otherwise, let $h_i' = 0$.

Repeating the above procedure for n times, f in Eq. (3.38) is estimated as $f = \sqrt{3}\pi \cdot \frac{1}{n}\sum_{i=1}^{n} h_i'$. Table 3.3 gives the results of the importance sampling method for different values of simulation runs. Compared with Table 3.2, it can be seen that the importance sampling method yields the desired result faster than the naïve Monte Carlo simulation.

3.5.3 Latin Hypercube Sampling

As has been mentioned before, the naïve Monte Carlo simulation provides a powerful tool for alternatively estimating a multiple-fold integral, and is especially competitive for the case of a high-dimensional problem. However, a large amount of simulation replications is essentially required in many cases to achieve a desired simulation accuracy. As a useful approach to improve the precision of the simulation-based estimate, especially for the case where one of the random inputs is particularly important, the *Latin hypercube sampling* method has witnessed its wide application in practical engineering, which generates a near-random sample of the inputs based on the stratification of each random variable simultaneously [11].

Recall Eq. (3.121) in Sect. 3.5.1, where we used a simulation-based approach to estimate $\mathbb{E}(h(\mathbf{X}))$ with $\frac{1}{n}\sum_{i=1}^{n} h_i$ (c.f. Eq. (3.122)). We assume that each element of $\mathbf{X}$, X_i ($i = 1, 2, \ldots r$), is a continuous random variable and is independent of each other. The stratification of the distribution of X_i is done by dividing the CDF of X_i into n non-overlapping and identical sections, as illustrated in Fig. 3.12a. Correspondingly, the horizontal axis (i.e., the domain of X_i) is also divided into n equiprobable intervals (not necessarily equal length). We randomly generate a sample of X_i for each interval, denoted by $x_{i1}, x_{i2}, \ldots x_{in}$, respectively, by letting

$$x_{ik} = F_{X_i}^{-1}\left(\frac{k - 1 + u_{ik}}{n}\right) \tag{3.139}$$

for $k = 1, 2, \ldots n$, where u_{ik} is a uniform sample over $[0, 1]$ and F_{X_i} is the CDF of X_i (when n is large enough, we can also simply assign $x_{ik} = F_{X_i}^{-1}\left(\frac{k-\frac{1}{2}}{n}\right)$, i.e., use the probability midpoint of each interval). In such a manner, a sequence of samples is obtained for each X_i, $i = 1, 2, \ldots r$. We write these $r \times n$ samples in a matrix form, as shown in Fig. 3.12c. Now, for the ith row (i.e., the samples of X_i), we randomly permutate the elements from $x_1, x_2, \ldots x_n$ to $x_{i\zeta_{i1}}, x_{i\zeta_{i2}}, \ldots x_{i\zeta_{in}}$, where $\{\zeta_{i1}, \zeta_{i2}, \ldots \zeta_{in}\}$ is a random permutation of $\{1, 2, \ldots n\}$, as shown in Fig. 3.12d. With this, the kth column ($k = 1, 2, \ldots n$) of the matrix in Fig. 3.12d, denoted by $\mathbf{x}_k$, contains samples of $X_1, X_2, \ldots X_r$ respectively.

It can be seen from Fig. 3.12 that the Latin hypercube sampling method forces the samples of each variable coming from different intervals, and thus ensures that the entire range of each random input is fully covered.

Similar to the vectors $\mathbf{x}_1, \mathbf{x}_2, \ldots \mathbf{x}_n$ in Fig. 3.12d, if we repeat the random permutation of the matrix in Fig. 3.12c for q times, then we will obtain totally $n \times q$ samples of $\mathbf{X}$, denoted by $\mathbf{x}_1, \mathbf{x}_2, \ldots \mathbf{x}_n, \mathbf{x}_{n+1}, \mathbf{x}_{n+2}, \ldots \mathbf{x}_{2n}, \mathbf{x}_{(q-1)n+1}, \mathbf{x}_{(q-1)n+2}, \ldots \mathbf{x}_{qn}$. With this, $\mathbb{E}(h(\mathbf{X}))$ in Eq. (3.121) is approximated by

$$\mathbb{E}(h(\mathbf{X})) = \widetilde{P}_{\text{LH},q} = \frac{1}{nq}\sum_{i=1}^{nq} h(\mathbf{x}_i) \tag{3.140}$$

Specifically, if $q = 1$ in Eq. (3.140), then

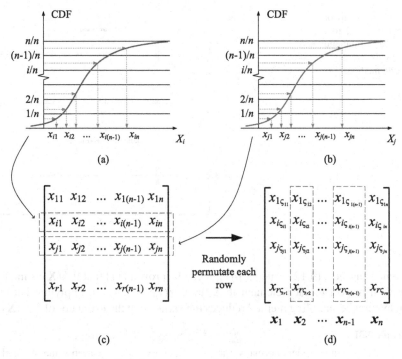

Fig. 3.12 Illustration of the Latin hypercube sampling method

$$\mathbb{E}(h(\mathbf{X})) = \widetilde{P}_{\text{LH}} = \frac{1}{n}\sum_{i=1}^{n} h(\mathbf{x}_i) \tag{3.141}$$

It can be shown that $\widetilde{P}_{\text{LH}}$ in Eq. (3.141) is an unbiased estimator of $\mathbb{E}(h(\mathbf{X}))$. To do so, we first rewrite Eq. (3.141) as follows,

$$\mathbb{E}(h(\mathbf{X})) = \widetilde{P}_{\text{LH}} = \frac{1}{n}\sum_{i=1}^{n} h_1(\mathbf{y}_i) \tag{3.142}$$

where $h_1(\mathbf{y}_i) = h_1(y_{i1}, y_{i2}, \ldots y_{in}) = h(F_{X_1}^{-1}(x_{i1}), F_{X_2}^{-1}(x_{i2}), \ldots F_{X_r}^{-1}(x_{in}))$. With Eq. (3.139) and Fig. 3.12d, $y_{ik} = \frac{\zeta_{ik}-1+u_{ik}}{n}$ for $k = 1, 2, \ldots n$, where $\zeta_{ik} \in \{1, 2, \ldots n\}$. Conditional on $\zeta_{ik} = \zeta \in \{1, 2, \ldots n\}$, y_{ik} is uniformly distributed over $\left[\frac{\zeta-1}{n}, \frac{\zeta}{n}\right]$. Also, using the law of total probability, y_{ik} follows a uniform distribution over $[0, 1]$. Furthermore, since the elements in $\mathbf{y}_i$ is independent of each other, one has

$$\mathbb{E}(\widetilde{P}_{\text{LH}}) = \frac{1}{n}\mathbb{E}\left(\sum_{i=1}^{n} h_1(\mathbf{y}_i)\right) = \mathbb{E}(h_1(\mathbf{y}_i)) = \mathbb{E}(h(\mathbf{x}_i)) = \mathbb{E}(h(\mathbf{X})) \tag{3.143}$$

Fig. 3.13 Comparison between the brute-force Monte Carlo simulation and the Latin hypercube sampling methods

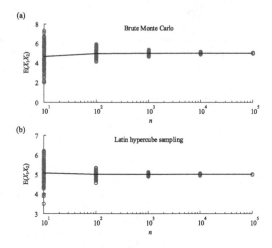

Comparing Eqs. (3.122) and (3.141), it can be proven [11] that if $h(\mathbf{X})$ is monotonic with respect to each element in $\mathbf{X}$, then $\mathbb{V}(\widetilde{P}_{LH}) \leq \mathbb{V}(\widetilde{P})$, implying that the Latin hypercube sampling method reduces the variance of the estimator of $\mathbb{E}(h(\mathbf{X}))$.

Example 3.31

Consider two independent normal variables X_1 and X_2. X_1 has a mean value of 1 and a standard deviation of 0.5; X_2 has a mean value of 5 and a standard deviation of 3. Compute the mean value of $X_1 X_2$ using both the brute-force Monte Carlo simulation and the Latin hypercube sampling methods, and compare their simulation efficiencies.

Solution

The mean value of $X_1 X_2$ can be calculated directly by $\mathbb{E}(X_1 X_2) = \mathbb{E}(X_1)\mathbb{E}(X_2) = 5$. Now, we use both simulation methods to approximate $\mathbb{E}(X_1 X_2)$. Figure 3.13 presents the simulation results associated with the two methods, using Eqs. (3.122) and (3.141). The number of simulation runs, n, equals 10^1 through 10^5. For each n, the results of 100 trials are presented in Fig. 3.13a, b. It can be seen that both methods yield an estimate of $\mathbb{E}(X_1 X_2) = 5$ as n is large enough. Also, the variance of the estimator $\widetilde{P}_{LH}$ is smaller than that of $\widetilde{P}$, indicating that the simulation efficiency is improved by the Latin hypercube sampling.

3.5.4   Subset Simulation

When the probability of an event is small enough, the *subset simulation* method provides a powerful tool for estimating such a small probability [2, 3].

In order to estimate the probability of an event F, we introduce a sequence of events $F_1, F_2, \ldots F_m = F$ such that $F_n \subset F_{m-1} \subset F_{m-2} \subset \ldots F_2 \subset F_1$. With this, it is easy to verify that $F_i = \bigcap_{j=1}^{i} F_j$ for $i = 1, 2, \ldots m$. Thus,

$$
\begin{aligned}
\mathbb{P}(F) = \mathbb{P}(F_m) &= \mathbb{P}(F_m \cap F_{m-1}) = \mathbb{P}(F_m | F_{m-1}) \cdot \mathbb{P}(F_{m-1}) \\
&= \mathbb{P}(F_m | F_{m-1}) \mathbb{P}(F_{m-1} | F_{m-2}) \cdot \mathbb{P}(F_{m-2}) = \ldots \\
&= \mathbb{P}(F_1) \prod_{i=1}^{m-1} \mathbb{P}(F_{i+1} | F_i)
\end{aligned}
\tag{3.144}
$$

According to Eq. (3.144), the probability $\mathbb{P}(F)$ equals the product of $\mathbb{P}(F_1)$ and the conditional probabilities $\mathbb{P}(F_{i+1} | F_i)$, $i = 1, 2, \ldots m - 1$. With this regard, even for an extremely small $\mathbb{P}(F)$, we can properly choose the sequence $\{F_i\}$ so that the conditional probability $\mathbb{P}(F_{i+1} | F_i)$ is relatively large for the sake of estimating via simulation-based approaches.

Remark

To illustrate the basic idea of subset simulation, simply consider two rare events A and B, and the probability of $\mathbb{P}(B) = \mathbb{P}(A) \cdot \mathbb{P}(A | B)$. If $\mathbb{P}(A) = \mathbb{P}(A | B) = 10^{-3}$, then $\mathbb{P}(B) = 10^{-6}$. Using the naïve Monte Carlo simulation to estimate $\mathbb{P}(B)$, the minimum number of simulation runs, $n_{\min}$, is $\sim 10^6$ according to Eq. (3.132). However, employing the subset simulation to estimate $\mathbb{P}(A)$ and $\mathbb{P}(A | B)$ first, the number of replications is, roughly, $\sim 2 \times 10^3$.

Specifically (and representatively), we consider the event F as $C < d$, where C is a random variable (e.g., the deteriorated structural resistance) and d is a deterministic value (threshold). Let $d_1 > d_2 > \cdots > d_{m-1} > d_m = d$, and define event F_i as $C < d_i$ for $i = 1, 2, \ldots m$. Clearly, $F_m \subset F_{m-1} \subset F_{m-2} \subset \ldots F_2 \subset F_1$ holds. Similar to Eq. (3.144), one has

$$
\mathbb{P}(F) = \mathbb{P}(C < d_m) = \mathbb{P}(C < d_1) \prod_{i=1}^{m-1} \mathbb{P}(C < d_{i+1} | C < d_i)
\tag{3.145}
$$

Example 3.32

In Eq. (3.145), we assume that C is a normal random variable with a mean value of 100 and a standard deviation of 20. Let $d = 30$ and $d_i = 10 \times (10 - i)$ for $i = 1, 2, \ldots 7$. Calculate $\mathbb{P}(C < d_{i+1} | C < d_i)$ for $i = 1, 2, \ldots 6$, and estimate $\mathbb{P}(C < d)$ using Eq. (3.145).

Solution

Let $F_C(x)$ denote the CDF of C. Conditional on $C < d_i$, the probability of $C < d_{i+1}$ is determined by

$$
\mathbb{P}(C < d_{i+1} | C < d_i) = \frac{\mathbb{P}(C < d_{i+1} \cap C < d_i)}{\mathbb{P}(C < d_i)} = \frac{F_C(d_{i+1})}{F_C(d_i)}
\tag{3.146}
$$

Table 3.4 presents the numerical solutions of $\mathbb{P}(C < d_{i+1}|C < d_i)$, as well as $\mathbb{P}(C < d)$ calculated according to Eq. (3.145). It can be seen that each conditional probability is fairly large, compared with the probability of $\mathbb{P}(C < d)$, so that it can be estimated via simulation-based approaches efficiently.

Regarding the implementation of (3.145), there are two important questions to be answered: (1) How to properly choose the sequence of $\{d_1, d_2, \ldots d_m\}$? (2) How to efficiently compute the conditional probability $\mathbb{P}(F_{i+1}|F_i)$?

We will answer the two questions by writing C as $C(\mathbf{X})$ in Eq. (3.145) to emphasize that it is a function of random factors $\mathbf{X} = \{X_1, X_2, \ldots X_r\}$. In fact, Eq. (3.145) is equivalent to

$$\mathbb{P}(F) = \mathbb{P}(C(\mathbf{X}) < d) = \underbrace{\int \cdots \int}_{r-\text{fold}} \mathbb{I}(C(\mathbf{x}) < d) f_{\mathbf{X}}(\mathbf{x}) d\mathbf{x} \qquad (3.147)$$

with $f_{\mathbf{X}}(\mathbf{x})$ being the joint PDF of $\mathbf{X}$, which takes a similar form of Eq. (3.121).

Now, for Question (1), in order to properly determine the sequence $\{d_1, d_2, \ldots d_m\}$, one can use the uniform conditional probabilities [3], that is,

$$\mathbb{P}(C < d_1) \approx \mathbb{P}(C < d_2|C < d_1) \approx \cdots \approx \mathbb{P}(C < d_m|C < d_{m-1}) \approx p_0 \quad (3.148)$$

where $p_0 \in (0, 1)$ is the "level probability", which is suggested to take a value of around 0.1 [2] or $0.1 \sim 0.3$ [23]. However, it is expected that $1/p_0$ is a positive integer for the sake of subsequent analysis.

With Eq. (3.148), we can first generate n samples of $\mathbf{X}$, $\mathbf{x}_1, \mathbf{x}_2, \ldots \mathbf{x}_n$, and order them in an ascending order in terms of the magnitude of $C(\mathbf{x}_i)$, based on which the value of d_1 is simply the p_0-percentile of the ordered samples. Subsequently, for $i = 1, 2, \ldots$, in an iterative manner, we simulate n samples of $\mathbf{X}$ conditional on $\mathbf{X} \in F_i$, $\mathbf{x}_1^{(i)}, \mathbf{x}_2^{(i)}, \ldots \mathbf{x}_n^{(i)}$ (details are described subsequently), and order them in an ascending order in terms of the magnitude of $C(\mathbf{x}_i^{(i)})$, with which the value of d_i is assigned as the p_0-percentile of the ordered samples $\{\mathbf{x}_i^{(i)}\}$. Repeating this process until the p_0-percentile of the ordered samples is smaller than d (correspondingly, $m = i$, and we assume that d is located at the q-percentile, $q \in (0, 1)$). With this, the probability of F, $\mathbb{P}(F) = \mathbb{P}(C(\mathbf{X}) < d)$, is estimated by

Table 3.4 Numerical solution of $\mathbb{P}(C < d_{i+1}|C < d_i)$ and $\mathbb{P}(C < d)$

$\mathbb{P}(C < d_1)$	$\mathbb{P}(C < d_2|C < d_1)$	$\mathbb{P}(C < d_3|C < d_2)$	$\mathbb{P}(C < d_4|C < d_3)$
0.3085	0.5142	0.4211	0.3405
$\mathbb{P}(C < d_5|C < d_4)$	$\mathbb{P}(C < d_6|C < d_5)$	$\mathbb{P}(C < d_7|C < d_6)$	$\mathbb{P}(C < d)$
0.2729	0.2174	0.1723	2.326×10^{-4}

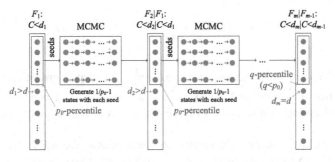

Fig. 3.14 Illustration of the subset simulation method for estimating $\mathbb{P}(F)$ in Eq. (3.145)

$$\mathbb{P}(C(\mathbf{X}) < d) = \widetilde{P}_{\text{SS}} = p_0^{m-1} q \tag{3.149}$$

The subset simulation algorithm is illustrated in Fig. 3.14. Clearly, a vital step in Fig. 3.14 is to simulate samples of $\mathbf{X}$ conditional on $\mathbf{X} \in F_j$, which is indeed closely related to Question (2). Basically, one can use the brute-force Monte Carlo simulation to estimate $\mathbb{P}(F_1)$ easily (c.f. Sect. 3.5.1). However, the simulation of conditional samples of $\mathbf{X}$ on $\mathbf{X} \in F_j$ is relatively difficult and inefficient if using the brute-force Monte Carlo simulation, since on average it needs $1/\mathbb{P}(F_i)$ simulation runs to generate a sample within the region of F_i. Alternatively, the MCMC simulation method can be use to improve the efficiency of estimating $\mathbb{P}(F_{i+1}|F_i)$ via simulation-based approaches [2, 15].

In Sect. 3.4.4, we mentioned that with the MCMC simulation method, in order to generate a sample of random variable X with a PDF of $f_X(x)$, one can generate a sample sequence of a Markov chain $\{X_1, X_2, \ldots X_n\}$ with a stationary PDF of $f_X(x)$, and treat the sample of X_n as that of X as n is large enough. Specifically, if X_1 has a PDF of $f_X(x)$, then each element of the sample sequence $\{X_1, X_2, \ldots X_n\}$ also has a PDF of $f_X(x)$. That being the case, $X_1, X_2, \ldots X_n$ are identically distributed (but not independent) and X_1 is called a "seed" of the sample sequence.

Employing the MCMC simulation method, we can simulate samples of X conditional on $X \in F_i$ as follows. With the $n \times p_0$ samples in F_i readily obtained (c.f. Fig. 3.14), we treat each sample as a "seed" to further simulate $\frac{1}{p_0} - 1$ samples (states) that have a common PDF of $f_{\mathbf{X}|F_i}(\mathbf{x})$ (note that each "seed" already has a PDF of $f_{\mathbf{X}|F_i}(\mathbf{x})$). Technically, with a sample $\mathbf{x}_j \in F_i$, the procedure to generate $\mathbf{x}_{j+1} \in F_i$ is as follows using MCMC simulation method.

(1) Simulate a sample $\mathbf{x}^*$ from the PDF $q(\mathbf{x}^*|\mathbf{x}_j)$, and a uniform variable u over $[0, 1]$.

(2) If $u \leq \min\left[1, \frac{f_{\mathbf{X}}(\mathbf{x}^*)}{f_{\mathbf{X}}(\mathbf{x}_j)} \cdot \frac{q(\mathbf{x}_j|\mathbf{x}^*)}{q(\mathbf{x}^*|\mathbf{x}_j)}\right]$ and $\mathbf{x}^* \in F_i$, let $\mathbf{x}_{i+1} = \mathbf{x}^*$; otherwise, let $\mathbf{x}_{i+j+1} = \mathbf{x}_j$.

As such, we can obtain totally $np_0 \times \left(\frac{1}{p_0} - 1\right) + np_0 = n$ samples of $\mathbf{X} \in F_i$, denoted by $\mathbf{x}_1^{(i)}, \mathbf{x}_2^{(i)}, \ldots \mathbf{x}_n^{(i)}$. Finally, the probability of F_m conditional on F_{m-1} is estimated by

$$\mathbb{P}(F_m \mid F_{m-1}) = q = \frac{1}{n} \sum_{j=1}^{n} \mathbb{I}(C(\mathbf{x}_j^{(m-1)}) < d) \qquad (3.150)$$

It is noticed that in Eq. (3.150), since each $\mathbf{x}_j^{(m-1)}$ is dependent on each other if generated from the same seed, $\mathbb{I}(C(\mathbf{x}_j^{(m-1)}) < d)$ may also be dependent mutually. While this dependency affects the variance of q, it has no impact on the mean value of q.

Finally, we briefly discuss the statistics of the estimator $\widetilde{P}_{SS}$ in Eq. (3.149). First, $\mathbb{E}(\widetilde{P}_{SS}) \neq \mathbb{P}(C(\mathbf{X}) < d)$ due to the correlation between the the estimators of different conditional probabilities. As such, $\widetilde{P}_{SS}$ is a biased estimator, with an order of $\frac{1}{n}$. Next, comparing the efficiencies of brute-force Monte Carlo simulation and the subset simulation method, the COV of $\widetilde{P}_{SS}$ in Eq. (3.149) is in proportion to $\left(\ln \frac{1}{\mathbb{P}(F)}\right)^{\xi}$, where $\xi \in \left[1, \frac{3}{2}\right]$ [3]. Recall that the COV of the estimator is proportional to $\frac{1}{\sqrt{\mathbb{P}(F)}}$ when $\mathbb{P}(F)$ is small enough using the naïve Monte Carlo simulation (c.f. Eq. (3.122)). Thus, it can be seen that the COV of $\widetilde{P}_{SS}$ increases at a slower rate than that of brute-force Monte Carlo simulation as $\mathbb{P}(F)$ diminishes.

Problems

3.1 Disk Line Picking Problem. Randomly pick up two points on a unit disk (with radius 1), and let D be the distance between the two points. What is the mean value of D?

3.2 A random variable X follows a Rayleigh distribution with scale parameter σ. Using the inverse transformation method, develop an algorithm to generate a sample for X, x.

3.3 For a Weibull random variable X with a scale parameter of u and a shape parameter of α, develop an algorithm to generate a sample for X using the inverse transformation method.

3.4 For a Cauchy random variable X whose PDF is as in Eq. (2.380), develop an algorithm to generate a sample for X using the inverse transformation method.

3.5 Given a positive λ, we simulate a sequence of independent uniform $(0,1)$ samples $u_1, u_2, \ldots$, and find such an integer n that

$$n = \min\left(k : \prod_{i=1}^{k} u_i < \exp(-\lambda)\right) - 1 \qquad (3.151)$$

Argue that n is a sample of Poisson random variable with mean λ [17].

3.6 The PDF of random variable X is as follows,

$$f_X(x) = \frac{1}{\sqrt{2\pi^{\frac{3}{2}}}} \int_{-\infty}^{\infty} \exp\left[-\frac{(x-y)^2}{2}\right] \cdot \frac{1}{y^2+1} dy, \quad -\infty < x < \infty \quad (3.152)$$

Establish a procedure to sample X using the composition method.

3.7 The PDF of random variable X is as follows with $\theta \geq 0$,

$$f_X(x) = \frac{x}{\theta+2} \exp\left(-\frac{x^2}{2(\theta+1)}\right) + (\theta+2)x \exp\left(-(\theta+2)x\right), \quad x \geq 0$$

$$(3.153)$$

Establish a procedure to sample X using the composition method.

3.8 If the PDF of random variable X is

$$f_X(x) = \frac{4}{41}(x^3 + 3x^2 + 6x), \quad x \in [0, 1] \quad (3.154)$$

establish a procedure to generate a realization of X [*Hint. Note that* $1 + x + \frac{x^2}{2} + \frac{x^3}{6} < e^x$ *when* $x \geq 0$.].

3.9 For a Laplace variable X whose PDF is as in Eq. (2.25), establish a procedure to generate a realization of X using the acceptance-rejection method.

3.10 Consider two continuous random variables X and Y with marginal CDFs of $F_X(x)$ and $F_Y(y)$ respectively, and a joint CDF of $C(F_X(x), F_Y(y))$, where C is the Farlie-Gumbel-Morgenstern copula, $C(u, v) = uv + \theta uv(1 - u)(1 - v)$ with $\theta \in [-1, 1]$. Develop an algorithm to simulate a sample pair for (X, Y), (x, y).

3.11 For two correlated random variables X_1, X_2 with an identical COV of 0.5, let ρ_X be their linear correlation coefficient. When using the Nataf transformation method to generate a sample pair of X_1, X_2, let Y_1, Y_2 be the corresponding standard normal variables (that is, $\Phi(Y_i) = F_{X_i}(X_i)$ for $i = 1, 2$, where F_{X_i} is the CDF of X_i), and ρ_Y the linear correlation coefficient for Y_1, Y_2. Plot the relationship between ρ_X and ρ_Y for the following two cases respectively.
(1) Both X_1 and X_2 follow an Extreme Type I distribution.
(2) X_1 is a normal variable while X_2 follows an Extreme Type I distribution.

3.12 Consider three identically lognormally distributed variables X_1, X_2 and X_3 with a mean value of 2 and a standard deviation of 0.5. Let ρ_{ij} be the correlation coefficient of X_i, X_j for $i, j \in \{1, 2, 3\}$. When $\rho_{ij} = \exp(-|i - j|)$, how to sample a realization for X_1, X_2, X_3 based on the Nataf transformation method? State the detailed steps.

3.13 For two correlated Poisson variables X_1 and X_2 with mean values of 1 and 2 respectively, let ρ_{12} be their correlation coefficient. When it is applicable to use Eq. (3.62) to generate a sample pair for X_1, X_2, what is the range for ρ_{12}?

3.14 With a reference period of $[0, T]$, the non-stationary Poisson process is used to model the occurrence of a sequence of repeated events. For the following two cases, establish a method to simulate a sequence of occurrence times of the events.
(1) The occurrence rate is $\lambda(t) = 2 + \sin(at)$, where $a > 0$ is a time-invariant parameter.
(2) The occurrence rate is $\lambda(t) = \exp(bt)$, where $b > 0$ is a time-invariant parameter.

3.15 With a reference period of $[0, T]$, the non-stationary Poisson process is used to model the occurrence of a sequence of repeated events. If the occurrence rate is

$$\lambda(t) = \begin{cases} 2 + \sin(at), & t \in [0, T_1] \\ \exp(bt), & t \in [T_1, T] \end{cases} \tag{3.155}$$

where $0 < T_1 < T$, a and b are two positive constants, establish a method to simulate a sequence of occurrence times of the events.

3.16 Recall Example 2.47. For a reference period of 50 years (i.e., $T = 50$ years), assume that the damage loss conditional on the occurrence of one cyclone event has a (normalized) mean value of 1 and a COV of 0.4. The cyclone occurrence is modelled as a stationary Poisson process with a mean occurrence rate of 0.5/year. Use a simulation-based approach to verify the mean value and variance of D_c as obtained in Example 2.47.

3.17 In Problem 3.16, if the cyclone occurrence is modelled as a non-stationary Poisson process with a mean occurrence rate of $\lambda(t) = 0.5(1 + 0.01t)$, where t is in years, what are the simulation-based mean value and variance of D_c? Compare the results with those in Problem 3.16.

3.18 Let N be a non-negative integer random variable with a mean value of μ_N, and $X_1, X_2, \ldots X_N$ a sequence of statistically independent and identically distributed random variables (independent of N) with a mean value of μ_X. Show that

$$\mathbb{E}\left(\sum_{i=1}^{N} X_i\right) = \mu_N \mu_X \tag{3.156}$$

Equation (3.156) is known as the Wald's equation.

3.19 The cables used in a suspension bridge have a lognormally distributed service life with a mean value of 40 years and a COV of 0.15.
(1) In order to guarantee the serviceability of the bridge in a long service term, at which rate does each cable have to be replaced by a new one?
(2) Use a simulation-based method to verify the result in (1).

3.20 In Problem 3.19, we additionally consider the impact of earthquakes on the cable serviceability. Suppose that the occurrence rate of earthquakes is a stationary Poisson process with a rate of 0.02/year, and that the cable is immediately replaced after each earthquake.

(1) In order to guarantee the serviceability of the bridge in a long service term, at which rate does each cable have to be replaced by a new one?
(2) Use a simulation-based method to verify the result in (1).

3.21 A structure can be categorized into one of the following four states: state 1 no damage; state 2 minor damage; state 3 moderate damage; state 4 severe/total damage. Let $X_1, X_2, \ldots$ be a stochastic process that represents the yearly state of the structure. Modeling the sequence $X_1, X_2, \ldots$ as a Markov chain, the transition matrix is as follows, where P_{ij} means the probability of the structure being classified as state j provided that the previous state is i for $i, j \in \{1, 2, 3, 4\}$.

$$\mathbf{P} = [P_{ij}] = \begin{bmatrix} 0.90 & 0.05 & 0.05 & 0 \\ 0 & 0.90 & 0.06 & 0.04 \\ 0 & 0 & 0.80 & 0.20 \\ 0 & 0 & 0 & 1 \end{bmatrix} \tag{3.157}$$

(1) If the structure is in state 1 at initial time, what is the probability of state 3 after five (5) years?
(2) If the structure is in state 2 at initial time, what is the probability of state 4 after ten (10) years?
(3) Show that after many years, the structure will be in state 4 with probability 1.

3.22 Consider the serviceability of a lining structure subjected to water seepage, as discussed in Problem 2.32. Assume that the hydraulic conductivity K and the water pressure p are two independent random variables. The structural performance is deemed as "satisfactory" if the water seepage depth at time t, $\chi(t)$, does not exceed a predefined threshold χ_{cr}. Establish a procedure to compute the probability of satisfactory structural performance at time t_0.

3.23 In Problem 3.22, if $\chi_{cr} = 0.15$ m, the hydraulic conductivity K is lognormally distributed having a mean value of 5.67×10^{-13} m/s and a COV of 0.3, and the water pressure is a Gamma variable with a mean value of 0.06 MPa and a COV of 0.2, what is the probability of satisfactory structural performance at the end of 50 years? Note that ω_0 is the water density (997 kg/m^3).

3.24 Reconsider the portal frame in Example 2.7. Develop a simulation procedure to estimate the probability that V occurs only conditional on the failure mode of "beam".

3.25 Estimate the following integral using Monte Carlo simulation method,

$$f = \int_0^\infty \int_0^z \int_0^y x y z^2 \exp\left(-\frac{x^2}{2} - y - z\right) dx\,dy\,dz \tag{3.158}$$

3.26 In Problem 3.25, numerically show that the use of importance sampling will improve the calculation efficiency.

3.27 Suppose that a normal random variable X has a mean value of 2 and a standard deviation of 1. Using a simulation-based approach to estimate $\mathbb{E}(X^2)$, show that the Latin hypercube sampling performs better than the brute-force Monte Carlo simulation in terms of convergence rate.

3.28 Suppose that X is a Gamma variable with a mean of 3 and a standard deviation of 1; Y is a lognormal variable with a mean of 2 and a standard deviation of 0.7. Use the subset simulation method to estimate the probability $\mathbb{P}[\exp(-X) \cdot Y > 1.5]$.

References

1. Andrieu C, De Freitas N, Doucet A, Jordan MI (2003) An introduction to mcmc for machine learning. Mach Learn 50(1–2):5–43. https://doi.org/10.1023/A:1020281327116
2. Au SK, Beck JL (2001) Estimation of small failure probabilities in high dimensions by subset simulation. Probab Eng Mech 16(4):263–277. https://doi.org/10.1016/S0266-8920(01)00019-4
3. Au SK, Wang Y (2014) Engineering risk assessment with subset simulation. Wiley
4. Avramidis AN, Channouf N, L'Ecuyer P (2009) Efficient correlation matching for fitting discrete multivariate distributions with arbitrary marginals and normal-copula dependence. INFORMS J Comput 21(1):88–106. https://doi.org/10.1287/ijoc.1080.0281
5. Badger L (1994) Lazzarini's lucky approximation of π. Math Mag 67(2):83–91
6. Der Kiureghian A, Liu PL (1986) Structural reliability under incomplete probability information. J Eng Mech 112(1):85–104. https://doi.org/10.1061/(ASCE)0733-9399(1986)112:1(85)
7. Ghitany M, Atieh B, Nadarajah S (2008) Lindley distribution and its application. Math Comput Simul 78(4):493–506. https://doi.org/10.1016/j.matcom.2007.06.007
8. Kahle W, Mercier S, Paroissin C (2016) Gamma processes. Wiley, Chap 2:49–148. https://doi.org/10.1002/9781119307488.ch2
9. Li Q, Wang C, Ellingwood BR (2015) Time-dependent reliability of aging structures in the presence of non-stationary loads and degradation. Struct Saf 52:132–141. https://doi.org/10.1016/j.strusafe.2014.10.003
10. Mazhdrakov M, Benov D, Valkanov N (2018) The Monte Carlo method: engineering applications. ACMO Academic Press
11. McKay MD, Beckman RJ, Conover WJ (2000) A comparison of three methods for selecting values of input variables in the analysis of output from a computer code. Technometrics 42(1):55–61. https://doi.org/10.1080/00401706.2000.10485979
12. Melchers RE, Beck AT (2018) Structural reliability analysis and prediction, 3rd edn. Wiley. https://doi.org/10.1002/9781119266105
13. Mori Y, Ellingwood BR (1993) Reliability-based service-life assessment of aging concrete structures. J Struct Eng 119(5):1600–1621. https://doi.org/10.1061/(ASCE)0733-9445(1993)119:5(1600)
14. Nakagawa T (2011) Stochastic processes: With applications to reliability theory. Springer Science & Business Media, Berli. https://doi.org/10.1007/978-0-85729-274-2
15. Papaioannou I, Betz W, Zwirglmaier K, Straub D (2015) MCMC algorithms for subset simulation. Probab Eng Mech 41:89–103. https://doi.org/10.1016/j.probengmech.2015.06.006
16. van Ravenzwaaij D, Cassey P, Brown SD (2018) A simple introduction to markov chain monte-carlo sampling. Psychon Bull Rev 25(1):143–154. https://doi.org/10.3758/s13423-016-1015-8
17. Ross SM (2010) Introduction to probability models, 10th edn. Academic
18. Shin K, Pasupathy R (2010) An algorithm for fast generation of bivariate poisson random vectors. INFORMS J Comput 22(1):81–92. https://doi.org/10.1287/ijoc.1090.0332

19. Van Ophem H (1999) A general method to estimate correlated discrete random variables. Econ Theory 15(2):228–237
20. Wang C, Zhang H (2018) Roles of load temporal correlation and deterioration-load dependency in structural time-dependent reliability. Comput Struct 194:48–59. https://doi.org/10.1016/j. compstruc.2017.09.001
21. Xiao Q (2017) Generating correlated random vector involving discrete variables. Commun Stat-Theory Methods 46(4):1594–1605. https://doi.org/10.1080/03610926.2015.1024860
22. Zio E (2013) The Monte Carlo simulation method for system reliability and risk analysis. Springer, Berlin. https://doi.org/10.1007/978-1-4471-4588-2
23. Zuev KM, Beck JL, Au SK, Katafygiotis LS (2012) Bayesian post-processor and other enhancements of subset simulation for estimating failure probabilities in high dimensions. Comput Struct 92:283–296. https://doi.org/10.1016/j.compstruc.2011.10.017

Chapter 4
Structural Reliability Assessment

Abstract This chapter discusses the method and application in practical engineering of structural reliability assessment. Reliability theory is the basis for measuring structural safety level under a probability-based framework. This chapter starts from the definition of structural reliability and reliability index, followed by the computation methods of reliability index, including first-order second-moment method, first-order reliability method, simulation-based method and moment-based method. Subsequently, some simple models for system reliability are presented, namely series system, parallel system and k-out-of-n system. The probability-based modelling of structural resistance and external loads are also addressed, followed by the probability-based limit state design approaches. The reliability theory can be further used to optimize structural design and conduct life-cycle cost analysis. Finally, the evaluation of structural reliability in the presence of imprecisely-informed random variables is considered.

4.1 Reliability and Reliability Index

For a structural member or system, the resistance, R, is expected to be greater than the load effect, S, so as to ensure the safety of the structure. The structural resistance may take different forms depending on the structural target performance of interest, e.g., the maximum moment-bearing capacity at the midpoint of a simply supported beam, the ultimate tensile stress of a steel bar, or the maximum displacement at the top of a structure. Correspondingly, the load effect S would have the same unit as R.

It is an important topic in practical engineering to estimate the possibility of structural failure (i.e., the probability of S exceeding R) under a probability-based framework, considering that both R and S are random variables with uncertainties arising from structural geometry, material properties, load magnitude (e.g., wind load effect is dependent on the wind speed) and others. We can employ the probability models (c.f. Chap. 2) to quantitatively measure the uncertainties associated with both R and S, given that the statistical information about resistance and load is observed or known.

© The Author(s), under exclusive license to Springer Nature Switzerland AG 2021   165
C. Wang, *Structural Reliability and Time-Dependent Reliability*,
Springer Series in Reliability Engineering,
https://doi.org/10.1007/978-3-030-62505-4_4

Fig. 4.1 Finding the probability of failure by comparing R and S

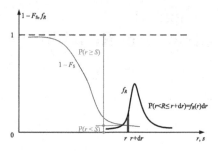

By definition, the structural reliability, denoted by $\mathbb{L}$, equals $1 - \mathbb{P}_f$, where $\mathbb{P}_f$ is the probability of failure. Using the law of total probability, it follows,

$$\mathbb{P}_f = 1 - \mathbb{L} = \mathbb{P}(R < S) = \int_0^\infty \int_0^s f_{R,S}(r, s)\mathrm{d}r\mathrm{d}s \tag{4.1}$$

where $f_{R,S}(r, s)$ is the joint PDF of R and S. Furthermore, if we assume that R and S are statistically independent, by referring to Fig. 4.1, one has

$$\mathbb{P}_f = \int_0^\infty F_R(s) f_S(s)\mathrm{d}s = \int_0^\infty [1 - F_S(r)] f_R(r)\mathrm{d}r \tag{4.2}$$

where F_R and F_S are the CDFs of R and S; f_R and f_S are the PDFs of R and S, respectively.

The expression of the failure probability $\mathbb{P}_f$ in Eq. (4.1) can be alternatively obtained by considering the ratio of R/S. Since both R and S are practically positive, it follows,

$$\mathbb{P}_f = \mathbb{P}(R < S) = \mathbb{P}\left(\frac{R}{S} < 1\right) = F_{R/S}(1) \tag{4.3}$$

where $F_{R/S}$ is the CDF of R/S. Note that

$$\frac{\mathrm{d}F_{R/S}(x)}{\mathrm{d}x} = f_{R/S}(x) = \int_0^\infty \tau \cdot f_{R,S}(x\tau, \tau)\mathrm{d}\tau \tag{4.4}$$

It is easy to see that Eqs. (4.1) and (4.3) give the same estimate of the failure probability.

Under the assumption of independent R and S, if both R and S are normally distributed, then their difference, $Z = R - S$, is also a normal variable. The mean value and variance of Z are respectively $\mu_Z = \mu_R - \mu_S$ and $\sigma_Z^2 = \sigma_R^2 + \sigma_S^2$, where $\mu_R, \mu_S, \sigma_R, \sigma_S$ are the mean values and standard deviations of R and S respectively. With this, the failure probability in Eq. (4.1) is simply

$$\mathbb{P}_f = \mathbb{P}(Z \leq 0) = \Phi\left(\frac{0 - \mu_Z}{\sigma_Z}\right) = 1 - \Phi\left(\frac{\mu_Z}{\sigma_Z}\right) = 1 - \Phi\left(\frac{\mu_R - \mu_S}{\sqrt{\sigma_R^2 + \sigma_S^2}}\right) \quad (4.5)$$

If treating Z as a function of R and S, then $Z = R - S$ is referred to as a "limit state function". The case of $Z < 0$ corresponds to structural failure; on the contrary, the structure survives if $Z \geq 0$. Clearly, the probability of structural failure (c.f. Eq. (4.5)) equal that of $Z < 0$.

Typically the limit state can be classified into two types [9]. The first is called *ultimate limit state*, which is related to the structural collapse of part or whole of a structure. The second, *serviceability limit state*, deals with the disruption of structural normal function due to damage or deterioration.

The structural *reliability index*, as a function of the failure probability, is defined as follows,

$$\beta = \Phi^{-1}(1 - \mathbb{P}_f) \quad (4.6)$$

which is a widely-used indicator to represent the safety level of a structure (used exchangeably with the probability of failure). For visualization purpose, Fig. 4.2a illustrates the relationship between the failure probability $\mathbb{P}_f$ and the reliability index β. Furthermore, with Eq. (4.5), when both R and S follow a normal distribution, one has

$$\beta = \frac{\mu_R - \mu_S}{\sqrt{\sigma_R^2 + \sigma_S^2}} \quad (4.7)$$

Equation (4.7) implies that the reliability index β equals the distance from origin to the mean value of Z' in a normalized space ($Z' = $ normalized $Z = \frac{Z - \mu_Z}{\sigma_Z}$), as illustrated in Fig. 4.2b.

Post to the assessment of the reliability index (or equivalently, the structural failure probability), one can compare with a predefined target reliability index to comment on the satisfaction of structural performance. Table 4.1 presents the target reliability indices adopted in ASCE/SEI 7-10 [4] (associated with an ultimate limit state) for different risk categories (the risk category of buildings and other structures, which

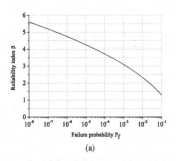

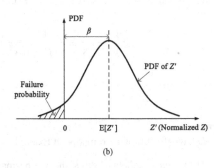

(a) (b)

Fig. 4.2 Relationship between the failure probability $\mathbb{P}_f$ and the reliability index β

Table 4.1 Target reliability index for different engineering cases on a 50-year-service-life basis

Basis	Risk category*			
	I	II	III	IV
Failure that is not sudden and does not lead to wide-spread progression of damage	2.5	3.0	3.25	3.5
Failure that is either sudden or leads to wide-spread progression of damage	3.0	3.5	3.75	4.0
Failure that is sudden and results in wide spread progression of damage	3.5	4.0	4.25	4.5

Note *The description of different risk categories can be found in Table 1.5-1 of ASCE/SEI 7-10 [4]

are determined according to their posed consequences to human life/surrounding community in the event of failure) and failure modes of engineered structures.

Example 4.1

In the presence of the limit state function of $Z = R - S$, we define a *central factor of safety*, θ, as $\frac{\mu_R}{\mu_S}$. Let v_R and v_S denote the COVs of R and S respectively.
(1) Derive the failure probability $\mathbb{P}_f$ in terms of θ, v_R and v_S. (2) Calculate $\mathbb{P}_f$ with $\theta = 2.5$, $v_R = 0.15$ and $v_S = 0.3$.
Solution

(1) Let σ_R and σ_S denote the standard deviations of R and S respectively. According to Eq. (4.5),

$$\mathbb{P}_f = 1 - \Phi\left(\frac{\mu_R - \mu_S}{\sqrt{\sigma_R^2 + \sigma_S^2}}\right)$$

$$= 1 - \Phi\left(\frac{(\theta - 1)\mu_S}{\sqrt{(\theta\mu_S v_R)^2 + (\mu_S v_S)^2}}\right) = 1 - \Phi\left(\frac{\theta - 1}{\sqrt{(\theta v_R)^2 + v_S^2}}\right) \quad (4.8)$$

(2) When $\theta = 2.5$, $v_R = 0.1$ and $v_S = 0.25$,

$$\mathbb{P}_f = 1 - \Phi\left(\frac{2.5 - 1}{\sqrt{(2.5 \times 0.1)^2 + 0.3^2}}\right) = 1 - \Phi(\underbrace{3.124}_{\beta}) = 8.936 \times 10^{-4} \quad (4.9)$$

Remark

We have the following remarks on Example 4.1.

(1) We have modelled both R and S as normal variables. Practically, R and S are expected to be definitely positive variables. Note that a normal variable, theoretically, may take

a negative value. This fact, nonetheless, does not affect the assumption of normality of R and S if the probability of a negative value is negligible. For example, the failure probability $\mathbb{P}_f$ in Eq. (4.9) is alternatively estimated by

$$\mathbb{P}_f = \mathbb{P}(R < S \cap S \leq 0) + \mathbb{P}(R < S \cap S > 0)$$

$$= \int_{-\infty}^{0} \Phi\left(\frac{\frac{S}{\mu_R} - 1}{0.15}\right) \cdot \frac{1}{\sigma_S} \phi\left(\frac{\frac{S}{\mu_S} - 1}{0.3}\right) ds + \int_{0}^{\infty} \Phi\left(\frac{\frac{S}{\mu_R} - 1}{0.15}\right) \cdot \frac{1}{\sigma_S} \phi\left(\frac{\frac{S}{\mu_S} - 1}{0.3}\right) ds$$

$$= 2.25 \times 10^{-15} + 8.936 \times 10^{-4} = 8.936 \times 10^{-4}$$

$$(4.10)$$

which implies that $\mathbb{P}(R < S \cap S \leq 0)$ is negligible compared with $\mathbb{P}_f$.

(2) Since $\beta = \dfrac{\theta - 1}{\sqrt{(\theta v_R)^2 + v_S^2}}$, one can also treat θ as a function of β, v_R and v_S as follows,

$$\theta = f(\beta, v_R, v_S) = \frac{1 + \sqrt{1 - (1 - \beta^2 v_R^2)(1 - \beta^2 v_S^2)}}{1 - \beta^2 v_R^2} \tag{4.11}$$

(3) We can assign a characteristic value of R or S (at a specific percentile) as the nominal resistance or nominal load. For example, if the nominal resistance r_n is at the 5% percentile of R, and the nominal load s_n at the 90% percentile of S, then

$$r_n = \mu_R - 1.645\sigma_R, \quad s_n = \mu_S + 1.282\sigma_S \tag{4.12}$$

With this,

$$\frac{r_n}{s_n} = \frac{\mu_R - 1.645\sigma_R}{\mu_S + 1.282\sigma_S} = \frac{1 - 1.645 v_R}{1 + 1.282 v_S}\theta \tag{4.13}$$

For the specific values of θ, v_R and v_S in Question (2), we have $r_n = 1.36 s_n$.

Example 4.2

A nominal load can be the load that has a mean recurrence interval of n years, or the mean maximum load during a service period of T years, or the load associated with other formulations [9].

For a location at Miami-Dade County, Florida, historical records show that the annual maximum wind speed (3-s gust) follows a Weibull distribution with a mean value of 24.3 m/s and a COV of 0.584 [7, 30]. Let the nominal wind speed v_n be that with an exceeding probability of 7% in 50 years. Compute v_n.

Solution

The annual maximum wind speed V (m/s) follows a Weibull distribution, with which its CDF is given as follows (c.f. Eq. (2.249)),

$$F_V(v) = 1 - \exp\left[-\left(\frac{v}{u}\right)^\alpha\right], \quad v \geq 0 \tag{4.14}$$

where u and α are the scale and shape parameters, respectively. Now, with

$$\mathbb{E}(V) = u\Gamma\left(1 + \frac{1}{\alpha}\right) = 24.3, \quad \mathbb{V}(V) = u^2\left[\Gamma\left(1 + \frac{2}{\alpha}\right) - \Gamma^2\left(1 + \frac{1}{\alpha}\right)\right] = (24.3 \times 0.584)^2$$
$$(4.15)$$

the two parameters u and α are found as 27.301 and 1.769 respectively.

Since v_n has an exceeding probability of 7% in 50 years, if the annual exceeding probability is p, it follows,

$$(1 - p)^{50} = 1 - 7\% \Rightarrow p = 0.00145 \tag{4.16}$$

This means that the nominal wind speed has a return period of $\frac{1}{p} \approx 700$ years. Furthermore,

$$F_V(v_n) = 1 - p \Rightarrow v_n = u\,(-\ln p)^{\frac{1}{\alpha}} = 78.9 \tag{4.17}$$

Thus, the nominal 3-s gust wind speed is 78.9 m/s having an exceeding probability of 7% in 50 years.

The structural resistance R follows a Weibull distribution with a location parameter of u_R and a shape parameter of α_R. The load effect, S, which is independent of R, has an Extreme Type II distribution with a scale parameter of u_S and a shape parameter of k_S. Let $Y = \frac{R}{S}$ be the overall factor of safety. Show that the CDF of Y is

$$F_Y(x) = \int_0^\infty \exp(-\tau)\left[1 - \exp\left(-(xu_S/u_R)^{\alpha_R}\tau^{-\alpha_R/k_S}\right)\right]d\tau \tag{4.18}$$

Solution

First, by definition,

$$F_Y(x) = \mathbb{P}(Y \le x) = \mathbb{P}(R \le xS) = \int_{-\infty}^\infty F_R(sx)f_S(s)ds \tag{4.19}$$

where F_R is the CDF of R and f_S is the PDF of S.

According to Eq. (2.240),

$$f_S(s) = k_S u_S^{k_S} s^{-k_S-1} \exp\left[-\left(\frac{s}{u_S}\right)^{-k_S}\right], \quad s \ge 0 \tag{4.20}$$

With Eq. (2.249),

$$F_R(r) = 1 - \exp\left[-\left(\frac{r}{u_R}\right)^{\alpha_R}\right], \quad r \ge 0 \tag{4.21}$$

Thus,

$$F_Y(x) = \int_{-\infty}^{\infty} \left\{ 1 - \exp\left[-\left(\frac{sx}{u_R} \right)^{\alpha_R} \right] \right\} \cdot k_S u_S^{k_S} s^{-k_S-1} \exp\left[-\left(\frac{s}{u_S} \right)^{-k_S} \right] ds \tag{4.22}$$

We let $\tau = \left(\frac{s}{u_S} \right)^{-k_S}$, with which $s = u_S \tau^{-\frac{1}{k_S}}$ and $ds = -\frac{u_S}{k_S} \tau^{-\frac{1}{k_S}-1} d\tau$. Thus,

$$F_Y(x) = \int_0^{\infty} \exp(-\tau) \left[1 - \exp\left(-\left(\frac{x u_S}{u_R} \right)^{\alpha_R} \tau^{-\alpha_R/k_S} \right) \right] d\tau \tag{4.23}$$

which completes the proof.

Example 4.4

Given a target reliability index of $\beta = 3.0$ for a reference period of 50 years (c.f. Table 4.1), calculate the corresponding annual failure probability $\mathbb{P}_{f,\text{annual}}$.

Solution

The reliability index of $\beta = 3.0$ yields a failure probability of $\mathbb{P}_{f,\,50\text{ year}} = 1 - \Phi(3.0) = 1.35 \times 10^{-3}$ for a 50-year service period. Furthermore, note that

$$1 - \mathbb{P}_{f,\,50\text{ year}} = \left(1 - \mathbb{P}_{f,\text{annual}} \right)^{50} \approx 1 - 50 \mathbb{P}_{f,\text{annual}} \tag{4.24}$$

which further gives

$$\mathbb{P}_{f,\text{annual}} = \frac{1}{50} \mathbb{P}_{f,\,50\text{ year}} = 2.7 \times 10^{-5} \tag{4.25}$$

If both R and S follow a lognormal distribution, then $Y = \frac{R}{S}$ is also lognormally distributed since $\ln Y = \ln \frac{R}{S} = \ln R - \ln S$ is a normal random variable. Let λ_R, λ_S, $\varepsilon_R, \varepsilon_S$ denote the mean value and standard deviation of $\ln R$ and $\ln S$ respectively (c.f. Eq. (2.201)). With this, $\ln Y$ has a mean value of $\lambda_R - \lambda_S$ and a standard deviation of $\sqrt{\varepsilon_R^2 + \varepsilon_S^2}$. Furthermore, with Eq. (4.3),

$$\mathbb{P}_f = \mathbb{P}\left(\ln \frac{R}{S} \le \ln 1 \right) = \mathbb{P}(\ln Y \le 0) = \Phi\left(\frac{0 - (\lambda_R - \lambda_S)}{\sqrt{\varepsilon_R^2 + \varepsilon_S^2}} \right) = 1 - \Phi\left(\frac{\lambda_R - \lambda_S}{\sqrt{\varepsilon_R^2 + \varepsilon_S^2}} \right) \tag{4.26}$$

Now, if the mean value and COV of R and S are μ_R, μ_S, ν_R and ν_S respectively, then, according to Eq. (2.203), one has

$$\varepsilon_R^2 = \ln(v_R^2 + 1), \quad \varepsilon_S^2 = \ln(v_S^2 + 1) \tag{4.27}$$

$$\lambda_R = \ln \mu_R - \frac{1}{2}\varepsilon_R^2, \quad \lambda_S = \ln \mu_S - \frac{1}{2}\varepsilon_S^2 \tag{4.28}$$

With this, Eq. (4.26) becomes

$$
\begin{aligned}
\mathbb{P}_f &= 1 - \Phi\left(\frac{\lambda_R - \lambda_S}{\sqrt{\varepsilon_R^2 + \varepsilon_S^2}}\right) \\
&= 1 - \Phi\left(\frac{\ln \frac{\mu_R \sqrt{v_S^2+1}}{\mu_S \sqrt{v_R^2+1}}}{\sqrt{\ln[(v_R^2 + 1)(v_S^2 + 1)]}}\right) = 1 - \Phi\left(\frac{\ln \frac{\theta \sqrt{v_S^2+1}}{\sqrt{v_R^2+1}}}{\sqrt{\ln[(v_R^2 + 1)(v_S^2 + 1)]}}\right)
\end{aligned}
\tag{4.29}
$$

where $\theta = \frac{\mu_R}{\mu_S}$.

Specifically, if both v_R and v_S are small enough, Eq. (4.29) becomes

$$\mathbb{P}_f \approx 1 - \Phi\left(\frac{\ln \theta}{\sqrt{v_R^2 + v_S^2}}\right) = 1 - \Phi\left(\frac{\ln \mu_R - \ln \mu_S}{\sqrt{v_R^2 + v_S^2}}\right) \tag{4.30}$$

Example 4.5

Reconsider Question (2) of Example 4.1. Supposing that both R and S are lognormally distributed, recalculate the failure probability $\mathbb{P}_f$ with Eq. (4.29).

Solution

According to Eq. (4.29),

$$\mathbb{P}_f = 1 - \Phi\left(\frac{\ln \frac{2.5\sqrt{0.3^2+1}}{\sqrt{0.15^2+1}}}{\sqrt{\ln[(0.15^2 + 1)(0.3^2 + 1)]}}\right) = 1 - \Phi\underbrace{(2.880)}_{\beta} = 1.990 \times 10^{-3} \tag{4.31}$$

Comparing the difference between the failure probabilities given by Eqs. (4.9) and (4.31), it can be seen that the failure probability is sensitive to the choice of distribution types of random inputs. This is known as the *tail sensitivity problem*.

Note that in Eq. (4.2), we have considered R and S as independent random variables. More generally, considering the potential correlation between R and S, Fig. 4.3 illustrates the joint distribution of (R, S). In such a case, we assume that the joint CDF of R and S is $F_{R,S}(r, s) = C(F_R(r), F_S(s))$, where C is a copula function (c.f. Sect. 2.3.3). Since $\mathbb{P}(V \le v | U = u) = \frac{\partial C(u,v)}{\partial u}$ holds according to Eq. (3.29), where

Fig. 4.3 Illustration of the joint PDF of R and S

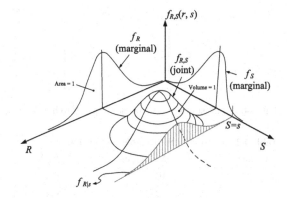

U and V are two uniform variables over $[0, 1]$ with a joint CDF of $C(u, v)$, one has

$$\mathbb{P}(R \leq r | S = s) = \mathbb{P}(F_R(R) \leq F_R(r) | F_S(S) = F_S(s))$$
$$= \frac{\partial F_{R,S}(r, s)}{\partial s} \cdot \frac{ds}{dF_S(s)} = \frac{1}{f_S(s)} \frac{\partial F_{R,S}(r, s)}{\partial s} \qquad (4.32)$$

Furthermore, using the law of total probability, it follows,

$$\mathbb{P}_f = \mathbb{P}(R \leq S) = \int \mathbb{P}(R \leq s | S = s) \cdot f_S(s)ds = \int \frac{\partial F_{R,S}(s, s)}{\partial s} ds \qquad (4.33)$$

Example 4.6

Consider a portal frame subjected to horizontal load $F = 4$ kN, as shown in Fig. 4.4a. The height of the frame is $l = 1$ m. The column materials are of the same mechanical properties for both AB and CD and the ultimate stress is $\sigma_{max} = 20$ MPa. The beam is assumed to be of infinite stiffness. The cross section of beams AB and CD are rectangular and the plan view is given in Fig. 4.4b. The geometric dimensions are summarized in Table 4.2. Consider the safety of column AB. Let R be the resistance (moment-bearing capacity) of the cross section at A (i.e., the bottom of column AB), and S be the moment at A due to the horizontal load F. The failure is defined as the occurrence of the plastic hinge at the bottom of column AB.

Using a simulation-based approach, find the linear correlation coefficient between R and S, and compute the failure probability considering/not considering the correlation between R and S.

Solution

First, using the knowledge of structural mechanics, we have

$$R = \sigma_{max} \cdot \frac{1}{6}b_1 h_1^2, \quad S = \frac{I_1}{I_1 + I_2} Fl = \frac{b_1 h_1^3}{b_1 h_1^3 + b_2 h_2^3} Fl \qquad (4.34)$$

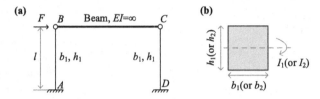

Fig. 4.4 An illustrative portal frame subjected to horizontal load and resistance-load correlation

Table 4.2 Statistics of the variables in Fig. 4.4b

Variable	Mean (cm)	COV	Distribution type
b_1	10	0.2	Normal
b_2	10	0.2	Normal
h_1	12	0.2	Normal
h_2	12	0.2	Normal

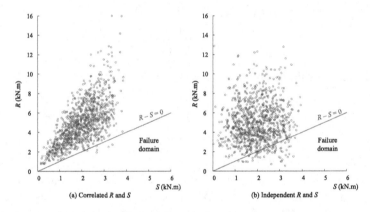

Fig. 4.5 Scatters of sample pairs of R and S

Using a simulation-based approach, we simulate 1,000,000 sample pairs of (R, S), with which the linear correlation coefficient between R and S is found as 0.6649. Figure 4.5a shows the scatters of sample pairs of (R, S).

Furthermore, with the sample pairs of (R, S), the failure probability is 0.0092. If ignoring the R-S correlation, the sample scatters of (R, S) are illustrated in Fig. 4.5b, and the failure probability is 0.0783. Clearly, a reasonable estimate of structural failure probability depends on the correct modelling of resistance-load correlation.

Recall that in Example 4.6, the failure probability is estimated as the probability of $R - S < 0$. Note that

$$R - S < 0 \Leftrightarrow b_1 h_1^2 + b_2 h_2^2 \cdot \frac{h_2}{h_1} < \frac{6Fl}{\sigma_{\max}} \qquad (4.35)$$

Fig. 4.6 Scatters of sample pairs of (h_1, h_2)

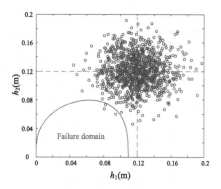

Specifically, if $b_1 = b_2 = b = $ a deterministic value cm, then the failure probability is alternatively estimated as

$$\mathbb{P}_f = \mathbb{P}\left(f(h_1, h_2) < f_0\right) \tag{4.36}$$

where $f(h_1, h_2) = h_1^2 + h_2^2 \cdot \frac{h_2}{h_1}$, and $f_0 = \frac{6Fl}{b\sigma_{max}}$. The scatters of sample pairs of (h_1, h_2) are shown in Fig. 4.6, assuming $b = 10$ cm, with which the failure probability is determined as 0.0049. Equation (4.36) implies that in some cases, the structural resistance and applied load effect (i.e., R and S) are not necessarily defined in an explicit form. Nonetheless, the failure probability can be estimated once the limit state function regarding the random inputs is available.

Example 4.7

For a column subjected to the combination effects of axial force and bending moment, the moment-axial force interaction diagram is illustrated in Fig. 4.7. The solid curve denotes the limit state of structural failure, consisting of two underlying mechanisms: compression-controlled failure (segment ab) and tension-controlled failure (segment bc). Now, consider a column subjected to axial force P. The relationship between M (kN·m) and P (kN) is given as follows,

$$P = \begin{cases} 10(M - 200), & P \le 500 \text{ (tension-controlled failure)} \\ 10(300 - M), & P > 500 \text{ (compression-controlled failure)} \end{cases} \tag{4.37}$$

The axial force P has an eccentricity of e (i.e., P is applied at a distance of e from the centroid), with which the bending moment due to the load eccentricity is $M = Pe$. Suppose that P is a normal variable with a mean value of 800 kN and a COV of 0.25, and that the load eccentricity e also follows a normal distribution with a mean value of 0.1 m and a COV of 0.2. Compute the probability of failure.

Solution

We have the following two methods to calculate the failure probability (i.e., the force pair (M, P) falls outside the safety region).

Fig. 4.7 Bending moment-axial force interaction diagram

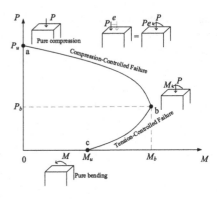

Method 1. Analytical solution. Note that

$$1 - \mathbb{P}_f = \mathbb{P}(3000 - 10M \geq P \cap 10M - 2000 \leq P)$$
$$= \mathbb{P}(3000 - 10Pe \geq P \cap 10Pe - 2000 \leq P)$$
$$= \int_{-\infty}^{0.1} \left[F_P \left(\frac{3000}{10x + 1} \right) - F_P \left(\frac{2000}{10x - 1} \right) \right] f_e(x) \mathrm{d}x$$
$$+ \int_{0.1}^{\infty} F_P \left[\min \left(\frac{3000}{10x + 1}, \frac{2000}{10x - 1} \right) \right] f_e(x) \mathrm{d}x \qquad (4.38)$$
$$= 0.99848$$

where F_P is the CDF of P, and f_e is the PDF of e. With this, $\mathbb{P}_f = 0.00152$.

Method 2. Simulation-based approach. The basic idea is to first generate, independently, n samples of P and n samples of e, and then check the number of sampled pairs (Pe, P) that fall outside the safety region, denoted by m. The failure probability is then approximated by $\frac{m}{n}$. Results show that the probability of failure equals $\mathbb{P}_f = 0.00149$ with 1,000,000 simulation runs. This is consistent with the result given by Method 1.

4.2   Computation of Structural Reliability

4.2.1   First-Order Second-Moment Reliability Method

We have discussed, in Sect. 4.1, the structural failure probability (as well as the reliability index) with a limit state function of $G = R - S$. The explicit expression of $\mathbb{P}_f$ is derived for the case where R and S both follow a normal distribution (c.f. Eq. (4.5)) or a lognormal distribution (c.f. Eq. (4.29)). For other distribution types of R and S, one can use Eq. (4.2) (or Eq. (4.33) if considering the potential correlation between R and S) to calculate the failure probability by solving a one-fold integral.

Note that both the structural resistance and load effect are functions of random inputs such as the structural geometry, material properties and others, denoted by $R = R(\mathbf{X})$ and $S = S(\mathbf{X})$, where $\mathbf{X}$ is a random vector representing the uncertain quantities involved in the reliability problem (c.f. Eq. (4.34)). However, in practical engineering, it is often the case that the distribution types of some random inputs are unknown but only the limited statistical information such as the mean value and variance are available. Correspondingly, the PDFs of structural resistance and/or load effect are also unaccessible. For example, with only a few observed samples of a random variable X, the mean value of X can be assigned as the average of the available samples and the standard deviation of X can be roughly estimated as $\frac{1}{4}$ times the difference between the maxima and minima of the samples, if the observed values are believed to vary within a range of mean ± 2 standard deviation of X (with a 95% confidence level if assuming a normal X). However, the specific distribution type of X depends on further probabilistic information.

Furthermore, it has been previously demonstrated that the failure probability may be fairly sensitive to the choice of distribution type of random variables in the presence of imprecise probabilistic information (i.e., the tail sensitivity problem, c.f. Example 4.5). With this regard, the reliability index (and correspondingly the probability of failure) cannot be evaluated exactly with Eq. (4.2) or Eq. (4.33). Alternatively, it can be computed based on the first-order and second-order moments of random inputs, as will be discussed in the following.

4.2.1.1 Linear Limit State Function

With a limit state function being

$$G(R, S) = R - S = 0 \tag{4.39}$$

we convert the original space of (S, R) into a *normalized space* (or a *reduced space*), denoted by (s, r). That is,

$$s = \frac{S - \mu_S}{\sigma_S}, \quad r = \frac{r - \mu_R}{\sigma_R} \tag{4.40}$$

where $\mu_R, \mu_S, \sigma_R, \sigma_S$ are the mean values and standard deviations of R and S respectively. With this, the limit state function in the normalized space is illustrated in Fig. 4.8a, taking a form of

$$g(r, s) = \sigma_R r - \sigma_S s + (\mu_R - \mu_S) \tag{4.41}$$

With Eq. (4.41), it is noticed that the shortest distance from the origin $(0, 0)$ to the line $g(r, s) = 0$ is

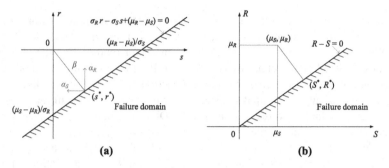

Fig. 4.8 Geometric representation of the reliability index β with a limit state of $R - S = 0$

$$\text{shortest distance} = \frac{\mu_R - \mu_S}{\sqrt{\sigma_R^2 + \sigma_S^2}} \qquad (4.42)$$

which takes a similar form compared with the reliability index in Eq. (4.7). Motivated by this, we define the shortest distance from origin (0, 0) to the line $g(r, s) = 0$ (c.f. Eq. (4.42)) as the reliability index, denoted by β_{HL}, regardless of the distribution types of R and S. We call β_{HL} the *Hasofer–Lind reliability index*. While β_{HL} may be less accurate compared with that in Eq. (4.7), it has been widely used in practical engineering because it does not require the specific distribution types of the random inputs and provides a reasonable estimate of structural safety level. Nonetheless, one should keep in mind that β_{HL} is a *notional* measure of structural reliability and is expected to be interpreted in a comparative sense.

When the limit state function is simply $G = R - S$, $\beta_{HL} = \frac{\mu_R - \mu_S}{\sqrt{\sigma_R^2 + \sigma_S^2}}$. If, further, both R and S are normal variables, β_{HL} would be exactly the reliability index in Eq. (4.6).

In Fig. 4.8a, let (s^*, r^*) be the point which lies on the line $g(r, s) = 0$ and yields the shortest distance from origin to $g(r, s) = 0$, referred to as the *checking point*. The normal vector of $g(r, s) = 0$ at (s^*, r^*) is denoted by $\boldsymbol{\alpha} = (\alpha_S, \alpha_R)$. Clearly, $\boldsymbol{\alpha}$, with a magnitude of 1, is parallel to $\left(\frac{\partial g}{\partial s}, \frac{\partial g}{\partial r} \right)$. Thus,

$$\alpha_S = \frac{\frac{\partial g}{\partial s}}{\sqrt{\left(\frac{\partial g}{\partial s} \right)^2 + \left(\frac{\partial g}{\partial r} \right)^2}} = \frac{-\sigma_S}{\sqrt{\sigma_S^2 + \sigma_R^2}}, \quad \alpha_R = \frac{\frac{\partial g}{\partial r}}{\sqrt{\left(\frac{\partial g}{\partial s} \right)^2 + \left(\frac{\partial g}{\partial r} \right)^2}} = \frac{\sigma_R}{\sqrt{\sigma_S^2 + \sigma_R^2}}$$

$$(4.43)$$

Furthermore, to determine (s^*, r^*), we have

$$(0, 0) - (s^*, r^*) = \beta_{HL} \boldsymbol{\alpha} \Rightarrow \begin{cases} s^* = -\alpha_S \beta_{HL}, \\ r^* = -\alpha_R \beta_{HL} \end{cases} \qquad (4.44)$$

Correspondingly, in the original space, the checking point (S^*, R^*) is as follows (c.f. Fig. 4.8b),

$$S^* = \mu_S + \sigma_S s^* = \mu_S - \sigma_S \alpha_S \beta_{HL}, \quad R^* = \mu_R + \sigma_R r^* = \mu_R - \sigma_R \alpha_R \beta_{HL} \quad (4.45)$$

Example 4.8

Consider a limit state function of $G = R - S$, with which the structural failure is deemed to occur when $G < 0$. The resistance R has a mean value of 2.5 and a COV of 0.15; the load effect S has a mean value of 1 and a COV of 0.3. Find the checking point (s^*, r^*) in the normalized space and that in the original space.

Solution

First, the limit state function in the normalized space is $g(r, s) = 0.375r - 0.3s + 1.5 = 0$, with which

$$\beta_{HL} = \frac{1.5}{\sqrt{0.3^2 + 0.375^2}} = 3.1235 \quad (4.46)$$

and

$$\alpha_S = \frac{-0.3}{\sqrt{0.3^2 + 0.375^2}} = -0.6247, \quad \alpha_R = \frac{0.375}{\sqrt{0.3^2 + 0.375^2}} = 0.7809 \quad (4.47)$$

Thus,

$$s^* = -\alpha_S \beta_{HL} = 0.6247 \times 3.1235 = 1.9512,$$
$$r^* = -\alpha_R \beta_{HL} = -0.7809 \times 3.1235 = -2.4391 \quad (4.48)$$

Furthermore, according to Eq. (4.45),

$$S^* = \mu_S + \sigma_S s^* = 1 + 0.3 \times 1.9512 = 1.5854,$$
$$R^* = \mu_R + \sigma_R r^* = 2.5 - 0.375 \times 2.4391 = 1.5854 \quad (4.49)$$

Example 4.9

If the limit state function is $G = R - S = 0$, then it is expected that $G(R^*, S^*) = R^* - S^* = 0$. Show that in Eq. (4.45), $R^* = S^*$.

Solution

It is easy to see that

$$R^* - S^* = \mu_R - \mu_S + (\sigma_S \alpha_S \beta_{HL} - \sigma_R \alpha_R \beta_{HL}) = \mu_R - \mu_S - \beta_{HL}\sqrt{\sigma_S^2 + \sigma_R^2} = 0 \quad (4.50)$$

which completes the proof.

Example 4.10

Reconsider Example 4.8. We define the resistance factor φ_R and the load factor γ_S as $\varphi_R = \frac{R^*}{r_n}$ and $\varphi_S = \frac{S^*}{s_n}$, where r_n and s_n are the nominal hesitance and load effect respectively, defined in Eq. (4.12). Compute φ_R and φ_S.

Solution

With Eqs. (4.12) and (4.45),

$$
\begin{aligned}
\varphi_R &= \frac{R^*}{r_n} = \frac{\mu_R - \sigma_R \alpha_R \beta_{\text{HL}}}{\mu_R - 1.645\sigma_R} = \frac{1.5854}{2.5 - 1.645 \times 0.375} = 0.84 \\
\varphi_S &= \frac{S^*}{s_n} = \frac{\mu_S - \sigma_S \alpha_S \beta_{\text{HL}}}{\mu_S + 1.282\sigma_S} = \frac{1.5854}{1 + 1.282 \times 0.3} = 1.15
\end{aligned}
\tag{4.51}
$$

Remark

(1) Equation (4.42) is supported by the fact that in an x-y coordinate system, the shortest distance from a point (x_0, y_0) to a line $ax + by + c = 0$ is $\frac{|ax_0 + by_0 + c|}{\sqrt{a^2 + b^2}}$, where a, b, c are three constants.

(2) As a general approach, we can also use the method of Lagrange multipliers to find the shortest distance from the origin to $g(r, s) = 0$ (c.f. Eq. (4.42)). We introduce a new variable λ (the Lagrange multiplier) to construct the Lagrange function as follows,

$$
L(r, s, \lambda) = \sqrt{r^2 + s^2} + \lambda g(r, s) = \sqrt{r^2 + s^2} + \lambda \left[\sigma_R r - \sigma_S s + (\mu_R - \mu_S) \right]
\tag{4.52}
$$

Let $\frac{\partial L}{\partial \bullet} = 0$, where $\bullet = \lambda, r$ or s, with which

$$
\begin{cases}
\dfrac{\partial L}{\partial \lambda} = 0 \Rightarrow \sigma_R r - \sigma_S s + (\mu_R - \mu_S) = 0 \\[2mm]
\dfrac{\partial L}{\partial r} = 0 \Rightarrow \dfrac{r}{\sqrt{r^2 + s^2}} + \lambda \sigma_R = 0 \\[2mm]
\dfrac{\partial L}{\partial s} = 0 \Rightarrow \dfrac{s}{\sqrt{r^2 + s^2}} - \lambda \sigma_S = 0
\end{cases}
\tag{4.53}
$$

yielding

$$
\lambda = \frac{1}{\sqrt{\sigma_S^2 + \sigma_R^2}}, \quad r = \frac{-\sigma_R (\mu_R - \mu_S)}{\sigma_S^2 + \sigma_R^2}, \quad s = \frac{\sigma_S (\mu_R - \mu_S)}{\sigma_S^2 + \sigma_R^2}
\tag{4.54}
$$

which is consistent with Eq. (4.44).

We can extend Eq. (4.42) to the case of an n-dimensional problem. Consider a limit state function $G = a_0 + \sum_{i=1}^{n} a_i X_i$, where X_i are random variables ($i = 1, 2, \ldots, n$), and a_i ($i = 0, 1, 2, \ldots, n$) are constants. We first transform G into that in the normalized space by letting $x_i = \frac{X_i - \mu_i}{\sigma_i}$, where μ_i and σ_i are the mean value and standard deviation of X_i respectively. With this,

$$g(x_1, x_2, \ldots, x_n) = \left(a_0 + \sum_{i=1}^{n} a_i \mu_i\right) + \sum_{i=1}^{n} a_i \sigma_i x_i \tag{4.55}$$

Again, we define the reliability index as the shortest distance from the origin to $g(x_1, x_2, \ldots, x_n) = 0$, denoted by β_{HL}. Thus,

$$\beta_{HL} = \frac{a_0 + \sum_{i=1}^{n} a_i \mu_i}{\sqrt{\sum_{i=1}^{n} (a_i \sigma_i)^2}} \tag{4.56}$$

Let $(x_1^*, x_2^*, \ldots, x_n^*)$ denote the checking point in the normalized space. The normal vector $\boldsymbol{\alpha} = (\alpha_1, \alpha_2, \ldots, \alpha_n)$ at the checking point is parallel to $\left(\frac{\partial g}{\partial x_1}, \frac{\partial g}{\partial x_2}, \ldots, \frac{\partial g}{\partial x_n}\right) = (a_1 \sigma_1, a_2 \sigma_2, \ldots, a_n \sigma_n)$ and has a magnitude of 1. Thus,

$$\alpha_i = \frac{a_i \sigma_i}{\sqrt{\sum_{i=1}^{n} (a_i \sigma_i)^2}}, \quad i = 1, 2, \ldots, n \tag{4.57}$$

Furthermore,

$$(0, 0, \ldots, 0) - (x_1^*, x_2^*, \ldots, x_n^*) = \beta_{HL} \boldsymbol{\alpha} \Rightarrow x_i^* = -\frac{\beta_{HL} a_i \sigma_i}{\sqrt{\sum_{i=1}^{n} (a_i \sigma_i)^2}}, \quad i = 1, 2, \ldots, n \tag{4.58}$$

Correspondingly, in the original space, the checking point $(X_1^*, X_2^*, \ldots, X_n^*)$ is determined by

$$X_i^* = \mu_i - \frac{\beta_{HL} a_i \sigma_i^2}{\sqrt{\sum_{i=1}^{n} (a_i \sigma_i)^2}}, \quad i = 1, 2, \ldots, n \tag{4.59}$$

It is easy to verify that with Eq. (4.59), $G(X_1^*, X_2^*, \ldots, X_n^*) = 0$. This is because

$$G(X_1^*, X_2^*, \ldots, X_n^*) = a_0 + \sum_{i=1}^{n} a_i X_i^* = a_0 + \sum_{i=1}^{n} a_i \mu_i - \sum_{i=1}^{n} a_i \frac{\beta_{HL} a_i \sigma_i^2}{\sqrt{\sum_{i=1}^{n} (a_i \sigma_i)^2}}$$

$$= a_0 + \sum_{i=1}^{n} a_i \mu_i - \frac{a_0 + \sum_{i=1}^{n} a_i \mu_i}{\sqrt{\sum_{i=1}^{n} (a_i \sigma_i)^2}} \cdot \frac{\sum_{i=1}^{n} a_i^2 \sigma_i^2}{\sqrt{\sum_{i=1}^{n} (a_i \sigma_i)^2}} = 0 \tag{4.60}$$

Table 4.3 Statistical information of the random variables R, S_1 and S_2

Parameter	Mean	COV
R	4.0	0.1
S_1	1.5	0.3
S_2	0.5	0.3

Example 4.11

Consider a limit state function of $G = R - S_1 - S_2$, where R is the structural resistance, S_1 and S_2 are two load effects that are applied to the structure simultaneously. The mean value and COV of each random variable are summarized in Table 4.3. Defining the nominal resistance and nominal loads (for both S_1 and S_2) with Eq. (4.12), calculate the resistance factor φ_R and the load factors γ_{S_1} and γ_{S_2} (c.f. Example 4.10).

Solution

We first calculate the checking point (S_1^*, S_2^*, R^*) according to Eq. (4.59). With Eq. (4.56),

$$\beta_{HL} = \frac{4 - 1.5 - 0.5}{\sqrt{0.4^2 + (-0.45)^2 + (-0.15)^2}} = 3.223 \tag{4.61}$$

Thus,

$$R^* = 4.0 - \frac{3.223 \times 0.4^2}{\sqrt{0.4^2 + (-0.45)^2 + (-0.15)^2}} = 3.169$$

$$S_1^* = 1.5 + \frac{3.223 \times (-0.45)^2}{\sqrt{0.4^2 + (-0.45)^2 + (-0.15)^2}} = 2.552 \tag{4.62}$$

$$S_2^* = 0.5 + \frac{3.223 \times (-0.15)^2}{\sqrt{0.4^2 + (-0.45)^2 + (-0.15)^2}} = 0.617$$

According to Eq. (4.12),

$$r_n = 4.0 - 1.645 \times 0.4 = 3.342, \quad s_{n1} = 1.5 + 1.282 \times 0.45 = 2.077,$$

$$s_{n2} = 0.5 + 1.282 \times 0.15 = 0.692 \tag{4.63}$$

Similar to Eq. (4.51),

$$\varphi_R = \frac{R^*}{r_n} = 0.95, \quad \varphi_{S_1} = \frac{S_1^*}{s_{n1}} = 1.23, \quad \varphi_{S_2} = \frac{S_2^*}{s_{n2}} = 0.89 \tag{4.64}$$

4.2.1.2   Nonlinear Limit State Function

Note that the limit state functions in Eqs. (4.41) and (4.55) are linear. Now we further consider the generalized case where the limit state function is nonlinear. Mathematically, in the presence of n random inputs $X_1, X_2, \ldots, X_n$, the limit state function is expressed as $G = G(\mathbf{X}) = G(X_1, X_2, \ldots, X_n)$. In the normalized space,

$$g = g(\mathbf{x}) = g(x_1, x_2, \ldots, x_n) = G(\mu_1 + \sigma_1 x_1, \mu_2 + \sigma_2 x_2, \ldots, \mu_n + \sigma_n x_n) \tag{4.65}$$

where μ_i and σ_i are the mean value and standard deviation of x_i respectively.

A straightforward idea is that we can linearize $g(\mathbf{x})$ and use Eq. (4.56) to estimate the reliability index. This can be done by taking the first-order Taylor expansion of $g(\mathbf{x})$ [22], that is,

$$g(\mathbf{x}) = g(x_1, x_2, \ldots, x_n) \approx \tilde{g}(\mathbf{x}) = g(x_1', x_2', \ldots, x_n') + \sum_{i=1}^{n}(x_i - x_i')\frac{\partial g(\mathbf{x}')}{\partial x_i} \tag{4.66}$$

where $\mathbf{x}' = \{x_1', x_2', \ldots, x_n'\}$ is the point where the Taylor expansion is performed. If we assign the elements in $\mathbf{x}'$ as the mean values of each x_i (which equals 0), then Eq. (4.66) becomes

$$g(\mathbf{x}) \approx \tilde{g}(\mathbf{x}) = g(0, 0, \ldots, 0) + \sum_{i=1}^{n}(x_i - 0)\frac{\partial g(0, 0, \ldots, 0)}{\partial x_i} \tag{4.67}$$

With Eq. (4.56), the reliability index, β_{HL}, is obtained as follows,

$$\beta_{\text{HL}} = \frac{g(0, 0, \ldots, 0)}{\sqrt{\sum_{i=1}^{n}\left(\frac{\partial g(0,0,\ldots,0)}{\partial x_i}\right)^2}} \tag{4.68}$$

that is, the reliability index is the shortest distance from the origin to $\tilde{g}(\mathbf{x}) = 0$, as shown in Fig. 4.9.

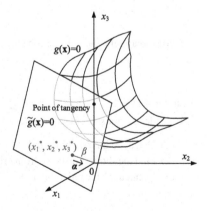

Fig. 4.9 The shortest distance from origin to the linearized surface $\tilde{g}(\mathbf{x}) = 0$: a 3-D example

By noting that $\frac{\partial g(0,0,\ldots,0)}{\partial x_i} = \sigma_i \cdot \frac{\partial G(\mu_1,\mu_2,\ldots,\mu_n)}{\partial X_i}$, Eq. (4.68) can be alternatively written as follows,

$$\beta_{HL} = \frac{G(\mu_1, \mu_2, \ldots, \mu_n)}{\sqrt{\sum_{i=1}^{n} \left(\sigma_i \cdot \frac{\partial G(\mu_1,\mu_2,\ldots,\mu_n)}{\partial X_i}\right)^2}} \tag{4.69}$$

The Hasofer–Lind reliability index in Eq. (4.68) or Eq. (4.69) has taken use of the first-order expansion of the limit state function and the first-order and second-order moments of the random variables. As such, Eq. (4.68) or Eq. (4.69) is known as the *first-order second-moment reliability method* (FOSM reliability method).

Example 4.12

Consider a steel bar subjected to tension force F having a mean value of 10 kN and a COV of 0.3. The ultimate stress of the steel has a mean value of 230 MPa and a COV of 0.2. The mean value and COV of the diameter of the steel cross section are 14 mm and 0.2 respectively. Calculate the failure probability using Eq. (4.69).

Solution

We let Y and D denote the ultimate stress and cross section diameter of the steel bar respectively. The limit state function is nonlinear, taking a form of

$$G(Y, D, F) = Y \cdot \pi \left(\frac{D}{2}\right)^2 - F = \frac{\pi}{4} Y D^2 - F \tag{4.70}$$

The failure occurs if $G < 0$. In the following, we use the units of m for D, kN for F and kPa for Y. Note that $\frac{\partial G}{\partial Y} = \frac{\pi}{4} D^2$, $\frac{\partial G}{\partial D} = \frac{\pi}{2} Y D$, $\frac{\partial G}{\partial F} = -1$, according to Eq. (4.69),

$$\beta_{HL} = \frac{G(230,000, 0.014, 10)}{\sqrt{\left(46,000 \cdot \frac{\partial G(230000,0.014,10)}{\partial Y}\right)^2 + \left(0.0028 \cdot \frac{\partial G(230000,0.014,10)}{\partial D}\right)^2 + \left(3 \cdot \frac{\partial G(230000,0.014,10)}{\partial F}\right)^2}}$$

$$= \frac{\frac{\pi}{4} \times 230,000 \times 0.014^2 - 10}{\sqrt{\left(46,000 \cdot \frac{\pi}{4} \times 0.014^2\right)^2 + \left(0.0028 \cdot \frac{\pi}{2} \times 230,000 \times 0.014\right)^2 + (3 \times (-1))^2}}$$

$$= 1.5765$$

$$\tag{4.71}$$

Correspondingly, the failure probability is 0.0575.

Example 4.13

Consider a limit state function $g(r, s) = r - \ln\left(s + \frac{1}{2}\right) + 4$ in a normalized space.

(1) Calculate the reliability index β_{HL} with Eq. (4.68).
(2) Find the shortest distance from origin to $g = 0$, and the corresponding point (s_1, r_1).
(3) Recalculate the reliability index by performing the first-order expansion of $g(r, s) = 0$ at (s_1, r_1).

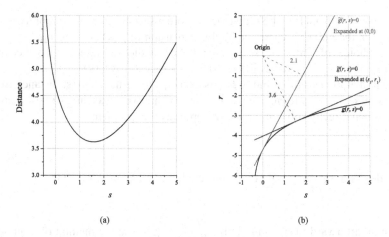

Fig. 4.10 Finding the reliability index in Example 4.13

Solution

(1) First, note that $\frac{\partial g(0,0)}{\partial r} = 1$ and $\frac{\partial g(0,0)}{\partial s} = -2$. Thus, the linearized limit state function at $(0, 0)$ is $\tilde{g}(r, s) = r - 2s + 4 + \ln 2$. According to Eq. (4.68),

$$\beta_{HL} = \frac{g(0, 0)}{\sqrt{\left(\frac{\partial g(0,0)}{\partial r}\right)^2 + \left(\frac{\partial g(0,0)}{\partial s}\right)^2}} = \frac{4 + \ln 2}{\sqrt{1^2 + (-2)^2}} = 2.099 \qquad (4.72)$$

(2) The shortest distance from origin to $g(r, s) = 0$ is found numerically as 3.6296 (c.f. Fig. 4.10a). Correspondingly, $s_1 = 1.575$ and $r_1 = -3.270$.

(3) Performing the first-order expansion of $g(r, s) = 0$ at (s_1, r_1), according to Eq. (4.66) (c.f. Fig. 4.10b),

$$
\begin{aligned}
\tilde{g}(r, s) &= g(r_1, s_1) + (r - r_1)\frac{\partial g(r_1, s_1)}{\partial r} + (s - s_1)\frac{\partial g(r_1, s_1)}{\partial s} \\
&= r_1 - \ln\left(s_1 + \frac{1}{2}\right) + 4 + (r - r_1) - \frac{2(s - s_1)}{2s_1 + 1} \\
&= r - 0.4819s + 4.0291
\end{aligned}
\qquad (4.73)
$$

Thus,

$$\beta_{HL} = \frac{4.0291}{\sqrt{1^2 + (-0.4819)^2}} = 3.6296 \qquad (4.74)$$

which is consistent with that in Question (2).

It can be seen from Eq. (4.67) that the first-order second-moment reliability index depends on the point $\mathbf{x}'$ where the Taylor expansion of $g(\mathbf{x})$ is performed (that is, different choices of $\mathbf{x}'$ may yield different reliability indices). Moreover, the checking point $(x_1^*, x_2^*, \ldots, x_n^*)$ satisfies $\widetilde{g}(\mathbf{x}) = 0$ but not necessarily $g(\mathbf{x}) = 0$.

Recall that the reliability index β_{HL} is expected to be the shortest distance from origin to $g(\mathbf{x}) = 0$ according to its definition. To improve the accuracy of estimating β_{HL}, one can directly solve the following optimization problem,

$$\beta_{\text{HL}} = \text{minimum} \sqrt{\sum_{i=1}^{n} x_i^2}, \quad \text{subjected to } g(x_1, x_2, \ldots, x_n) = 0 \qquad (4.75)$$

A straightforward ideas to minimize $\sqrt{\sum_{i=1}^{n} x_i^2}$ to use the method of Lagrange multipliers (c.f. Eq. (4.52)). We introduce a Lagrange multiplier λ and construct the Lagrange function as follows,

$$L(x_1, x_2, \ldots, x_n, \lambda) = \sqrt{\sum_{i=1}^{n} x_i^2} + \lambda g(x_1, x_2, \ldots, x_n) \qquad (4.76)$$

Letting $\frac{\partial L}{\partial x_i} = 0$ $(i = 1, 2, \ldots, n)$ in Eq. (4.76) gives $\frac{x_i}{\sqrt{\sum_{i=1}^{n} x_i^2}} + \lambda \frac{\partial g}{\partial x_i} = 0$. Thus, if the desired checking point is $\mathbf{x}^* = (x_1^*, x_2^*, \ldots, x_n^*)$, then

$$\begin{cases} g(x_1^*, x_2^*, \ldots, x_n^*) = 0 \\ \dfrac{x_i^*}{\sqrt{\sum_{i=1}^{n} (x_i^*)^2}} + \lambda \dfrac{\partial g(x_1^*, x_2^*, \ldots, x_n^*)}{\partial x_i} = 0, \quad i = 1, 2, \ldots, n \end{cases} \qquad (4.77)$$

Equation (4.77) indicates that if we have successfully determined the desired checking point $\mathbf{x}^* = (x_1^*, x_2^*, \ldots, x_n^*)$, then the vector $\mathbf{x}^*$ is perpendicular to the plane $\widetilde{g}(\mathbf{x}) = 0$ which is the first-order expansion of $g(\mathbf{x}) = 0$ performed at $\mathbf{x}^*$. The proof can be found in Example 4.17.

Example 4.14

Recall the limit state function in Example 4.13. Find the reliability index using Eq. (4.77).
Solution
Applying Eq. (4.77) gives

$$g(r^*, s^*) = 0 \Rightarrow r^* - \ln\left(s^* + \frac{1}{2}\right) + 4 \qquad (4.78)$$

$$\frac{r^*}{\sqrt{(r^*)^2 + (s^*)^2}} + \lambda = 0, \quad \frac{s^*}{\sqrt{(r^*)^2 + (s^*)^2}} - \frac{\lambda}{s^* + \frac{1}{2}} = 0 \qquad (4.79)$$

Fig. 4.11 Numerically finding the checking point of load, s^*

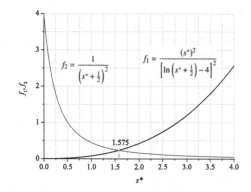

Thus,

$$\frac{(s^*)^2}{\left[\ln\left(s^* + \frac{1}{2}\right) - 4\right]^2} = \frac{1}{\left(s^* + \frac{1}{2}\right)^2} \tag{4.80}$$

with which s^* can be solved numerically as 1.575 (c.f. Fig. 4.11). Furthermore, $r^* = \ln\left(s^* + \frac{1}{2}\right) - 4 = -3.270$. Thus, the reliability index β_{HL} is estimated as $\sqrt{(-3.270)^2 + 1.575^2} = 3.6295$.

Solving $\mathbf{x}^*$ and λ numerically according to Eq. (4.77) may be unfortunately difficult in many cases due to the implicity of Eq. (4.77). In such a case, an iteration-based approach can be used alternatively to help find the reliability index β_{HL}. Suppose that $\mathbf{x}^{(j-1)}$ is the $(j-1)$th approximation of the optimized checking point ($j = 1, 2, \ldots$), based on which we will find the jth approximation, $\mathbf{x}^{(j)}$. Performing the first-order Taylor expansion of $g(\mathbf{x}) = 0$ at $\mathbf{x}^{(j-1)}$ gives

$$g(\mathbf{x}) \approx \tilde{g}(\mathbf{x}) = g(\mathbf{x}^{(j-1)}) + (\mathbf{x} - \mathbf{x}^{(j-1)}) \cdot \left[\frac{\partial g(\mathbf{x}^{(j-1)})}{\partial x_1}, \frac{\partial g(\mathbf{x}^{(j-1)})}{\partial x_2}, \ldots, \frac{\partial g(\mathbf{x}^{(j-1)})}{\partial x_n}\right]^T$$

$$= g(\mathbf{x}^{(j-1)}) + \mathbf{x}\mathbf{g}'^T - \mathbf{x}^{(j-1)}\mathbf{g}'^T \tag{4.81}$$

where $\mathbf{g}' = \left[\frac{\partial g(\mathbf{x}^{(j-1)})}{\partial x_1}, \frac{\partial g(\mathbf{x}^{(j-1)})}{\partial x_2}, \ldots, \frac{\partial g(\mathbf{x}^{(j-1)})}{\partial x_n}\right]$. Now, the basic idea is to assign $\mathbf{x}^{(j)}$ as the checking point of the linearized limit state function in Eq. (4.81). Let $\beta_{\text{HL}}^{(j)}$ denote the reliability index associated with the jth iteration. According to Eq. (4.68),

$$\beta_{\text{HL}}^{(j)} = \frac{g(\mathbf{x}^{(j-1)}) - \mathbf{x}^{(j-1)}\mathbf{g}'^T}{\sqrt{\mathbf{g}'\mathbf{g}'^T}} = \frac{g(\mathbf{x}^{(j-1)}) - \mathbf{x}^{(j-1)}\mathbf{g}'^T}{|\mathbf{g}'|} \tag{4.82}$$

With the normal vector being

Fig. 4.12 Illustration of the iteration-based algorithm to search for reliability index: the R-S case

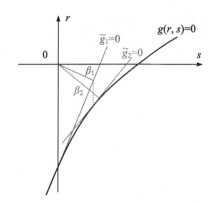

$$\boldsymbol{\alpha}^{(j)} = \frac{\mathbf{g}'}{|\mathbf{g}'|} \tag{4.83}$$

one has

$$\mathbf{x}^{(j)} = -\beta_{\text{HL}}^{(j)}\boldsymbol{\alpha}^{(j)} \tag{4.84}$$

Thus, the iteration-based procedure is as follows, referred to as *Method 1*.

(1) Preliminary: set the initial checking point $\mathbf{x}^{(0)}$ (e.g., the mean value of each random input).
(2) For the jth iteration ($j = 1, 2, \ldots$), calculate $\beta_{\text{HL}}^{(j)}$ according to Eq. (4.82), and compute $\mathbf{x}^{(j)}$ with Eq. (4.84).
(3) The procedure ends if such a j is found that $|\beta_{\text{HL}}^{(j)} - \beta_{\text{HL}}^{(j-1)}| \leq \varepsilon$, where ε is a predefined threshold (e.g., 0.01).

Finally, the reliability index is the converged value of $\beta_{\text{HL}}^{(j)}$.

The aforementioned process of searching for the shortest distance is, again, based on the first-order and second-order moments of each random variable, and the first-order expansion of the nonlinear surface $g(\mathbf{x}) = 0$. Figure 4.12 illustrates the basic concept of the iteration-based algorithm.

One can use an alternative derivation to find the reliability index iteratively. Since $\mathbf{x}^{(j)}$ satisfies $\widetilde{g}(\mathbf{x}) = 0$ (c.f. Eq. (4.81)), that is,

$$g(\mathbf{x}^{(j-1)}) + \mathbf{x}^{(j)}\mathbf{g}'^{\mathsf{T}} - \mathbf{x}^{(j-1)}\mathbf{g}'^{\mathsf{T}} = 0 \tag{4.85}$$

it follows,

$$\mathbf{x}^{(j)}\mathbf{g}'^{\mathsf{T}} = \mathbf{x}^{(j-1)}\mathbf{g}'^{\mathsf{T}} - g(\mathbf{x}^{(j-1)}) \tag{4.86}$$

Since $\boldsymbol{\alpha}^{(j)} = \frac{\mathbf{g}'}{|\mathbf{g}'|}$, $\mathbf{x}^{(j-1)} = -\beta_{\text{HL}}^{(j-1)}\boldsymbol{\alpha}^{(j-1)}$, one has

$$|\mathbf{g}'|\mathbf{x}^{(j)}\boldsymbol{\alpha}^{(j)\mathsf{T}} = -\beta_{\text{HL}}^{(j-1)}\boldsymbol{\alpha}^{(j-1)} \cdot |\mathbf{g}'|\boldsymbol{\alpha}^{(j)\mathsf{T}} - g(\mathbf{x}^{(j-1)}) \tag{4.87}$$

We additionally assume that $\boldsymbol{\alpha}^{(j)} \approx \boldsymbol{\alpha}^{(j-1)}$, with which $\mathbf{x}^{(j)} = k\boldsymbol{\alpha}^{(j-1)}$ as $\mathbf{x}^{(j)} /\!/ \boldsymbol{\alpha}^{(j)}$, where k is a constant to be determined. Thus,

$$k\boldsymbol{\alpha}^{(j-1)}\boldsymbol{\alpha}^{(j-1)\mathsf{T}} = -\beta_{\text{HL}}^{(j-1)} - \frac{g(\mathbf{x}^{(j-1)})}{|\mathbf{g}'|} \Rightarrow k = -\beta_{\text{HL}}^{(j-1)} - \frac{g(\mathbf{x}^{(j-1)})}{|\mathbf{g}'|} \quad (4.88)$$

and

$$\mathbf{x}^{(j)} = -\boldsymbol{\alpha}^{(j-1)} \left(\beta_{\text{HL}}^{(j-1)} + \frac{g(\mathbf{x}^{(j-1)})}{|\mathbf{g}'|} \right) \quad (4.89)$$

where

$$\beta_{\text{HL}}^{(j-1)} = \sqrt{\sum_{i=1}^{n} \left(x_i^{(j-1)} \right)^2} \quad (4.90)$$

With Eq. (4.89), the algorithm for finding the reliability index β_{HL} is as follows, referred to as *Method 2*.

(1) Preliminary: set the initial checking point $\mathbf{x}^{(0)}$ (e.g., the mean value of each random input).
(2) For the jth iteration ($j = 1, 2, \ldots$), calculate $\beta_{\text{HL}}^{(j-1)}$ according to Eq. (4.90), and compute $\mathbf{x}^{(j)}$ with Eq. (4.89).
(3) The procedure ends if such a j is found that $|\beta_{\text{HL}}^{(j)} - \beta_{\text{HL}}^{(j-1)}| \le \varepsilon$, where ε is a predefined threshold (e.g., 0.01).

Finally, $\beta_{\text{HL}}^{(j)}$ stabilizes at the reliability index β_{HL} as desired.

Example 4.15

Recall Example 4.13. Find the reliability index β_{HL}. Set $\varepsilon = 0.01$.
Solution
We first assign $\mathbf{x}^{(0)} = (0, 0)$. If we use *Method 1*, the result is as follows, yielding a reliability index of 3.6239.

Iteration No.(j)	$\beta_{\text{HL}}^{(j)}$	$\mathbf{x}^{(j)}$
1	2.0988	(−0.9386, 1.8773)
2	3.6168	(−3.3338, 1.4024)
3	3.6239	(−3.2077, 1.6862)

Alternatively, if one uses *Method 2*, the result is as follows. The reliability index is found as 3.6296. Clearly, the results given by the two methods are consistent with each other.

Iteration No.(j)	$\beta_{\text{HL}}^{(j)}$	$\mathbf{x}^{(j)}$
1	0	(−0.9386, 1.8773)
2	2.0988	(−3.8000, 1.5985)
3	4.1225	(−3.2805, 1.5633)
4	3.6339	(−3.2662, 1.5830)
5	3.6296	(−3.2721, 1.5708)

Example 4.16

Recall Example 4.12. Find the reliability index β_{HL}. Set $\varepsilon = 0.01$.

Solution

The limit state function in the normalized space is as follows,

$$g(y, d, f) = \frac{\pi}{4}(46{,}000y + 230{,}000) \times (0.0028d + 0.014)^2 - (3f + 10) \quad (4.91)$$

We assign $\mathbf{x}^{(0)} = (y, d, f)^{(0)} = (0, 0, 0)$. Using the procedure where $\beta_{\mathrm{HL}}^{(j)}$ is calculated according to Eq. (4.82) and $\mathbf{x}^{(j)}$ is computed with Eq. (4.84), the result is as follows, yielding a reliability index of 2.0963.

Iteration No.(j)	$\beta_{\mathrm{HL}}^{(j)}$	$\mathbf{x}^{(j)}$
1	1.5765	$(-0.6927, -1.3854, 0.2935)$
2	2.0672	$(-0.7632, -1.8189, 0.6187)$
3	2.0966	$(-0.6916, -1.8422, 0.7238)$
4	2.0963	$(-0.6775, -1.8487, 0.7196)$

Example 4.17

In a normalized space, let $\mathbf{x}^* = \{x_1^*, x_2^*, \ldots, x_n^*\}$ be the point on the surface $g(\mathbf{x}) = 0$ which leads to the shortest distance from origin to $g(\mathbf{x}) = 0$. The first-order Taylor expansion of $g(\mathbf{x})$ at $\mathbf{x}^*$ is denoted by $\widetilde{g}(\mathbf{x}) = 0$. Show that the vector $\mathbf{x}^*$ is perpendicular to the plane $\widetilde{g}(\mathbf{x}) = 0$.

Solution

First, according to Eq. (4.66),

$$\widetilde{g}(\mathbf{x}) = \widetilde{g}(x_1, x_2, \ldots, x_n) = g(x_1^*, x_2^*, \ldots, x_n^*) + \sum_{i=1}^{n}(x_i - x_i^*)\frac{\partial g(\mathbf{x}^*)}{\partial x_i} \quad (4.92)$$

Let $\mathbf{x}^c = \{x_1^c, x_2^c, \ldots, x_n^c\}$ be the checking point on the plane $\widetilde{g}(\mathbf{x}) = 0$, and we will show that $\mathbf{x}^c = \mathbf{x}^*$. To this end, we only need to prove that $\mathbf{x}^c // \mathbf{x}^*$, since both $\mathbf{x}^c$ and $\mathbf{x}^*$ are on the plane $\widetilde{g}(\mathbf{x}) = 0$ and $\mathbf{x}^c$ is perpendicular to $\widetilde{g}(\mathbf{x}) = 0$. With Eq. (4.58), $\mathbf{x}^c // \mathbf{g}'$, where $\mathbf{g}' = \left[\frac{\partial g(\mathbf{x}^*)}{\partial x_1}, \frac{\partial g(\mathbf{x}^*)}{\partial x_2}, \ldots, \frac{\partial g(\mathbf{x}^*)}{\partial x_n}\right]$. Furthermore, according to Eq. (4.77), $\frac{\mathbf{x}^*}{|\mathbf{x}^*|} = -\lambda \mathbf{g}' \Rightarrow \mathbf{x}^* // \mathbf{g}'$. Thus, $\mathbf{x}^c // \mathbf{g}' // \mathbf{x}^*$, which completes the proof.

Example 4.18

Consider a limit state function of $G = R - 2S_1 S_2 - S_2^2$, where the statistical information of R, S_1 and S_2 can be found in Table 4.3. If S_1 and S_2 are correlated with a linear correlation coefficient of ρ ($0 \le \rho \le 1$), find numerically the relationship between the reliability index β_{HL} and ρ.

Solution

We first transform the two correlated variables S_1 and S_2 into two independent variable Q_1, Q_2 by letting

$$\mathbf{Q} = \begin{bmatrix} Q_1 \\ Q_2 \end{bmatrix} = \mathbf{AS} \tag{4.93}$$

where $\mathbf{S} = \begin{bmatrix} S_1 \\ S_2 \end{bmatrix}$ and $\mathbf{A}$ is a transform matrix to be determined. With this, $\mathbf{S} = \mathbf{A}^{-1}\mathbf{Q}$ and

$$\mu_{\mathbf{Q}} = \mathbf{A}\mu_{\mathbf{S}}, \quad \mathbf{C}_{\mathbf{Q}} = \mathbf{AC}_{\mathbf{S}}\mathbf{A}^{\mathsf{T}} \tag{4.94}$$

The covariance matrix $\mathbf{C}_{\mathbf{Q}}$ for $\mathbf{Q}$ is expected to be a diagonal matrix. To achieve this, we let the rows of $\mathbf{A}$ be the (transpose of the) eigenvectors of $\mathbf{C}_{\mathbf{S}}$. Let ψ_1, ψ_2 be the two eigenvectors of $\mathbf{C}_{\mathbf{S}}$ and λ_1, λ_2 the two corresponding eigenvalues. With this, it follows,

$$\mathbf{C}_{\mathbf{S}}\psi_i = \lambda_i \psi_i \Rightarrow \det\left(\mathbf{C}_{\mathbf{S}} - \lambda_i \mathbf{I}\right) = 0, \quad i = 1, 2 \tag{4.95}$$

Note that the covariance matrix of $\mathbf{S}$ is

$$\mathbf{C}_{\mathbf{S}} = \begin{bmatrix} 0.45^2 & 0.45 \times 0.15\rho \\ 0.45 \times 0.15\rho & 0.15^2 \end{bmatrix} = \begin{bmatrix} 0.2025 & 0.0675\rho \\ 0.0675\rho & 0.0225 \end{bmatrix} \tag{4.96}$$

Thus,

$$\begin{vmatrix} 0.2025 - \lambda_i & 0.0675\rho \\ 0.0675\rho & 0.0225 - \lambda_i \end{vmatrix} = 0 \Rightarrow (0.2025 - \lambda_i)(0.0225 - \lambda_i) = (0.0675\rho)^2 \tag{4.97}$$

with which λ_i is solved as follows,

$$\lambda_{1,2} = \frac{0.225 \pm \sqrt{0.0324 + 0.018225\rho^2}}{2} \tag{4.98}$$

Now, with λ_i ready, one can find the eigenvectors of $\mathbf{C}_{\mathbf{S}}$ as

$$\psi_i = \begin{bmatrix} 0.0675k_i\rho \\ k_i(\lambda_i - 0.2025) \end{bmatrix}, \quad i = 1, 2 \tag{4.99}$$

where k_i is a constant so that ψ_i has a magnitude of 1. We let $\mathbf{A} = \begin{bmatrix} \psi_1^{\mathsf{T}} \\ \psi_2^{\mathsf{T}} \end{bmatrix}$, with which $\mu_{\mathbf{Q}}$ and $\mathbf{C}_{\mathbf{Q}}$ can be found with Eq. (4.94). Now, we are ready to calculate β_{HL} numerically given a value of ρ. Figure 4.13 shows the dependence of β_{HL} on the correlation between S_1 and S_2, ρ. It can be seen that a greater correlation between S_1 and S_2 leads to a smaller reliability index.

Fig. 4.13 Dependence of
β_{HL} on the correlation
between S_1 and S_2

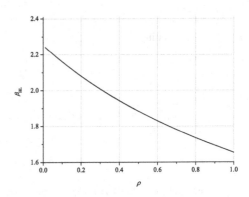

Remark

When using FOSM reliability method, only the mean value and variance of the random variables are used regardless of their distribution type. Also, all the random inputs have been assumed to follow a normal distribution. In fact, for a variable X defined in $(-\infty, \infty)$, if its mean value and variance are known as μ_X and σ_X^2 respectively, then the assumption of normality for X yields the greatest differential entropy, implying that a normal distribution leads to the minimal prior constraint on the probabilistic information of X beyond the first-order and second-order moments. This observation is proven as follows.

Let $f(x)$ be an arbitrary distribution function of X satisfying the following constraints,

$$\begin{cases} \displaystyle\int_{-\infty}^{\infty} f(x)\mathrm{d}x = 1 \\[2mm] \displaystyle\int_{-\infty}^{\infty} xf(x)\mathrm{d}x = \mu_X \\[2mm] \displaystyle\int_{-\infty}^{\infty} (x - \mu_X)^2 f(x)\mathrm{d}x = \sigma_X^2 \end{cases} \tag{4.100}$$

The differential entropy is

$$H(X) = -\int_{-\infty}^{\infty} f(x)\ln f(x)\mathrm{d}x \tag{4.101}$$

We use the method of Lagrange multipliers to find the minima of $H(X)$ by introducing three multipliers λ_1, λ_2 and λ_3 to construct the Lagrange function as follows,

$$L(f, \lambda_1, \lambda_1, \lambda_3) = -\int_{-\infty}^{\infty} f(x)\ln f(x)\mathrm{d}x + \lambda_1 \left(\int_{-\infty}^{\infty} f(x)\mathrm{d}x - 1 \right)$$
$$+ \lambda_2 \left(\int_{-\infty}^{\infty} xf(x)\mathrm{d}x - \mu_X \right) + \lambda_3 \left(\int_{-\infty}^{\infty} (x - \mu_X)^2 f(x)\mathrm{d}x - \sigma_X^2 \right) \tag{4.102}$$

We let $\frac{\delta L}{\delta f} = 0$, with which

$$\ln f(x) + 1 - \lambda_1 - \lambda_2 x - \lambda_3 (x - \mu_X)^2 = 0 \tag{4.103}$$

Thus,

$$f(x) = \exp\left(\lambda_1 + \lambda_2 x + \lambda_3 (x - \mu_X)^2 - 1\right) \tag{4.104}$$

Using the constraints in Eq. (4.100) yields $f(x) = \frac{1}{\sqrt{2\pi}\sigma_X} \exp\left(-\frac{(x-\mu_X)^2}{\sigma_X}\right)$, which is the PDF of a normal distribution.

4.2.2 First-Order Reliability Method

Recall that when using FOSM reliability method to estimate the Hasofer–Lind reliability index (and subsequently, the probability of failure), only the mean value and variance of each random variable have been utilized. When the distribution types of the random inputs are unknown, taking use of the distribution information would improve the accuracy of reliability assessment, especially for the cases of non-normal random inputs (c.f. Sect. 2.2 for many non-normal distribution types).

One straightforward approach to use the probability distribution information is to estimate the reliability by solving a multi-fold integral (as discussed in Sect. 3.5.1, as well as Eq. (4.117) in the following). Due to the complexity of numerically computing a multidimensional integral, one can, alternatively, first transform the non-normal variables into normal variables and then employ the techniques in FOSM (c.f. Sect. 4.2.1) to find the reliability index. In such a case, the first-order Taylor expansion of the limit state function would be used (as before), and we call this approach the *first-order reliability method* (FORM).

The transformation from a non-normal variable X_i to a normal variable X_i' can be achieved by finding the mean value and standard deviation of X_i' such that the CDFs and PDFs of the two variables evaluated at the checking point of X_i, x_i^*, are respectively identical [24]. The justification is that the structural failure is most likely to occur around the checking point. That being the case, let $F_{X_i}(x)$ and $f_{X_i}(x)$ denote the CDF and PDF of X_i respectively. The mean value and standard deviation of X_i', μ_i' and σ_i', are determined according to

$$F_{X_i}(x_i^*) = \Phi\left(\frac{x_i^* - \mu_i'}{\sigma_i'}\right), \quad f_{X_i}(x_i^*) = \frac{1}{\sigma_i'}\phi\left(\frac{x_i^* - \mu_i'}{\sigma_i'}\right) \tag{4.105}$$

which further gives

$$\sigma_i' = \frac{\phi\left(\Phi^{-1}[F_{X_i}(x_i^*)]\right)}{f_{X_i}(x_i^*)}, \quad \mu_i' = x_i^* - \sigma_i'\Phi^{-1}[F_{X_i}(x_i^*)] \tag{4.106}$$

Now, having converted X_i into a normal variable, one can subsequently use the techniques in FOSM to evaluate the reliability index by first considering the limit state function in a reduced space. The transform method is called the *normal tail transform* since the statistics of X_i' are determined by considering the tail behaviour of the normal distribution. It is noticed, however, that both μ_i' and σ_i' are dependent on the checking point of X_i and thus should be recalculated in each iteration.

One can also use the Rosenblatt transformation method or the Nataf transformation method to accomplish the transformation from a non-normal variable to a standard normal variable (c.f. Sect. 3.3), when the correlation between different variables is considered. The former works when the joint PDF of the variables is known, and the latter is applicable when only the marginal distributions and the correlation matrix of the random variables are given. Specially, when the variables are independent, both Rosenblatt and Nataf transformation methods gives

$$y_i = \Phi^{-1}[F_i(X_i)], \quad i = 1, 2, \ldots \tag{4.107}$$

where y_i is a standard normal variable. With this, the limit state function in a normalized space is as follows,

$$g(y_1, y_2, \ldots, y_n) = G\left(F_1^{-1}[\Phi(y_1)], F_2^{-1}[\Phi(y_2)], \ldots, F_n^{-1}[\Phi(y_n)]\right) \tag{4.108}$$

where $G(X_1, X_2, \ldots, X_n)$ is the limit state function in the original space.

We now show the consistency between Eqs. (4.105) and (4.107). At the checking point x_i^*, the first-order Taylor expansion of Eq. (4.107) is

$$y_i = y_i(X_i) \approx \Phi^{-1}[F_i(x_i^*)] + \frac{\mathrm{d}y_i}{\mathrm{d}X_i}(X_i - x_i^*) \tag{4.109}$$

Since $\Phi(y_i) = F_i(X_i) \Rightarrow \phi(y_i)\mathrm{d}y_i = f_{X_i}(X_i)\mathrm{d}X_i$, it follows,

$$y_i \approx \Phi^{-1}[F_i(x_i^*)] + \frac{\phi(y_i)}{f_{X_i}(X_i)}(X_i - x_i^*) = \frac{X_i - \mu_i'}{\sigma_i'} \tag{4.110}$$

which implies that y_i obtained from Eq. (4.107) can be treated as the normalized form of a normal variable X_i' having a mean value of μ_i' and a standard deviation of σ_i'.

Example 4.19

Reconsider Example 4.12. If, additionally, F has a Weibull distribution, the ultimate stress has a lognormal distribution and the cross section diameter has a Gamma distribution, find the probability of failure using FORM.

Solution

We as before let Y and D denote the ultimate stress and cross section diameter of the steel bar respectively. Taking use of the probability information of each variable, the limit state function in the reduced space takes the form of

$$g(x_1, x_2, x_3) = \frac{\pi}{4} F_Y^{-1}[\Phi(x_1)] \cdot \{F_D^{-1}[\Phi(x_2)]\}^2 - F_F^{-1}[\Phi(x_3)] \tag{4.111}$$

where $F_\bullet$ is the CDF of $\bullet$ ($\bullet = Y, D, F$). Note that

$$\begin{aligned}
\frac{\partial g}{\partial x_1} &= \frac{\pi}{4} \{F_D^{-1}[\Phi(x_2)]\}^2 \cdot \frac{\phi(x_1)}{f_Y\{F_Y^{-1}[\Phi(x_1)]\}} \\
\frac{\partial g}{\partial x_2} &= \frac{\pi}{4} F_Y^{-1}[\Phi(x_1)] \cdot 2\{F_D^{-1}[\Phi(x_2)]\} \cdot \frac{\phi(x_2)}{f_D\{F_D^{-1}[\Phi(x_2)]\}} \\
\frac{\partial g}{\partial x_3} &= -\frac{\phi(x_3)}{f_F\{F_F^{-1}[\Phi(x_3)]\}}
\end{aligned} \tag{4.112}$$

where $f_\bullet$ is the PDF of $\bullet$ ($\bullet = Y, D, F$).
Using the iteration-based approach in Sect. 4.2.1, one can find the reliability index (with *Method 1* herein) to be 2.2477 with four (4) iterations as follows, setting $\varepsilon = 0.01$.

Iteration No.(j)	$\beta_{HL}^{(j)}$	$\mathbf{x}^{(j)}$
1	1.5360	$(-0.6646, -1.3499, 0.3089)$
2	2.1835	$(-0.8367, -1.8637, 0.7708)$
3	2.2473	$(-0.8106, -1.8720, 0.9427)$
4	2.2477	$(-0.8125, -1.8775, 0.9311)$

Comparing the reliability index obtained herein ($\beta_{HL} = 2.2477$) with that in Example 4.16 ($\beta_{HL} = 2.0963$), it can be seen that the use of probability information affects the reliability assessment significantly.

Example 4.20

Reconsider Example 4.18. If R has a lognormal distribution; both S_1 and S_2 follows an Extreme Type I distribution, and the correlation coefficient between S_1 and S_2 is 0.5, calculate the probability of failure.

Solution

We first transform S_1 and S_2 into two correlated standard normal variables Y_1 and Y_2. The correlation between Y_1 and Y_2, ρ', equals

$$\rho' = \rho(1.064 - 0.069\rho + 0.005\rho^2) = 0.5154 \tag{4.113}$$

with which the correlation matrix of (Y_1, Y_2) is $\boldsymbol{\rho}' = \begin{bmatrix} 1 & 0.5154 \\ 0.5154 & 1 \end{bmatrix}$. The Cholesky

decomposition of $\boldsymbol{\rho}'$ gives $\mathbf{A} = \begin{bmatrix} 1 & 0 \\ 0.5144 & 0.8576 \end{bmatrix}$ such that $\mathbf{AA}^T = \boldsymbol{\rho}'$ (c.f. Sect. 3.3).

With this, we can further convert Y_1, Y_2 into two independent standard normal variable U_1, U_2 by letting

$$\begin{bmatrix} Y_1 \\ Y_2 \end{bmatrix} = \begin{bmatrix} 1 & 0 \\ 0.5144 & 0.8576 \end{bmatrix} \begin{bmatrix} U_1 \\ U_2 \end{bmatrix} \Rightarrow \begin{cases} Y_1 = U_1 \\ Y_2 = 0.5144U_1 + 0.8576U_2 \end{cases} \tag{4.114}$$

Thus, the original limit state function is rewritten as follows in a reduced space,

$$g(x, u_1, u_2) = F_R^{-1}[\Phi(x)] - 2.5F_{S_1}^{-1}[\Phi(u_1)] \cdot F_{S_2}^{-1}[\Phi(0.5144u_1 + 0.8576u_2)]$$
$$- \{F_{S_2}^{-1}[\Phi(0.5144u_1 + 0.8576u_2)]\}^2 \tag{4.115}$$

where $F_\bullet$ is the CDF of $\bullet$ ($\bullet = R, S_1, S_2$). Furthermore, one has

$$\frac{\partial g}{\partial x} = \frac{\phi(x)}{f_R\{F_R^{-1}[\Phi(x)]\}}$$

$$\frac{\partial g}{\partial u_1} = -2.5 \left\{ \frac{\phi(u_1)F_{S_2}^{-1}[\Phi(0.5144u_1 + 0.8576u_2)]}{f_{S_1}\{F_{S_1}^{-1}[\Phi(u_1)]\}} + \frac{0.5144\phi(0.5144u_1 + 0.8576u_2)F_{S_1}^{-1}[\Phi(u_1)]}{f_{S_2}\{F_{S_2}^{-1}[\Phi(0.5144u_1 + 0.8576u_2)]\}} \right\}$$

$$- 2\{F_{S_2}^{-1}[\Phi(0.5144u_1 + 0.8576u_2)]\} \cdot \frac{0.5144\phi(0.5144u_1 + 0.8576u_2)}{f_{S_2}\{F_{S_2}^{-1}[\Phi(0.5144u_1 + 0.8576u_2)]\}}$$

$$\frac{\partial g}{\partial u_2} = -2.5F_{S_1}^{-1}[\Phi(u_1)] \cdot \frac{0.8576\phi(0.5144u_1 + 0.8576u_2)}{f_{S_2}\{F_{S_2}^{-1}[\Phi(0.5144u_1 + 0.8576u_2)]\}}$$

$$- 2\{F_{S_2}^{-1}[\Phi(0.5144u_1 + 0.8576u_2)]\} \cdot \frac{0.8576\phi(0.5144u_1 + 0.8576u_2)}{f_{S_2}\{F_{S_2}^{-1}[\Phi(0.5144u_1 + 0.8576u_2)]\}}$$

$$\tag{4.116}$$

where $f_\bullet$ is the PDF of $\bullet$ ($\bullet = R, S_1, S_2$).

Next, we use the iteration-based approach as developed in Sect. 4.2.1 to estimate the reliability index (with *Method 1* in the following). The result associated with each iteration is presented as follows with $\varepsilon = 0.01$, with which the reliability index β_{HL} is determined as 1.7418.

Iteration No.(j)	$\beta_{HL}^{(j)}$	$\mathbf{x}^{(j)}$
1	2.8017	$(-1.2994, 2.1187, 1.2933)$
2	1.8275	$(-0.2030, 1.6088, 0.8428)$
3	1.7423	$(-0.3139, 1.5046, 0.8204)$
4	1.7418	$(-0.3283, 1.4990, 0.8242)$

Comparing the reliability index obtained in Example 4.18 (1.8842 according to Fig. 4.13), the difference between the reliability assessment results incorporating the probabilistic information of random variables or not is evident.

4.2.3 Simulation-Based Reliability Assessment

As has been mentioned in Sect. 4.2.1, both the structural resistance R and the load effect S are functions of a random vector $\mathbf{X}$ that accounts for the uncertain quantities involved in the reliability problem, that is, $R = R(\mathbf{X})$ and $S = S(\mathbf{X})$. Correspondingly, the limit state function is expressed as $G = G(\mathbf{X}) = R(\mathbf{X}) - S(\mathbf{X})$.

We consider the case where $\mathbf{X}$ contains n elements and the probability distribution of each element is known. With this, the probability of structural failure can be estimated as follows,

$$\mathbb{P}_f = \mathbb{P}(G(\mathbf{X}) < 0) = \int \cdots \int_{G(\mathbf{X}) < 0} f_{\mathbf{X}}(\mathbf{x}) d\mathbf{x} \qquad (4.117)$$

where $f_{\mathbf{X}}(\mathbf{x})$ is the joint PDF of $\mathbf{X}$. In some cases it is difficult to directly calculate the n-fold integral in Eq. (4.117), especially when n is large. In such a case, the Monte Carlo simulation method provides an alternative technique for calculating the integral, which can be used to simulate a large number of experiments, where the observation of the results yields an estimate of the structural failure probability. Details of Monte Carlo simulation, as well as some advanced simulation techniques have been discussed earlier in Sect. 3.5. Herein, we only present a short summary of this method for the completeness of this section.

First, simulate a realization of $\mathbf{X}$, $\mathbf{x}_1$, and determine the sign of $G(\mathbf{x}_1)$ (i.e., whether $G(\mathbf{x}_1) < 0$ or not), denoted by $\mathbb{I}[G(\mathbf{x}_1) < 0]$. Repeating this experiment for m times, if the sample of $\mathbf{X}$ is $\mathbf{x}_i$ for the ith simulation run, then the failure probability in Eq. (4.117) is approximated by

$$\mathbb{P}_f \approx \widetilde{P} = \frac{1}{m} \sum_{i=1}^{m} \mathbb{I}[G(\mathbf{x}_i) < 0] \qquad (4.118)$$

Equation (4.118), guaranteed by the *strong law of large numbers* (c.f. Theorem 3.1), provides a simulation-based approach to estimate $\mathbb{P}_f$ in Eq. (4.117). Furthermore, with the central limit theorem (c.f., Sect. 2.2.4), $\widetilde{P}$ follows a normal distribution when n is large enough. The mean and variance of $\widetilde{P}$ are respectively determined by

$$\mathbb{E}(\widetilde{P}) = \mathbb{I}[G(\mathbf{x}) < 0], \quad \mathbb{V}(\widetilde{P}) = \frac{1}{m} \mathbb{V}(\mathbb{I}[G(\mathbf{x}) < 0]) \qquad (4.119)$$

As such, it can be seen that the variance of $\widetilde{P}$ decreases with the number of simulation runs m and is in proportion to m^{-1}. Equation (4.119) implies that $\widetilde{P}$ is an unbiased estimate of $\mathbb{P}_f$ and that to improve the simulation accuracy of Eq. (4.118), one can either use a sufficiently large m or modify the simulation procedure to reduce $\mathbb{V}(\mathbb{I}[G(\mathbf{x}) < 0])$. In terms of the latter, some simulation techniques can be used alternatively, e.g., the importance sampling method, Latin hypercube sampling method, subset simulation method, and others (c.f. Sect. 3.5).

Example 4.21

Recall Example 4.19. Suppose that each variable (Y, D and F) is independent mutually. Use Monte Carlo simulation method to find the failure probability and the reliability index β.

Solution

The general simulation procedure has been previously discussed in Sect. 3.5.1. The basic idea is to first sample n sample pairs of Y, D, F, and then check the number of sample pairs leading to $G < 0$, denoted by m. The failure probability is then approximated by $\frac{m}{n}$ as n is large enough. Results show that the probability of failure is 0.011 with $n = 1,000,000$ simulation runs, with which the reliability index is 2.29, fairly close to that obtained in Example 4.19.

The minimum number of simulation replications depends on both the magnitude of failure probability and the confidence level of the result. As has been discussed in Sect. 3.5.1, in order to guarantee that the simulation error is less than 20% with a confidence level of 95%, then the minimum number of simulation runs is 8.63×10^3. On the other hand, with 1,000,000 simulation runs (as adopted herein), the simulation error is less than 2.5% with a confidence level of 99%.

Example 4.22

Consider a roof structure [26, 35] subjected to a uniformly distributed vertical load q, as shown in Fig. 4.14a. The top cords and the compression bars are concrete, and the bottom cords and the tension bars are steel. The load q is transformed into three nodal loads at C, D and F in structural analysis, each being $P = \frac{1}{4}ql$ (c.f. Fig. 4.14b). The serviceability limit state is the maximum vertical displacement of the roof structure $U_{\max}$ exceeding a predefined threshold u_0. As such, the limit state function is given by

$$G = u_0 - U_{\max} = u_0 - \frac{ql^2}{2}\left(\frac{3.81}{A_c E_c} + \frac{1.13}{A_s E_s}\right) \qquad (4.120)$$

where A_c, E_c, A_s, E_s are the cross-sectional areas and elasticity modulus of concrete and steel respectively. The statistical information of the parameters in Eq. (4.120) is summarized in Table 4.4, and the permitted maximum displacement u_0 is set 0.03 m. Assume all the parameters are independent.

Using a simulation-based method, calculate the probability of unsatisfactory structural performance (i.e, the probability of $G < 0$).

Solution

We first generate n sample pairs of (q, l, A_s, A_c, E_s, E_c), and then check the number of sample pairs with which $G < 0$. Results show that the probability of $G < 0$ is 9.328×10^{-3} with $n = 10^7$. This result is associated with a simulation error of less than 1% with a confidence level of 99%.

Fig. 4.14 An illustrative roof structure

(a)

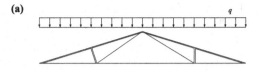

(b)

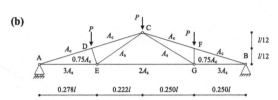

Table 4.4 Statistical information of the random inputs in Eq. (4.120)

Parameter	Mean	COV	Distribution type
q(N/m)	20000	0.07	Normal
l(m)	12	0.01	Normal
A_s (m^2)	9.82×10^{-4}	0.06	Normal
A_c (m^2)	400×10^{-4}	0.12	Normal
E_s (N/m^2)	10×10^{10}	0.06	Normal
E_c (N/m^2)	2×10^{10}	0.06	Normal

Example 4.23

We consider the reliability of an infinite slope example [15, 28], as shown in Fig. 4.15. Let H be the depth of soil above bedrock, α the slope inclination, γ the total unit weight, ϕ the effective stress friction angle, and c the cohesion. The limit state function is

$$G = \frac{\tau}{F} - 1 = \frac{c + \gamma H \cos^2 \alpha \tan \phi}{\gamma H \sin \alpha \cos \alpha} - 1 \tag{4.121}$$

where τ and F are the shear strength and the sliding force, respectively. The parameters c and ϕ both follow a lognormal distribution and are negatively correlated with a correlation coefficient of -0.5, while the remaining parameters are assumed to be deterministic. Let $\gamma = 17$ kN/m^3, $H = 5$ m, $\alpha = 30°$. The mean value and COV of c are 12 kPa and 0.4 respectively; the mean value and COV of ϕ are 30° and 0.2 respectively. Use a simulation-based approach to estimate the failure probability (i.e., the probability of $G < 0$), assuming a Gaussian copula for c and ϕ.

Solution

As before, we can first generate n sample pairs of (c, ϕ) and then check the number of sample pairs so that $G < 0$. The simulation of a sample pair for (c, ϕ) can be achieved with the Nataf transformation method (c.f. Sect. 3.3). One may also refer to Example 4.20 for the transformation from (c, ϕ) to two independent standard normal variables.

Fig. 4.15 An example of
slope reliability

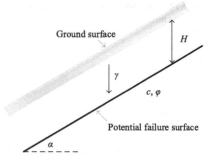

Results show that the probability of failure (i.e., the probability of $G < 0$) is 0.0303 with $n = 10^5$. This result is associated with a simulation error of less than 1% with a confidence level of 99%.

4.2.4   Moment-Based Reliability Assessment

Recall Eq. (4.117), which implies that the structural reliability can be estimated based on the probability information of random variables. In fact, as discussed in Sect. 2.1.6, a random variable's PDF can be fully determined by its moment information and vice versa. Motivated by this fact, one can also evaluate structural reliability by using the moment information of random variables. We have, in fact, used the first-order and second-order moments to estimate the reliability index in FOSM (c.f. Sect. 4.2.1). This section discusses the *moment-based reliability method* developed by [33], which takes an advantage over Eq. (4.117) when the calculation of moments is computationally easier (e.g., in a multidimensional problem).

4.2.4.1   One-Dimensional Problems

Consider the limit state function in Eq. (4.39). Let $s_{\max}$ be the maxima of S. With this, normalizing Eq. (4.39) with respect to $s_{\max}$ yields a new limit state function as follows, which is mathematically equivalent to Eq. (4.39).

$$G' = \mathcal{R} - \mathcal{S} = \frac{R}{s_{\max}} - \frac{S}{s_{\max}} \tag{4.122}$$

Similar to Eq. (4.2), the probability of failure $\mathbb{P}_f$ is estimated according to

$$\mathbb{P}_f = \int_0^1 \xi(x) f_{\mathcal{S}}(x) \mathrm{d}x \tag{4.123}$$

where $\xi(x)$ is the CDF of $R' = \frac{R}{s_{\max}}$, and $f_S(x)$ is the PDF of $S = \frac{S}{s_{\max}}$. Considering the Fourier expansion of $\xi(x)$, which gives

$$\xi(x) = \frac{a_0}{2} + \sum_{j=1}^{\infty} a_j \cos(jx\pi) \qquad (4.124)$$

where a_j is a Fourier coefficient, $a_j = 2\int_0^1 \xi(x)\cos(jx\pi)\mathrm{d}x$ for $j = 0, 1, 2, \ldots$. By noting that $\cos x = \dfrac{e^{\mathrm{i}x} + e^{-\mathrm{i}x}}{2}$ holds for any real number x with i being the unit of imaginary (i $= \sqrt{-1}$), one has

$$\xi(x) = \frac{a_0}{2} + \frac{1}{2}\sum_{j=1}^{\infty} a_j \left[\exp(\mathrm{i}\pi \cdot jx) + \exp(-\mathrm{i}\pi \cdot jx)\right] \qquad (4.125)$$

Substituting Eq. (4.125) to Eq. (4.123), it follows,

$$\mathbb{P}_\mathrm{f} = \frac{a_0}{2} + \frac{1}{2}\sum_{j=1}^{\infty} a_j \left[\phi(\mathrm{i}\pi \cdot j) + \phi(-\mathrm{i}\pi \cdot j)\right] \qquad (4.126)$$

where $\phi(t)$ is the MGF of S, which can be expressed in a polynomial form of $\phi(t) = \sum_{k=0}^{\infty} c_k t^k$, with each c_k being a constant.

The MGF satisfies that $\phi^{(j)}(0) = \mathbb{E}(S^j)$ for $j = 0, 1, 2, \ldots$, where $\phi^{(j)}$ is the jth order derivative of ϕ (c.f. Sect. 2.1.6). Thus, $\mathbb{E}(S^j) = \phi^{(j)}(0) = c_j \cdot j!$ holds for $j = 0, 1, 2, \ldots$.

Now we consider the first m orders of moment of S. It can be proven that there exist two coefficient sequences $\{\alpha_l, l = 1, 2, \ldots, m\}$, and $\{v_l > 0, l = 1, 2, \ldots, m\}$ satisfying $\mathbb{E}(S^j) = \sum_{l=1}^{m} \alpha_l \cdot v_l^j$ for $j = 0, 1, \ldots, m - 1$, which is alternatively written in a matrix form as follows,

$$\begin{bmatrix} 1 & 1 & \cdots & 1 \\ v_1 & v_2 & \cdots & v_m \\ v_1^2 & v_2^2 & \cdots & v_m^2 \\ \vdots & \vdots & \ddots & \vdots \\ v_1^{m-1} & v_2^{m-1} & \cdots & v_m^{m-1} \end{bmatrix} \cdot \begin{bmatrix} \alpha_1 \\ \alpha_2 \\ \alpha_3 \\ \vdots \\ \alpha_m \end{bmatrix} = \begin{bmatrix} 1 \\ \mathbb{E}(S) \\ \mathbb{E}(S^2) \\ \vdots \\ \mathbb{E}(S^{m-1}) \end{bmatrix} \qquad (4.127)$$

or $v \cdot \alpha = \mathbf{E}$ in short. The determinant of v, which is a Vandermonde matrix, is

$$\det(v) = \prod_{1 \le l < k \le m} (v_k - v_l) \qquad (4.128)$$

which is non-zero if $v_k \ne v_l$ for $\forall k \ne l$. Thus, the inverse of v exists and Eq. (4.127) holds for properly selected $\{\alpha_l\}$ and $\{v_l\}$.

With Eq. (4.127), $c_{2k} = \frac{E(S^{2k})}{(2k)!} = \frac{\sum_{l=1}^{m} \alpha_l \cdot v_l^{2k}}{(2k)!}$ for $k = 0, 1, 2, \ldots$. Furthermore, as m is large enough, the failure probability in Eq. (4.126) is estimated by

$$\mathbb{P}_f = \frac{a_0}{2} + \sum_{j=1}^{\infty} \left[a_j \sum_{k=0}^{\infty} \frac{\sum_{l=1}^{m} \alpha_l \cdot v_l^{2k}}{(2k)!} \cdot (j\pi)^{2k}(-1)^k \right] \qquad (4.129)$$

Note that the Taylor expansion of function $\cos x$ is $\sum_{k=0}^{\infty} \frac{x^{2k}}{(2k)!}(-1)^{2k}$ for a real number x, with which

$$\sum_{k=0}^{\infty} \frac{\sum_{l=1}^{m} \alpha_l \cdot v_l^{2k}}{(2k)!} \cdot (j\pi)^{2k}(-1)^k = \sum_{l=1}^{m} \alpha_l \cos(v_l \cdot j\pi) \qquad (4.130)$$

Thus, the failure probability in Eq. (4.129) is rewritten as follows,

$$\mathbb{P}_f = \frac{a_0}{2} + \sum_{l=1}^{m} \alpha_l \left[\sum_{j=1}^{\infty} a_j \cos(v_l \cdot j\pi) \right] = \frac{a_0}{2} + \sum_{l=1}^{m} \alpha_l \left[\xi(v_l) - \frac{a_0}{2} \right] \qquad (4.131)$$

Now, since $\sum_{l=1}^{m} \alpha_l = 1$, Eq. (4.131) further becomes $\mathbb{P}_f = \sum_{l=1}^{m} \alpha_l \xi(v_l)$ or, equivalently, in a matrix form,

$$\mathbb{P}_f = \left[\xi(v_1), \xi(v_2), \ldots, \xi(v_m) \right] \cdot \boldsymbol{\alpha} = \left[\xi(v_1), \xi(v_2), \ldots, \xi(v_m) \right] \cdot (\boldsymbol{v}^{-1}\mathbf{E}) \quad (4.132)$$

where $\boldsymbol{\alpha}$, $\boldsymbol{v}$ and $\mathbf{E}$ are those defined in Eq. (4.127). Equation (4.132) estimates the failure probability on the basis of the moment information of random variables and is independent of the probability distribution of S.

As a byproduct, let

$$\left[\gamma_1, \gamma_2, \ldots, \gamma_m \right] = \left[\xi(v_1), \xi(v_2), \ldots, \xi(v_m) \right] \cdot \boldsymbol{v}^{-1} \qquad (4.133)$$

with which Eq. (4.132) becomes $\mathbb{P}_f = \sum_{i=1}^{m} \gamma_i \mathbb{E}\left(S^{i-1} \right)$, indicating that the failure probability is the weighted sum of the moments of S.

Remark

Consider a case where the probability of failure is calculated according to Eq. (4.123) and the moment information on S is determined by n samples of S, $\underline{s}_1, \underline{s}_2, \ldots, \underline{s}_n$. The confidence level of the estimated failure probability, $\widetilde{\mathbb{P}}_f$ is estimated by first noting, with Eq. (4.133), that

$$\widetilde{\mathbb{P}}_f = \sum_{i=1}^{m} \left(\frac{\gamma_i}{n} \sum_{j=1}^{n} \underline{s}_j^{i-1} \right) \qquad (4.134)$$

As we have discussed before (c.f. Sect. 3.5), $\widetilde{\mathbb{P}}_f$ can be treated as a random variable taking into account the uncertainties associated with each sample. With this, the variance of $\widetilde{\mathbb{P}}_f$ is given by

$$
\mathbb{V}(\widetilde{\mathbb{P}}_f) = \mathbb{V}\left(\sum_{i=1}^{m}\left(\frac{\gamma_i}{n}\sum_{j=1}^{n}\underline{s}_j^{i-1}\right)\right) = \sum_{i=1}^{m}\sum_{j=1}^{m}\frac{\gamma_i\gamma_j}{n^2}\mathbb{C}\left(\sum_{p=1}^{n}\underline{s}_p^{i-1},\sum_{q=1}^{n}\underline{s}_q^{j-1}\right)
$$
$$
= \sum_{i=1}^{m}\sum_{j=1}^{m}\frac{\gamma_i\gamma_j}{n^2}\sum_{p=1}^{n}\sum_{q=1}^{n}\mathbb{C}\left(\underline{s}_p^{i-1},\underline{s}_q^{j-1}\right) = \sum_{i=1}^{m}\sum_{j=1}^{m}\frac{\gamma_i\gamma_j}{n^2}\sum_{p=1}^{n}\mathbb{C}\left(\underline{s}_p^{i-1},\underline{s}_p^{j-1}\right)
$$

(4.135)

By considering the identical distribution and statistical independence of each $\underline{s}_j$, one has

$$
\sum_{p=1}^{n}\mathbb{C}\left(\underline{s}_p^{i-1},\underline{s}_p^{j-1}\right) = n\left(\mathbb{E}(\mathcal{S}^{i+j-2}) - \mathbb{E}(\mathcal{S}^{i-1})\mathbb{E}(\mathcal{S}^{j-1})\right)
$$

(4.136)

Thus,

$$
\mathbb{V}(\widetilde{\mathbb{P}}_f) = \frac{1}{n}\sum_{i=1}^{m}\sum_{j=1}^{m}\gamma_i\gamma_j\left[\mathbb{E}(\mathcal{S}^{i+j-2}) - \mathbb{E}(\mathcal{S}^{i-1})\mathbb{E}(\mathcal{S}^{j-1})\right]
$$

(4.137)

which indicates that the variance of the estimated $\widetilde{\mathbb{P}}_f$ decreases with n and is in proportion to $\frac{1}{n}$. Recall that in Sect. 3.5.1, we mentioned when the naïve Monte Carlo simulation is used to estimate the structural failure probability, the simulated result with n samples also has a variance decreasing with n proportionally to $\frac{1}{n}$.

Example 4.24

Consider the limit state function in Eq. (4.39). Suppose that R is lognormally distributed with a mean value of 60 and a standard deviation of 10, and that the load effect S follows a Gamma distribution with a mean value of 30 and a standard deviation of 8. Calculate the probability of failure using Eq. (4.132) [33].

Solution

We first assign $s_{max} = 30 + 10 \times 8$ (i.e., mean value plus 10 times the standard deviation, which would be the maxima of S with negligible error). By referring to Eq. (2.218), the ith ($i = 1, 2, \ldots$) moment of $\mathcal{S} = \frac{S}{s_{max}}$ is estimated according to

$$
\mathbb{E}\left[\left(\frac{S}{s_{max}}\right)^i\right] = \frac{b^i}{s_{max}^i}\prod_{k=1}^{i}(a + i - k)
$$

(4.138)

where a and b are the shape and scale parameters of S respectively (c.f. Sect. 2.2.7). Letting $v_i = \frac{i}{m}$ for $i = 1, 2, \ldots, m$. Clearly, a greater m would increase the accuracy of Eq. (4.132). Results show that with a permissible error of 10^{-6}, the failure probability is determined as 0.012263 with $m = 21$. If calculating $\mathbb{P}_f$ directly with Eq. (4.2), the result is also 0.012263.

Remark

We herein present an alternative method to derive Eq. (4.132). First, we expand the function $\xi(x)$ in a polynomial form as follows,

$$\xi(x) = \sum_{i=0}^{m-1} \alpha_i x^i, \quad x \in [0, 1] \tag{4.139}$$

where the coefficients α_i's are to be determined. Substituting Eq. (4.139) into Eq. (4.123) gives

$$\mathbb{P}_f = \int_0^1 \left(\sum_{i=0}^{m-1} \alpha_i x^i \right) f_{\mathcal{S}}(x) dx = \sum_{i=0}^{m-1} \alpha_i \left(\int_0^1 x^i f_{\mathcal{S}}(x) dx \right) = \sum_{i=0}^{m-1} \alpha_i \mathbb{E}(\mathcal{S}^i) \tag{4.140}$$

To determine α_i's, we introduce m coefficients $v_i \in [0, 1]$ ($i = 1, 2, \ldots, m$), with which

$$\begin{bmatrix} \xi(v_1) \\ \xi(v_2) \\ \vdots \\ \xi(v_m) \end{bmatrix} = \begin{bmatrix} 1 & 1 & \cdots & 1 \\ v_1 & v_2 & \cdots & v_m \\ v_1^2 & v_2^2 & \cdots & v_m^2 \\ \vdots & \vdots & \ddots & \vdots \\ v_1^{m-1} & v_2^{m-1} & \cdots & v_m^{m-1} \end{bmatrix}^{\mathsf{T}} \begin{bmatrix} \alpha_0 \\ \alpha_1 \\ \vdots \\ \alpha_{m-1} \end{bmatrix} \tag{4.141}$$

Thus,

$$\begin{bmatrix} \alpha_0 & \alpha_1 & \ldots & \alpha_{m-1} \end{bmatrix} = \begin{bmatrix} \xi(v_1) & \xi(v_2) & \ldots & \xi(v_m) \end{bmatrix} \begin{bmatrix} 1 & 1 & \cdots & 1 \\ v_1 & v_2 & \cdots & v_m \\ v_1^2 & v_2^2 & \cdots & v_m^2 \\ \vdots & \vdots & \ddots & \vdots \\ v_1^{m-1} & v_2^{m-1} & \cdots & v_m^{m-1} \end{bmatrix}^{-1} \tag{4.142}$$

with which Eq. (4.140) becomes

$$\mathbb{P}_f = \sum_{i=0}^{m-1} \alpha_i \mathbb{E}(\mathcal{S}^i) = \begin{bmatrix} \xi(v_1) & \xi(v_2) & \ldots & \xi(v_n) \end{bmatrix} \begin{bmatrix} 1 & 1 & \cdots & 1 \\ v_1 & v_2 & \cdots & v_m \\ v_1^2 & v_2^2 & \cdots & v_m^2 \\ \vdots & \vdots & \ddots & \vdots \\ v_1^{m-1} & v_2^{m-1} & \cdots & v_m^{m-1} \end{bmatrix}^{-1} \begin{bmatrix} 1 \\ \mathbb{E}(\mathcal{S}) \\ \mathbb{E}(\mathcal{S}^2) \\ \vdots \\ \mathbb{E}(\mathcal{S}^{m-1}) \end{bmatrix} \tag{4.143}$$

which is consistent with Eq. (4.132).

4.2.4.2 Multi-dimensional Problems

In this section, we consider the case of a multiple-dimensional limit state function. The cases of both linear and nonlinear forms are considered.

First, we consider a linear limit state function G taking a form of

$$G = G(X_1, X_2, \ldots, X_n) = a_0 + \sum_{i=1}^{n} a_i X_i \qquad (4.144)$$

where a_i $(i = 0, 1, 2, \ldots, n)$ is a constant. Without loss of generality, we assume that $a_1 > 0$. With this, Eq. (4.144) is rewritten as follows,

$$G = \left[a_1 X_1 - \min\left(a_0' + \sum_{i=2}^{n} a_i' X_i \right) \right] - \left[a_0' + \sum_{i=2}^{n} a_i' X_i - \min\left(a_0' + \sum_{i=2}^{n} a_i' X_i \right) \right] \qquad (4.145)$$

where $a_i' = -a_i$ for $i = 0, 2, 3, \ldots, n$, and $a_0' + \min \sum_{i=2}^{n} a_i' X_i$ is the minima of $a_0' + \sum_{j=2}^{n} a_i' X_i$ (a deterministic value) so that both terms at the right hand of Eq. (4.145) would be non-negative. As before, we consider the normalized form of G with respect to $\max\left[a_0' + \sum_{i=2}^{n} a_i' X_i - \min\left(a_0' + \sum_{i=2}^{n} a_i' X_i \right) \right]$ as follows,

$$G' = \frac{a_1 X_1 - \min\left(a_0' + \sum_{i=2}^{n} a_i' X_i \right)}{\max\left[a_0' + \sum_{i=2}^{n} a_i' X_i - \min\left(a_0' + \sum_{i=2}^{n} a_i' X_i \right) \right]} - \frac{a_0' + \sum_{i=2}^{n} a_i' X_i - \min\left(a_0' + \sum_{i=2}^{n} a_i' X_i \right)}{\max\left[a_0' + \sum_{i=2}^{n} a_i' X_i - \min\left(a_0' + \sum_{i=2}^{n} a_i' X_i \right) \right]} \qquad (4.146)$$

or $G' = X_1^* - X_{\neq 1}^*$ in short. If treating X_1^* and $X_{\neq 1}^*$ as the generalized resistance and load effect, Eq. (4.146) indeed transforms the multiple-dimensional reliability problem into a one-dimensional case. With Eq. (4.132), one has

$$\mathbb{P}_f = \left[\xi(v_1), \xi(v_2), \ldots, \xi(v_m) \right] \cdot (v^{*-1} \mathbf{E}^*) \qquad (4.147)$$

where $\xi(x)$ is the CDF of X_1^*, $(v^*)_{ij} = v_j^{*i-1}$ and $(\mathbf{E}^*)_i = \mathbb{E}(X_{\neq 1}^{*i-1})$.

To estimate $\mathbf{E}^*$, note that for numbers $x_1, x_2, \ldots, x_p$, the multinomial theorem gives

$$\left(\sum_{i=1}^{p} x_i \right)^q = \sum_{\sum_{i=1}^{p} k_i = q} \frac{q!}{k_1! k_2! \ldots k_p!} \prod_{i=1}^{p} x_i^{k_i} \qquad (4.148)$$

where q is a positive integer. Thus, the jth order moment of $X_{\neq 1}^*$ is given by

$$\mathbb{E}(X_{\neq 1}^{*j}) = \frac{1}{m_a^j} \mathbb{E}\left[\left(\sum_{i=1}^{n} Z_i \right)^j \right] = \frac{1}{m_a^j} \sum \frac{j!}{k_1! k_2! \ldots k_n!} \prod_{i=1}^{n} \mathbb{E}\left(Z_i^{k_i} \right) \qquad (4.149)$$

where $m_a = \max\left[a_0' + \sum_{i=2}^{n} a_i' X_i - \min\left(a_0' + \sum_{i=2}^{n} a_i' X_i\right)\right]$, $Z_1 = a_0' - \min$ $\left(a_0' + \sum_{i=2}^{n} a_i' X_i\right)$, and $Z_i = a_i' X_i$ for $i = 2, 3, \ldots, n$.

The variables X_i's in Eq. (4.144) can, in a generalized case, take a form of a function of another random variable (e.g., $X_i = h(Y_i)$) or the product of several random variables. In such a case, one can still treat the limit state function as *linear* and use Eq. (4.147) to estimate the probability of failure. In fact, if $X_i = h(Y_i)$, then the jth order moment of X_i is

$$\mathbb{E}(X_i^j) = \mathbb{E}[h^j(Y_i)] = \int h^j(y) f_{Y_i}(y) dy \qquad (4.150)$$

where f_{Y_i} is the PDF of Y_i.

Furthermore, we consider the case of a nonlinear limit state function. The failure probability $\mathbb{P}_f$ is, in the presence of independent random inputs $\mathbf{X}=(X_1, X_2, \ldots, X_n)$, estimated as follows,

$$\mathbb{P}_f = \underbrace{\int \ldots \int}_{n\text{-fold}} \vartheta(x_1, x_2, \ldots, x_n) \prod_{i=1}^{n} f_{X_i}(x_i) d\mathbf{x} \qquad (4.151)$$

where $\vartheta(x_1, x_2, \ldots, x_n)$ is the failure probability conditional on $\mathbf{X}=\mathbf{x} = (x_1, x_2, \ldots, x_n)$, and $f_{X_i}(x)$ is the PDF of X_i, $i = 1, 2, \ldots, n$. To consider the normalized form of each X_i, let $Y_i = \frac{X_i - \underline{X}_i}{\overline{X}_i - \underline{X}_i}$, where $\overline{X}_i$ and $\underline{X}_i$ are the maxima and minima of X_i respectively. With this, Eq. (4.151) becomes

$$\mathbb{P}_f = \underbrace{\int_0^1 \ldots \int_0^1}_{n\text{-fold}} \varphi(y_1, y_2, \ldots, y_n) \prod_{i=1}^{n} f_{Y_i}(y_i) d\mathbf{y} \qquad (4.152)$$

where $f_{Y_i}(y)$ is the PDF of Y_i for $i = 1, 2, \ldots, n$, and

$$\varphi(y_1, y_2, \ldots, y_n) = \vartheta((\overline{X}_1 - \underline{X}_1)y_1 + \underline{X}_1, \ldots, (\overline{X}_n - \underline{X}_n)y_n + \underline{X}_n) \qquad (4.153)$$

We define a function

$$\varphi_1(y_1) = \underbrace{\int \ldots \int}_{(n-1)\text{-fold}} \varphi(y_1, y_2, \ldots, y_n) \prod_{i=2}^{n} f_{Y_i}(y_i) d\mathbf{y} \qquad (4.154)$$

with which Eq. (4.152) becomes

$$\mathbb{P}_f = \int \varphi_1(y_1) f_{Y_1}(y_1) dy_1 \qquad (4.155)$$

As before, there exists a sequence of $\{_1\nu_l > 0, l = 1, 2, \ldots, m\}$ such that

$$\mathbb{P}_f = \left[\varphi_1(_1\nu_1), \varphi_1(_1\nu_2), \ldots, \varphi_1(_1\nu_m)\right] \cdot (_1\boldsymbol{\nu}^{-1}\mathbf{E}_1) \qquad (4.156)$$

where $(_1\boldsymbol{\nu})_{ij} = {_1\nu_j^{i-1}}$ and $(\mathbf{E}_1)_j = \mathbb{E}(Y_1^{j-1})$. Next, to compute $\varphi_1(_1\nu_j)$ in Eq. (4.156), we define another function

$$\varphi_2(y_1, y_2) = \underbrace{\int \cdots \int}_{(n-2)\text{-fold}} \varphi(y_1, y_2, \ldots, y_n) \prod_{i=3}^{n} f_{Y_i}(y_i) \mathrm{d}\mathbf{y} \qquad (4.157)$$

with which

$$\varphi_1(y_1) = \int \varphi_2(y_1, y_2) f_{Y_2}(y_2) \mathrm{d}y_2 \qquad (4.158)$$

Again, with Eq. (4.132), there exists a sequence of $\{_2\nu_l > 0\}$ such that

$$\varphi_1(y_1) = \left[\varphi_2(x_1, {_2\nu_1}), \varphi_2(x_1, {_2\nu_2}), \ldots, \varphi_2(x_1, {_2\nu_m})\right] \cdot (_2\boldsymbol{\nu}^{-1}\mathbf{E}_2) \qquad (4.159)$$

in which $(_2\boldsymbol{\nu})_{ij} = {_2\nu_j^{i-1}}$ and $(\mathbf{E}_2)_j = \mathbb{E}(Y_2^{j-1})$.

Similar to Eqs. (4.154) and (4.157), we can define the function sequence φ_3, φ_4 through φ_n, and calculate φ_i with φ_{i+1} for $i = n-1, n-2, \ldots, 2, 1$ (note that $\varphi_n = \varphi$). The calculation of $\mathbb{P}_f$ is finalized once φ_1 is obtained.

Specifically, if $n = 2$, with Eqs. (4.156) and (4.159), the failure probability is given by

$$\mathbb{P}_f = (\mathbf{E}_1)^{\mathsf{T}} (_1\boldsymbol{\nu}^{-1})^{\mathsf{T}} \cdot \begin{bmatrix} \varphi_1(_1\nu_1) \\ \varphi_1(_1\nu_2) \\ \vdots \\ \varphi_1(_1\nu_m) \end{bmatrix}$$

$$= (\mathbf{E}_1)^{\mathsf{T}} (_1\boldsymbol{\nu}^{-1})^{\mathsf{T}} \begin{bmatrix} \varphi(_1\nu_1, {_2\nu_1}) & \varphi(_1\nu_1, {_2\nu_2}) & \cdots & \varphi(_1\nu_1, {_2\nu_m}) \\ \varphi(_1\nu_2, {_2\nu_1}) & \varphi(_1\nu_2, {_2\nu_2}) & \cdots & \varphi(_1\nu_2, {_2\nu_m}) \\ \vdots & \vdots & \ddots & \vdots \\ \varphi(_1\nu_m, {_2\nu_1}) & \varphi(_1\nu_m, {_2\nu_2}) & \cdots & \varphi(_1\nu_m, {_2\nu_m}) \end{bmatrix} (_2\boldsymbol{\nu}^{-1}\mathbf{E}_2) \qquad (4.160)$$

Example 4.25

Consider a limit state function of $G = R - S_1 - S_2 S_3$, where the statistical information on the random variables is presented in Table 4.5. The failure probability is obtained as 0.003718 via 1,000,000 runs of Monte Carlo simulation. Now, use the moment-based method in Sect. 4.2.4.2 to estimate the probability of failure [33].

Table 4.5 Statistical information of the variables in Example 4.25

Variable	Mean	Standard deviation	Distribution type
R	60	10	Normal
S_1	15	5	Gamma
S_2	1/2	$\sqrt{3}/6$	Uniform over [0, 1]
S_3	20	4	Weibull

Solution

First, we treat the limit state function in this example as *linear*, which can be rewritten as $G = R - S^*$, where $S^* = S_1 + S_2 S_3$. Let $s_{\max}^*$ and $s_{\min}^*$ denote the maxima and minima of S^*, respectively. We assign $s_{\min}^* = 0$ and $s_{\max}^* = \mu_{S_1} + 3\sigma_{S_1} + \mu_{S_3} + 3\sigma_{S_3}$. Normalizing G with respect to $s_{\max}^*$ gives $G' = \frac{R}{s_{\max}^*} - \frac{S^*}{s_{\max}^*}$.

With the multinomial theorem (c.f. Eq. (4.148)), the jth moment of $\frac{S^*}{s_{\max}^*}$ is estimated according to

$$
\mathbb{E}\left[\left(\frac{S^*}{s_{\max}^*}\right)^j\right] = \sum_{k=0}^{j} \left\{ \binom{j}{k} \mathbb{E}\left(\frac{S_1}{s_{\max}^*}\right)^k \mathbb{E}\left(S_2^{j-k}\right) \mathbb{E}\left(\frac{S_3}{s_{\max}^*}\right)^{j-k} \right\} \tag{4.161}
$$

We further let $v_i = \frac{i}{m}$ for $i = 1, 2, \ldots, m$. Results show that the failure probability converges to 0.003710 when $m = 12$, which agrees well with the Monte Carlo simulation solution.

We can alternatively treat the limit state function as *nonlinear* and re-calculate the failure probability with Eq. (4.151). Note that $\vartheta(s_1, s_2, s_3) = F_R(s_1 + s_2 s_3)$, where $F_R(s)$ is the CDF of R. Similar to Eq. (4.160), the failure probability is

$$
\mathbb{P}_f = (\mathbf{E}_1)^\mathsf{T} (_1 v^{-1})^\mathsf{T}
\begin{bmatrix}
\varphi_2(_1 v_1, _2 v_1) & \varphi_2(_1 v_1, _2 v_2) & \cdots & \varphi_2(_1 v_1, _2 v_m) \\
\varphi_2(_1 v_2, _2 v_1) & \varphi_2(_1 v_2, _2 v_2) & \cdots & \varphi_2(_1 v_2, _2 v_m) \\
\vdots & \vdots & \ddots & \vdots \\
\varphi_2(_1 v_m, _2 v_1) & \varphi_2(_1 v_m, _2 v_2) & \cdots & \varphi_2(_1 v_m, _2 v_m)
\end{bmatrix}
(_2 v^{-1} \mathbf{E}_2)
$$
$$\tag{4.162}$$

where

$$
\varphi_2(s_1, s_2) = \int \varphi(s_1, s_2, s) f_{S_3}(s)\,\mathrm{d}s = \left[\varphi(s_1, s_2, _1 v_1), \ldots, \varphi(s_1, s_2, _1 v_m) \right] \cdot (_3 v^{-1} \mathbf{E}_3) \tag{4.163}
$$

and $(\mathbf{E}_k)_j = \mathbb{E}\left[\left(\frac{S_k - \underline{S}_k}{\overline{S}_k - \underline{S}_k} \right)^{j-1} \right]$ for $k = 1, 2, 3$. Let $_1 v_i = _2 v_i = _3 v_i = \frac{i}{m}$ for $i = 1, 2, \ldots, m$, with which the failure probability converges to 0.003710 with $m = 10$, consistent with the previous results.

Note that each element in $\mathbf{X}$, X_i, has been assumed statistically independent in Eq. (4.151). If taking into account the correlation between each X_i, we can use the Nataf transformation method to convert $\mathbf{X}$ into dependent standard normal variable $(Y_1, Y_2, \ldots, Y_n)$ and then independent standard normal variables $(U_1, U_2, \ldots, U_n)$. We assume that the correlation matrix of $\{X_1, \ldots, X_n\}$ is $\boldsymbol{\rho} = [\rho_{ij}]_{n \times n}$, with which one can find the correlation matrix of $\mathbf{Y}$, $\boldsymbol{\rho}'$. Furthermore, it follows,

$$
\begin{bmatrix}
\Phi^{-1}[F_{X_1}(X_1)] \\
\Phi^{-1}[F_{X_2}(X_2)] \\
\vdots \\
\Phi^{-1}[F_{X_n}(X_n)]
\end{bmatrix}
= \mathbf{A}
\begin{bmatrix}
U_1 \\
U_2 \\
\vdots \\
U_n
\end{bmatrix}
=
\begin{bmatrix}
a_{11} & & & \\
a_{21} & a_{22} & & \\
\vdots & \vdots & \ddots & \\
a_{n1} & a_{n2} & \ldots & a_{nn}
\end{bmatrix}
\cdot
\begin{bmatrix}
U_1 \\
U_2 \\
\vdots \\
U_n
\end{bmatrix}
\tag{4.164}
$$

where $F_{X_i}()$ is the CDF of X_i, and $\mathbf{A}$ is a lower triangle matrix satisfying $\mathbf{A}\mathbf{A}^\mathsf{T} = \boldsymbol{\rho}'$. Thus, $X_i = F_{X_i}^{-1}\left[\Phi\left(\sum_{j=1}^{i} a_{ij}U_j\right)\right]$ for $i = 1, 2, \ldots, n$ and Eq. (4.151) becomes

$$
\mathbb{P}_\mathrm{f} = \underbrace{\int \cdots \int}_{n\text{-fold}} \vartheta \left(F_{X_1}^{-1}[\Phi(a_{11}u_1)], \ldots, F_{X_n}^{-1}\left[\Phi\left(\sum_{j=1}^{n} a_{nj}u_n\right)\right] \right) \cdot \prod_{i=1}^{n} \phi(u_i)\mathrm{d}\mathbf{u}
\tag{4.165}
$$

where $\phi()$ is the PDF of a standard normal variable. As such, the reliability problem incorporating correlated random variables can be transformed into that with independent variables only.

4.3 System Reliability

We have previously discussed the survival (or failure) of a single structural component. For a structural system, however, it may fail in many modes, depending on the complex interactions between the components involved. This section presents a concise introduction to system reliability, including three basic models: series system, parallel system and k-out-of-n system. In many cases, it is fairly difficult to calculate the system reliability directly. In such a case, the upper and lower bounds of the system reliability could be derived instead, providing a straightforward description on the range of the system safety level.

4.3.1 Series System

For a *series system* consisting of n components, let F_i denote the failure of member (or mode) i, $i = 1, 2, \ldots, n$. The system failure is defined by

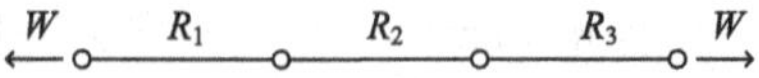

Fig. 4.16 Example of a series system

$$F_{\text{sys}} = F_1 \bigcup F_2 \bigcup \cdots \bigcup F_n \tag{4.166}$$

With this,

$$\mathbb{P}(F_{\text{sys}}) = \mathbb{P}\left(F_1 \bigcup F_2 \bigcup \cdots \bigcup F_n\right) = \mathbb{P}\left(\bigcup_{i=1}^{n} F_i\right) \tag{4.167}$$

Note that for $\forall i$, $F_i \subset F_1 \bigcup F_2 \bigcup \cdots \bigcup F_n$. Thus, we have the upper and lower bounds for $\mathbb{P}(F_{\text{sys}})$ as follows,

$$\max_{i=1}^{n} \mathbb{P}(F_i) \leq \mathbb{P}(F_{\text{sys}}) \leq \sum_{i=1}^{n} \mathbb{P}(F_i) \tag{4.168}$$

Specifically, if each F_i is independent of each other,

$$\mathbb{P}(F_{\text{sys}}) = 1 - \prod_{i=1}^{n}[1 - \mathbb{P}(F_i)] \tag{4.169}$$

which further is close to $\sum_{i=1}^{n} \mathbb{P}(F_i)$ for typical cases where each $\mathbb{P}(F_i)$ is small enough.

Example 4.26

Consider the series system in Fig. 4.16. Let W be deterministically 100 kN. Assume that both R_1 and R_2 are independently normal distributed, and $R_3 = \frac{1}{2}(R_1 + R_2)$. R_1 has a mean value of 400 kN and a COV of 0.3; R_2 has a mean value of 300 kN and a standard deviation of 0.25. Let mode i denote the failure of the ith element, $i = 1, 2, 3$.
(1) Calculate the failure probability associated with mode 1, 2 or 3. (2) Calculate the probability of system failure, and verify Eq. (4.168).
Solution

(1) Since $R_3 = \frac{1}{2}(R_1 + R_2)$, the mean value and standard deviation of R_3 are obtained respectively as 350 and 70.75 kN. Let p_i denote the failure probability associated with mode i, $i = 1, 2, 3$. With this,

$$p_1 = \mathbb{P}(R_1 < 100) = \Phi\left(\frac{100 - 400}{400 \times 0.3}\right) = 0.00621 \tag{4.170}$$

$$p_2 = \mathbb{P}(R_2 < 100) = \Phi\left(\frac{100 - 300}{300 \times 0.25}\right) = 0.00383 \tag{4.171}$$

$$p_3 = \mathbb{P}(R_3 < 100) = \Phi\left(\frac{100 - 350}{70.75}\right) = 0.00020 \tag{4.172}$$

(2) The probability of system failure is estimated by

$$\mathbb{P}(F_{sys}) = 1 - \mathbb{P}(R_1 \geq 100 \cap R_2 \geq 100 \cap R_3 \geq 100) = \mathbb{P}(R_1 \geq 100 \cap R_2 \geq 100)$$
$$= 1 - \mathbb{P}(R_1 \geq 100) \cdot \mathbb{P}(R_2 \geq 100)$$
$$= 1 - \left[1 - \Phi\left(\frac{100 - 400}{120}\right)\right] \cdot \left[1 - \Phi\left(\frac{100 - 300}{75}\right)\right] = 0.01002$$

$$(4.173)$$

Clearly, $\max_{i=1}^{n} p_i = 0.00621$ and $\sum_{i=1}^{n} p_i = 0.01024$. Thus, Eq. (4.168) holds.

Example 4.27

Consider a series system with two components. Let R_1, R_2, S_1 and S_2 denote the resistance and load effect of the two elements respectively, all following a normal distribution for simplicity. R_1 and R_2 are identically distributed with a mean value of 2.5 and a COV of 0.15; S_1 and S_2 are identically distributed with a mean value of 1 and a COV of 0.3. Let ρ_R and ρ_S denote the correlation coefficient of (R_1, R_2) and (S_1, S_2) respectively. Discuss the sensitivity of the system failure probability to ρ_R and ρ_S.

Solution

Since R_1, R_2, S_1 and S_2 all follow a normal distribution, we introduce two normal variables $\widetilde{R}$ and $\widetilde{S}$ such that

$$R_2 = \rho_R R_1 + \widetilde{R}, \quad S_2 = \rho_S S_1 + \widetilde{S} \tag{4.174}$$

Thus, $\widetilde{R}$ has a mean value of $2.5(1 - \rho_R)$ and a standard deviation of $0.375\sqrt{1 - \rho_R^2}$; $\widetilde{S}$ has a mean value of $1 - \rho_S$ and a standard deviation of $0.3\sqrt{1 - \rho_S^2}$. With this, we further define $\widetilde{G} = \widetilde{R} - \widetilde{S}$, which is also a normal variable having a mean value of $2.5(1 - \rho_R) - (1 - \rho_S)$ and a standard deviation of $\sqrt{0.375^2(1 - \rho_R^2) + 0.3^2(1 - \rho_S^2)}$. The failure probability of the 2-component series system is

$$\mathbb{P}(F_{sys}) = 1 - \mathbb{P}(R_1 > S_1 \cap R_2 > S_2) = 1 - \mathbb{P}\left(R_1 > S_1 \cap \rho_R R_1 + \widetilde{R} > \rho_S S_1 + \widetilde{S}\right)$$
$$= 1 - \int_{R_1} \int_{\widetilde{G}} F_{S_1}\left[\min\left(r_1, \frac{\rho_R r_1 + \widetilde{g}}{\rho_S}\right)\right] f_{\widetilde{G}}(\widetilde{g}) f_{R_1}(r_1) d\widetilde{g} dr_1$$

$$(4.175)$$

where $F_\bullet$ and $f_\bullet$ are the CDF of PDF of $\bullet$ respectively ($\bullet = R_1, S_1, \widetilde{G}$).

Figure 4.17 illustrates the dependence of system failure probability $\mathbb{P}(F_{sys})$ on ρ_R and ρ_S. It can be seen that the system failure probability is more sensitive to the correlation between R_1 and R_2.

Note that the two components in this example have a common failure probability, which is determined by

$$\mathbb{P}(F_i) = \mathbb{P}(R_1 < S_1) = \mathbb{P}(R_2 < S_2) = \int F_{R_1}(s) f_{S_1}(s) ds = 8.936 \times 10^{-4} \tag{4.176}$$

Fig. 4.17 Dependence of series system failure probability on ρ_R and ρ_S

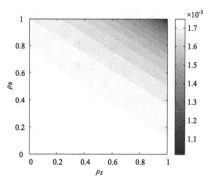

For the series system, specifically, if $\rho_R = \rho_S = 0$, then $\mathbb{P}(F_{sys}) = 1 - (1 - \mathbb{P}(F_i))^2 \approx 2\mathbb{P}(F_i) = 1.79 \times 10^{-3}$. On the other hand, if $\rho_R = \rho_S = 1$, then $\mathbb{P}(F_{sys}) = \mathbb{P}(F_i) = 8.936 \times 10^{-4}$. These two extreme cases indeed yield the lower and upper bounds of $\mathbb{P}(F_{sys})$ according to Eq. (4.168).

Remark

For an indeterministic system with multiple failure paths, let $\mathbf{X} = \{X_1, X_2, \ldots, X_n\}$ be the random vector representing the uncertain quantities of the structure (including both strength and applied loads), and D_i a subdomain of $(-\infty, \infty)^n$ which defines the set of failure mode i (i.e., the failure occurs in mode i only if $\mathbf{X} \in D_i, i = 1, 2, \ldots, m$). Clearly, $D_i \cap D_j = \varnothing$ for $\forall i \neq j$. If the system resistance associated with failure path i is R_i, then the system failure probability is estimated as follows subjected to a load effect of s,

$$\mathbb{P}(F_{sys}) = \sum_{i=1}^{m} \mathbb{P}(R_i \leq s \cap \mathbf{X} \in D_i) \tag{4.177}$$

Furthermore, if the PDF of load effect is $f_S(x)$, with the law of total probability, Eq. (4.177) becomes

$$\mathbb{P}(F_{sys}) = \sum_{i=1}^{m} \int \mathbb{P}(R_i \leq s \cap \mathbf{X} \in D_i) f_S(s) \mathrm{d}s \tag{4.178}$$

Equation (4.178) demonstrates that the estimate of the structural failure probability depends on the ability to identify D_i and R_i correctly.

Example 4.28

Consider the portal frame in Fig. 2.1a. Given the simultaneous actions of V and H, the frame may fail in one of the three modes: "beam" mode, "sway" mode and "combined" mode. We suppose that the span and height of the frame are 2 and 1 m respectively, with V acting at the middle point of the beam. Both V and H are independent normal variables.

V has a mean value of 300 kN and a COV of 0.2; the mean value and COV of H are 200 kN and 0.3. The moment capacities of beam and column are deterministically $M_b = 120$ kN·m and $M_c = 100$ kN·m respectively, suggesting a *strong beam-weak column design*. The three failure modes correspond to:

Beam $V > 2M_b + 2M_c \cap \frac{V}{H} > 1 + \frac{M_b}{M_c}$

Sway $H > 4M_c \cap \frac{V}{H} \leq \frac{M_b}{2M_c}$ (4.179)

Combined $V + H > 2M_b + 4M_c \cap \frac{M_b}{2M_c} < \frac{V}{H} \leq 1 + \frac{M_b}{M_c}$

Compute the system failure probability.

Solution

Note that it is given $M_b = 120$ kN·m and $M_c = 100$ kN·m, with which

Beam $V > 440 \cap \frac{V}{H} > 2.2$

Sway $H > 400 \cap \frac{V}{H} \leq 0.6$ (4.180)

Combined $V + H > 640 \cap 0.6 < \frac{V}{H} \leq 2.2$

The failure regions for the three failure modes, denoted by D_1, D_2, D_3 respectively, are illustrated in Fig. 4.18. Clearly, each D_i is mutually exclusive.

The probability of failure associated with "beam" mode is estimated by

$$\mathbb{P}(\text{beam mode}) = \mathbb{P}((H, V) \in D_1) = \mathbb{P}\left(V > 440 \cap \frac{V}{H} > 2.2\right)$$

$$= 1 - \int_H F_V \left(\max\left(440, 2.2h\right)\right) f_H(h)dh \qquad (4.181)$$

$$= 5.499 \times 10^{-3}$$

Similarly,

$$\mathbb{P}(\text{sway mode}) = \mathbb{P}((H, V) \in D_2) = \mathbb{P}\left(H > 400 \cap \frac{V}{H} \leq 0.6\right)$$

$$= 1 - \int_V F_H \left(\max\left(440, \frac{v}{0.6}\right)\right) f_V(v)dv \qquad (4.182)$$

$$= 1.019 \times 10^{-5}$$

and

$$\mathbb{P}(\text{combined mode}) = \mathbb{P}((H, V) \in D_3) = \mathbb{P}\left(V + H > 640 \cap 0.6 < \frac{V}{H} \leq 2.2\right)$$

$$= \int_{200}^{\infty} \left[F_V(2.2h) - F_V\left(\max\left(0.6h, 640 - h\right)\right)\right] f_H(h)dh$$

$$= 4.763 \times 10^{-2}$$

$$(4.183)$$

Fig. 4.18 Failure regions D_1, D_2, D_3 in Example 4.28

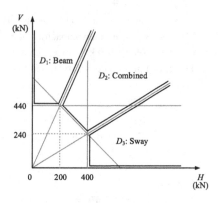

Thus, the system failure probability is $\mathbb{P}(F_{\text{sys}}) = \sum_{i=1}^{3} \mathbb{P}((H, V) \in D_i) = 5.499 \times 10^{-3} + 1.019 \times 10^{-5} + 4.763 \times 10^{-2} = 5.313 \times 10^{-2}$.

4.3.2 Parallel System

For a parallel or ductile model with n components, the system failure is determined by

$$F_{\text{sys}} = F_1 \bigcap F_2 \bigcap \cdots \bigcap F_n \tag{4.184}$$

where F_i is the failure of element (or mode) i as before. Since for $\forall i$, $F_1 \bigcap F_2 \bigcap \cdots \bigcap F_n \subset F_i$, it follows,

$$\mathbb{P}(F_{\text{sys}}) \leq \min_{i=1}^{n} \mathbb{P}(F_i) \tag{4.185}$$

which yields an upper bound of the failure probability of a parallel system. Furthermore, by noting that $\mathbb{P}(F_{\text{sys}}) = 1 - \mathbb{P}\left(\overline{F}_1 \bigcup \overline{F}_2 \bigcup \cdots \bigcup \overline{F}_n\right) = 1 - \mathbb{P}\left(\bigcup_{i=1}^{n} F_i\right)$, one has

$$\max\left[0, 1 - n + \sum_{i=1}^{n} \mathbb{P}(F_i)\right] \leq \mathbb{P}(F_{\text{sys}}) \tag{4.186}$$

which gives a lower bound of the parallel system.

Remark

One can compare the similarity of Eqs. (4.186) and (2.336) (its generalized form). For a series system, if we let the resistance of component i be R_i with a CDF of $F_{R_i}(x)$, and the load effect applied to the ith component be s_i, then $\mathbb{P}(F_i) = F_{R_i}(s_i)$ and $\mathbb{P}(F_{\text{sys}}) = F_{R_1, R_2, \ldots, R_n}(s_1, s_2, \ldots, s_n)$, where $F_{R_1, R_2, \ldots, R_n}$ is the joint CDF of $R_1, R_2, \ldots, R_n$.

Fig. 4.19 An example of parallel system

(a)

(b)

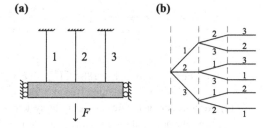

With this, a generalized form of Eq. (2.336) gives $\max \left[0, 1 - n + \sum_{i=1}^{n} F_{R_i}(s_i)\right] \leq F_{R_1, R_2, \ldots, R_n}(s_1, s_2, \ldots, s_n)$, which is consistent with Eq. (4.186).

The lower bound in Eq. (4.186) can be further tightened. First, note that the following relationship holds,

$$\mathbb{P}(F_{\text{sys}}) = \prod_{i=1}^{n} \mathbb{P}\left(F_i \left| \bigcap_{j=i+1}^{n} F_j\right.\right) \tag{4.187}$$

In many engineering problems, $\mathbb{P}(F_i | \bigcap_{j=i+1}^{n} F_j) > \mathbb{P}(F_i)$. Based on this, the lower bound of $\mathbb{P}(F_{\text{sys}})$ as in Eq. (4.186) is tightened as

$$\prod_{i=1}^{n} \mathbb{P}(F_i) \leq \mathbb{P}(F_{\text{sys}}) \tag{4.188}$$

Example 4.29

Consider a parallel system as shown in Fig. 4.19a. The capacities of each rope, R_1, R_2 and R_3, are identically normally distributed and statistically independent, having a mean value of 100 kN and a COV of 0.15. The load F also has a normal distribution with a mean value of 120 kN and a COV of 0.3. Suppose that the tensile load transformed to each rope is identically $\frac{F}{n}$ when n ropes are working ($n = 1, 2, 3$).

(1) Find the lower and upper bounds of the system failure probability according to Eqs. (4.185) and (4.188).
(2) Compute the system failure probability.

Solution

(1) We first calculate the probability of failure associated with each structural element (rope), denoted by $\mathbb{P}(F_i)$. According to Eq. (4.5),

$$\mathbb{P}(F_i) = 1 - \Phi\left(\frac{\mu_R - \mu_S}{\sqrt{\sigma_R^2 + \sigma_S^2}}\right) = 1 - \Phi\left(\frac{100 - \frac{120}{3}}{\sqrt{(100 \times 0.15)^2 + \left(\frac{120 \times 0.3}{3}\right)^2}}\right) = 8.936 \times 10^{-4}$$

(4.189)

With Eqs. (4.185) and (4.188), $\prod_{i=1}^{n} \mathbb{P}(F_i) \leq \mathbb{P}(F_{sys}) \leq \min_{i=1}^{n} \mathbb{P}(F_i)$. Thus, the upper and lower bounds are given as 8.936×10^{-4} and 7.136×10^{-10} respectively.

(2) We consider the failure of the parallel system using failure tree analysis. Figure 4.19b shows the potential failure paths of the system. Illustratively, we consider the path of $1 \rightarrow 2 \rightarrow 3$. First, according to Eq. (4.189), $\mathbb{P}(F_1) = 8.936 \times 10^{-4}$. Next,

$$\mathbb{P}(F_2|F_1) = 1 - \Phi\left(\frac{100 - \frac{120}{2}}{\sqrt{(100 \times 0.15)^2 + \left(\frac{120 \times 0.3}{2}\right)^2}}\right) = 0.044 \qquad (4.190)$$

and furthermore,

$$\mathbb{P}(F_3|F_1 F_2) = 1 - \Phi\left(\frac{100 - 120}{\sqrt{(100 \times 0.15)^2 + (120 \times 0.3)^2}}\right) = 0.696 \qquad (4.191)$$

Thus, $\mathbb{P}(1 \rightarrow 2 \rightarrow 3) = 8.936 \times 10^{-4} \times 0.044 \times 0.696 = 2.74 \times 10^{-5}$ and the system failure probability is $\mathbb{P}(F_{sys}) = 6\mathbb{P}(1 \rightarrow 2 \rightarrow 3) = 1.64 \times 10^{-4}$, which lies between the upper and lower bounds as indicated above.

Example 4.30

Reconsider Example 4.27. If the system is a parallel system, discuss the sensitivity of the system failure probability to ρ_R and ρ_S.

Solution

We as before introduce two normal variables $\widetilde{R}$ and $\widetilde{S}$ such that $R_2 = \rho_R R_1 + \widetilde{R}$ and $S_2 = \rho_S S_1 + \widetilde{S}$. The failure probability of the 2-component parallel system is

$$\mathbb{P}(F_{sys}) = \mathbb{P}(R_1 < S_1 \cap R_2 < S_2) = \mathbb{P}\left(R_1 < S_1 \cap \rho_R R_1 + \widetilde{R} < \rho_S S_1 + \widetilde{S}\right)$$

$$= 1 - \int_{R_1} \int_{\widetilde{G}} F_{S_1}\left[\max\left(r_1, \frac{\rho_R r_1 + \widetilde{g}}{\rho_S}\right)\right] f_{\widetilde{G}}(\widetilde{g}) f_{R_1}(r_1) \mathrm{d}\widetilde{g} \mathrm{d}r_1$$

(4.192)

where $\widetilde{G} = \widetilde{R} - \widetilde{S}$, $F_\bullet$ and $f_\bullet$ are the CDF of PDF of $\bullet$ respectively ($\bullet = R_1, S_1, \widetilde{G}$). The dependence of the parallel system reliability on ρ_R and ρ_S is presented in Fig. 4.20, from which it can be seen that the system failure probability is more sensitive to the correlation between R_1 and R_2.

Fig. 4.20 Dependence of parallel system failure probability on ρ_R and ρ_S

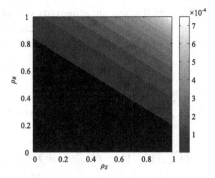

Since the failure probability for each element is $\mathbb{P}_f = 8.936 \times 10^{-4}$ according to Eq. (4.176), for the extreme case of $\rho_R = \rho_S = 0$, $\mathbb{P}(F_{\text{sys}}) = \mathbb{P}^2(F_i) = 7.99 \times 10^{-7}$. However, if $\rho_R = \rho_S = 1$, then $\mathbb{P}(F_{\text{sys}}) = \mathbb{P}(F_i) = 8.936 \times 10^{-4}$. These two extreme cases give the lower and upper bounds of $\mathbb{P}(F_{\text{sys}})$ according to Eqs. (4.185) and (4.188).

4.3.3 k-Out-of-n System

For a system with n components, if the system fails if at least k components fail, then the system is called a *k-out-of-n system*. Clearly, both a series system and a parallel system are specific cases of a k-out-of-n system, with $k = 1$ and $k = n$ respectively [8, 18, 25].

Suppose that each component has a failure probability of p and behaves independently. Let $F_{\text{sys}}(k, n, p)$ denote the probability of system failure, which is a function of k, n and p. By definition,

$$\mathbb{P}[F_{\text{sys}}(k, n, p)] = \sum_{i=k}^{n} \mathbb{P}(X = i) = \sum_{i=k}^{n} C_n^i p^i (1 - p)^{n-i} \qquad (4.193)$$

where X is the number of failed components, which follows a binomial distribution (c.f. Sect. 2.2.1.2).

Remark

We in fact have an equivalent definition of a k-out-of-n system: that is, a system with n components functions if at least k components work. With this regard, the cases of $k = n$ and $k = 1$ yield a series system and a parallel system respectively. Both definitions shall be used with clear demonstration.

We now discuss the sensitivity of $\mathbb{P}[F_{\text{sys}}(k, n, p)]$ in Eq. (4.193) to k, n and p.

First, since n is positive integer, we consider $\mathbb{P}[F_{\text{sys}}(k, n, p)] - \mathbb{P}[F_{\text{sys}}(k, n - 1, p)]$. Using the law of total probability, it follows [20]

$$
\mathbb{P}[F_{\text{sys}}(k, n, p)] = \mathbb{P}[F_{\text{sys}}(k - 1, n - 1, p)] \cdot p + \mathbb{P}[F_{\text{sys}}(k, n - 1, p)] \cdot (1 - p)
$$

$$
= p \sum_{i=k-1}^{n-1} C_{n-1}^i p^i (1 - p)^{n-1-i} + (1 - p) \sum_{i=k}^{n-1} C_{n-1}^i p^i (1 - p)^{n-1-i}
$$

$$
= p C_{n-1}^{k-1} p^{k-1} (1 - p)^{n-k} + p \sum_{i=k}^{n-1} C_{n-1}^i p^i (1 - p)^{n-1-i} + \sum_{i=k}^{n-1} C_{n-1}^i p^i (1 - p)^{n-1-i}
$$

$$
- p \sum_{i=k}^{n-1} C_{n-1}^i p^i (1 - p)^{n-1-i}
$$

$$
= p C_{n-1}^{k-1} p^{k-1} (1 - p)^{n-k} + \mathbb{P}[F_{\text{sys}}(k, n - 1, p)]
$$

$$
\tag{4.194}
$$

Thus,

$$
\mathbb{P}[F_{\text{sys}}(k, n, p)] - \mathbb{P}[F_{\text{sys}}(k, n - 1, p)] = C_{n-1}^k p^{k-1} (1 - p)^{n-k} \tag{4.195}
$$

The positivity of $C_{n-1}^k p^{k-1} (1 - p)^{n-k}$ in Eq. (4.195) implies that $\mathbb{P}[F_{\text{sys}}(k, n, p)]$ increases with a greater n.

Next, we consider $\mathbb{P}[F_{\text{sys}}(k, n, p)] - \mathbb{P}[F_{\text{sys}}(k-1, n, p)]$. According to Eq. (4.193), one has

$$
\mathbb{P}[F_{\text{sys}}(k, n, p)] - \mathbb{P}[F_{\text{sys}}(k - 1, n, p)] = \sum_{i=k}^{n} C_n^i p^i (1 - p)^{n-i} - \sum_{i=k-1}^{n} C_n^i p^i (1 - p)^{n-i}
$$

$$
= C_n^{k-1} p^{k-1} (1 - p)^{n-k+1} > 0
$$

$$
\tag{4.196}
$$

Finally, by noting that p is continuous in $[0, 1]$, we consider the derivative $\frac{d\mathbb{P}[F_{\text{sys}}(k,n,p)]}{dp}$, which is determined by

$$
\frac{d\mathbb{P}[F_{\text{sys}}(k, n, p)]}{dp} = \frac{d}{dp} \left(\sum_{i=k}^{n} C_n^i p^i (1 - p)^{n-i} \right)
$$

$$
= \sum_{i=k}^{n} C_n^i p^{i-1} (1 - p)^{n-i-1} [i(1 - p) - p(n - i)]
$$

$$
= k C_n^k p^{k-1} (1 - p)^{n-k} + \sum_{i=k+1}^{n} i C_n^i p^{i-1} (1 - p)^{n-i} - \sum_{i=k}^{n-1} (n - i) C_n^i p^i (1 - p)^{n-i-1}
$$

$$
\tag{4.197}
$$

Since $(i + 1)C_n^{i+1} = (n - i)C_n^i$, and

$$
\sum_{i=k+1}^{n} i C_n^i p^{i-1} (1 - p)^{n-i} = \sum_{i=k}^{n-1} (i + 1)C_n^{i+1} p^i (1 - p)^{n-i-1} = \sum_{i=k}^{n-1} (n - i)C_n^i p^i (1 - p)^{n-i-1}
$$

$$
\tag{4.198}
$$

it follows,

$$\frac{d\mathbb{P}[F_{sys}(k, n, p)]}{dp} = kC_n^k p^{k-1}(1 - p)^{n-k} \tag{4.199}$$

As before, the positive value of $kC_n^k p^{k-1}(1 - p)^{n-k}$ implies that the system failure probability is a monotonically increasing function of p.

Example 4.31

For a 4-out-of-6 system with a component failure probability of $p = 0.3$, calculate the system failure probability.

Solution

According to Eq. (4.193),

$$
\begin{aligned}
\mathbb{P}[F_{sys}(4, 6, 0.3)] &= \sum_{i=4}^{6} C_6^i 0.3^i (1 - p)^{6-i} \\
&= C_6^4 0.3^4 (1 - p)^{6-4} + C_6^5 0.3^5 (1 - p)^{6-5} + C_6^6 0.3^6 (1 - p)^{6-6} \\
&= 0.07
\end{aligned}
\tag{4.200}
$$

Thus, the system failure probability is 0.07.

Example 4.32

Reconsider the system in Fig. 4.19a, as discussed in Example 4.29. Now we assume that the system fails if at least two of the three ropes fail, and that the failure probability of each rope is identically p. Derive $\mathbb{P}[F_{sys}(k, n, p)]$ according to Eq. (4.199).

Solution

For the system in Fig. 4.19, $n = 3$ and $k = 2$. With Eq. (4.199),

$$\frac{d\mathbb{P}(F_{sys})}{dp} = 2C_3^2 p^{2-1}(1 - p)^{3-2} = 6p(1 - p) \tag{4.201}$$

with which

$$F_{sys} = \int 6p(1 - p)dp = 3p^2 - 2p^3 + c \tag{4.202}$$

where c is a constant to be determined. Note that a zero-value of p leads to F_{sys} being 0. Thus, $c = 0$ and $F_{sys} = 3p^2 - 2p^3$.

To verify this solution, we reconsider F_{sys} by noting that

$$F_{sys} = C_3^2 p^2 (1 - p)^1 + p^3 = 3p^2(1 - p) + p^3 = 3p^2 - 2p^3 \tag{4.203}$$

which is consistent with the previous result.

4.4 Reliability-Based Design and Safety Check

4.4.1 Modelling Structural Resistance

As has been discussed above, a structure's resistance is a random variable with uncertainties arising from the relevant parameters such as structural geometry, material properties, human error and others. Mathematically, one has

$$R = R(X_1, X_2, \ldots, X_n) \tag{4.204}$$

where R is structural resistance, and $\mathbf{X} = \{X_1, X_2, \ldots, X_n\}$ is a random vector reflecting the uncertain quantities considered in the probability-based analysis.

A resistance model is expected to reasonably represent the in situ conditions, which may differ significantly from the experimental tests in laboratories. In fact, analytical models used to predict the structural strength may not capture all the possible sources of uncertainty. For example, for a compact steel beam with an ultimate stress of F_{yn} and a plastic section modulus of Z_n, the nominal resistance (in terms of the maximum plastic moment that the cross-section can carry) is calculated as $R_n = F_{yn} Z_n$, which has not taken into account the uncertainties associated with material property, cross-sectional property and model bias.

To reduce the bias due to purely theoretical or experimental prediction models of structural strength, the following form is applicable for most cases,

$$R = P \cdot M \cdot F \cdot R_n \tag{4.205}$$

where R_n is the nominal resistance (deterministic), P is a random factor reflecting the difference between the actual in situ resistance and the predicted one (model bias), M is a factor reflecting the difference between the actual and nominal material properties, and F is to reflect the ratio of actual to nominal cross-sectional properties. With Eq. (4.205), it follows,

$$\mu_R = \mu_P \cdot \mu_M \cdot \mu_F \cdot R_n, \quad v_R = \sqrt{v_P^2 + v_M^2 + v_F^2} \tag{4.206}$$

where $\mu_\bullet$ and $v_\bullet$ denote the mean value and COV of $\bullet$ respectively. Furthermore, with Eq. (4.205), it is reasonable to use a lognormal distribution to model R. In some cases other distribution types such as normal or Weibull distribution may be also competitive. Table 4.6 lists some typical statistics for resistance of concrete and steel members.

Example 4.33

For a compact steel beam, the actual resistance R_{act} is calculated by

$$R_{act} = \frac{R_{act}}{Z F_y} \cdot Z F_y = \underbrace{\frac{R_{act}}{Z F_y}}_{P} \cdot \underbrace{(F Z_n) \cdot (M \cdot F_{yn})}_{MFR_n} \tag{4.207}$$

Table 4.6 Statistical information of resistances of typical concrete and steel members [9]

No.	Designation	μ_R/R_n	v_R
Concrete members			
1	Flexure, reinforced concrete, Grade 60	1.05	0.11
2	Flexure, reinforced concrete, Grade 40	1.14	0.14
3	Flexure, cast-in-place pre-tensioned beams	1.06	0.08
4	Flexure, cast-in-place post-tensioned beams	1.04	0.095
5	Short columns, compression failure, $f_c' = 21$ MPa	1.05	0.16
6	Short columns, tension failure, $f_c' = 21$ or 35 MPa	1.05	0.12
Steel members			
7	Tension members, limit state yielding	1.05	0.11
8	Tension members, limit state tensile strength	1.10	0.11
9	Compact beam, uniform moment	1.07	0.13
10	Beam-column	1.07	0.15
11	Plate girders, flexure	1.08	0.12
12	Axially loaded column	1.08	0.14

Now, given that $\mu_P = 1.02$, $v_P = 0.06$, $\mu_M = 1.05$, $v_M = 0.10$ and $\mu_F = 1.00$, $v_F = 0.05$ [9], calculate the mean value and COV of R_{act}.

Solution

According to Eq. (4.206),

$$\mu_{R_{act}} = \mu_P \cdot \mu_M \cdot \mu_F \cdot R_n = 1.02 \times 1.05 \times 1.00 R_n = 1.07 R_n$$

$$v_{R_{act}} = \sqrt{v_P^2 + v_M^2 + v_F^2} = \sqrt{0.06^2 + 0.10^2 + 0.05^2} = 0.13$$

(4.208)

4.4.2 Modelling Spatially Distributed Loads

Due to the uncertain characteristics, loads vary in space as well as in time. In terms of the former, external loads are often associated with spatial variation, and this variation will be discussed in this section. For the latter, the temporal variation of

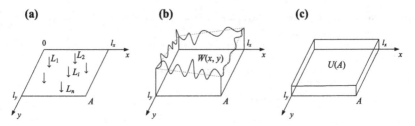

Fig. 4.21 Illustration of the spatially distributed loads

loads will be briefly discussed in Sect. 4.4.3 and further addressed in Chap. 5 in detail.

We consider a space with an area of A, as illustrated in Fig. 4.21a, b. The applied loads may be classified into two types: discrete and continuous. For both cases, we can use an *equivalent uniformly distributed load* (EUDL) to represent the spatially distributed load (c.f. Fig. 4.21c), which, by definition, will produce an identical structural action (e.g., moment, shear force, axial force and others) as the distributed load. A similar concept was used in [31] to represent the spatially distributed depth of water seepage for a lining structure.

Let $U(A)$ denote the EUDL, a function of area A, which is calculated as follows,

$$
U(A) = \begin{cases} \dfrac{1}{A} \sum_{i=1}^{n} L_i, & \text{discrete} \\[2ex] \dfrac{1}{A} \int\int W(x, y)\mathrm{d}x\mathrm{d}y, & \text{continuous} \end{cases} \tag{4.209}
$$

where n is the number of loads, L_i is the load magnitude of the ith load for the discrete case, and $W(x, y)$ is the load intensity at location (x, y) for the continuous case.

We first consider the discrete case. Assume that each L_i is identically distributed with a mean value of μ_L and a standard deviation of σ_L, $i = 1, 2, \ldots, n$. With this,

$$
\mathbb{E}[U(A)] = \frac{1}{A} \cdot n\mu_L \tag{4.210}
$$

which is independent of the area A. It is further assumed that $\frac{n}{A}$ is a constant, denoted by η.

Next, we discuss the variance of the EUDL. According to Eq. (4.209),

$$
\mathbb{V}[U(A)] = \frac{1}{A^2} \mathbb{V}\left(\sum_{i=1}^{n} L_i\right) = \frac{1}{A^2} \sum_{i=1}^{n}\sum_{j=1}^{n} \mathbb{C}(L_i, L_j) = \frac{1}{A^2}\left[n\sigma_L^2 + \sigma_L^2 \sum_{i=1}^{n}\sum_{j\neq i} \rho(L_i, L_j)\right] \tag{4.211}
$$

If $\rho(L_i, L_j) = 0$ for $i \neq j$, then

Fig. 4.22 Illustration of the nominal load based on the statistics of $U(A)$

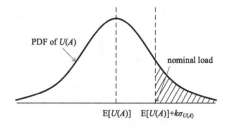

$$\mathbb{V}[U(A)] = \frac{n}{A^2}\sigma_L^2 = \frac{\eta}{A}\sigma_L^2 \qquad (4.212)$$

which implies that the variance of $U(A)$ decreases with A in proportion to $\frac{1}{A}$.

Remark

With Eqs. (4.210) and (4.212), if we define the nominal load, s_n, as a characteristic value of EUDL with a specific percentile (c.f. Eq. (4.12)), then s_n would be k times standard deviation above mean, as shown in Fig. 4.22. That is,

$$s_n = \eta\mu_L + \frac{k\sqrt{\eta}\sigma_L}{\sqrt{A}} \qquad (4.213)$$

where k depends on the percentile where the nominal load is defined. Thus, the nominal load would take a form of $\text{constant}_1 + \frac{\text{constant}_2}{\sqrt{A}}$. This expression, in fact, has been adopted in many design standards and codes of structures. For example, in the US standard ASCE/SEI 7-10, structural members with a value of $K_{LL}A_T$ being 37.16 m^2 or more are permitted to be designed with a reduced live load as follows,

$$L_n = L_0\left(0.25 + \frac{4.57}{\sqrt{K_{LL}A_T}}\right) \qquad (4.214)$$

where L_n is the (nominal) reduced live load supported by the member, L_0 is the unreduced design live load, K_{LL} is the live load element factor, and A_T is the tributary area (in m^2). Also, in the Australia/New Zealand standard AS/NZS 1170.1:2002, the live load reduction factor takes a form of $\psi_a = 0.3 + \frac{3}{\sqrt{A}}$ varying between 0.5 and 1, where A is the sum of areas supported by the member in m^2. While there are some additional limitations on the use of the reduced live load in both standards, the general formulation of the reduction factors is similar to that in Eq. (4.213).

We now discuss the statistics of EUDL for the continuous case. Let $W(x, y) = \overline{W} + \varepsilon(x, y)$, where $\overline{W}$ is the *spatial average* of $W(x, y)$ over the whole area A, and $\varepsilon(x, y)$ is the residual term at location (x, y), which has a mean value of zero. Assume that the mean value and standard deviation of $\overline{W}$ are $\mu_{\overline{W}}$ and $\sigma_{\overline{W}}$ respectively. Since $U(A) = \frac{1}{A}\int\int W(x, y)\mathrm{d}x\mathrm{d}y$, one has

$$\mathbb{E}[U(A)] = \mathbb{E}\left[\frac{1}{A}\int\int W(x, y)dxdy\right] = \frac{1}{A}\cdot\mathbb{E}\left[\int\int W(x, y)dxdy\right]$$
$$= \frac{1}{A}\cdot\int\int \mathbb{E}(\overline{W} + \varepsilon(x, y))dxdy = \frac{1}{A}\cdot\int\int \mathbb{E}(\overline{W})dxdy \qquad (4.215)$$
$$= \frac{1}{A}\cdot A\mu_{\overline{W}} = \mu_{\overline{W}}$$

which implies that the mean value of $U(A)$ is independent of the area A, consistent with the observation from the discrete case.

Next, consider the variance of the EUDL. By noting that $\overline{W}$ is independent of $\varepsilon(x, y)$, the variance of $U(A)$ is estimated by

$$V[U(A)] = V\left[\frac{1}{A}\int\int W(x, y)dxdy\right] = \frac{1}{A^2}\cdot V\left[\int\int W(x, y)dxdy\right]$$
$$= \frac{1}{A^2}\cdot V\left[\int\int \overline{W}dxdy\right] + \frac{1}{A^2}\cdot V\left[\int\int \varepsilon(x, y)dxdy\right] \qquad (4.216)$$

where

$$V\left[\int\int \overline{W}dxdy\right] = V\left[\overline{W}\int\int dxdy\right] = V(\overline{W}A) = A^2\sigma_{\overline{W}}^2 \qquad (4.217)$$

and

$$V\left[\int\int \varepsilon(x, y)dxdy\right] = \int\int\int\int \mathbb{C}[\varepsilon(x, y), \varepsilon(x', y')]dx'dy'dxdy \qquad (4.218)$$

Specifically, if $\mathbb{C}[\varepsilon(x, y), \varepsilon(x', y')] = \sigma_\varepsilon^2$ (i.e., $\varepsilon(x, y)$ is spatially fully correlated), then $V\left[\int\int \varepsilon(x, y)dxdy\right] = A^2\sigma_\varepsilon^2$.

Now, to estimate Eq. (4.218), we assume that $\mathbb{C}[\varepsilon(x, y), \varepsilon(x', y')]$ depends on the distance between (x, y) and (x', y') only, that is, $\varepsilon(x, y)$ is spatially homogeneous and isotropic. It is further assumed that $\mathbb{C}[\varepsilon(x, y), \varepsilon(x', y')]$ decreases with $r = \sqrt{(x - x')^2 + (y - y')^2}$ following a Gaussian function, that is,

$$\mathbb{C}[\varepsilon(x, y), \varepsilon(x', y')] = \sigma_\varepsilon^2 \exp\left(-\frac{r^2}{l_c^2}\right) \qquad (4.219)$$

where l_c is the effective length.

Substituting Eq. (4.219) into Eq. (4.218), one has

$$\iint \mathbb{C}[\varepsilon(x, y), \varepsilon(x', y')] dx' dy' = \sigma_\varepsilon^2 \int_0^{l_y} \int_0^{l_x} \exp\left(-\frac{r^2}{l_c^2}\right) dx' dy'$$

$$= \pi l_c^2 \sigma_\varepsilon^2 \int_0^{l_y} \int_0^{l_x} \frac{1}{\sqrt{2\pi} \cdot \frac{l_c}{\sqrt{2}}} \exp\left(-\frac{(x - x')^2}{l_c^2}\right) \cdot \frac{1}{\sqrt{2\pi} \cdot \frac{l_c}{\sqrt{2}}} \exp\left(-\frac{(y - y')^2}{l_c^2}\right) dx' dy'$$

$$= \pi l_c^2 \sigma_\varepsilon^2 \int_0^{l_x} \frac{1}{\sqrt{2\pi} \cdot \frac{l_c}{\sqrt{2}}} \exp\left(-\frac{(x' - x)^2}{2\left(\frac{l_c}{\sqrt{2}}\right)^2}\right) dx' \cdot \int_0^{l_y} \frac{1}{\sqrt{2\pi} \cdot \frac{l_c}{\sqrt{2}}} \exp\left(-\frac{(y' - y)^2}{2\left(\frac{l_c}{\sqrt{2}}\right)^2}\right) dy'$$

$$= \pi l_c^2 \sigma_\varepsilon^2 \left[\Phi\left(\frac{l_x - x}{\frac{l_c}{\sqrt{2}}}\right) - \Phi\left(\frac{-x}{\frac{l_c}{\sqrt{2}}}\right)\right] \cdot \left[\Phi\left(\frac{l_y - y}{\frac{l_c}{\sqrt{2}}}\right) - \Phi\left(\frac{-y}{\frac{l_c}{\sqrt{2}}}\right)\right]$$

(4.220)

and furthermore,

$$\iiiint \mathbb{C}[\varepsilon(x, y), \varepsilon(x', y')] dx' dy' dx dy$$

$$= \pi l_c^2 \sigma_\varepsilon^2 \int_0^{l_y} \int_0^{l_x} \left[\Phi\left(\frac{l_x - x}{\frac{l_c}{\sqrt{2}}}\right) - \Phi\left(\frac{-x}{\frac{l_c}{\sqrt{2}}}\right)\right] \cdot \left[\Phi\left(\frac{l_y - y}{\frac{l_c}{\sqrt{2}}}\right) - \Phi\left(\frac{-y}{\frac{l_c}{\sqrt{2}}}\right)\right] dx dy$$

(4.221)

Since the following two integrals hold,

$$\int_0^l \Phi\left(\frac{l - x}{\sigma}\right) dx = \frac{\sigma}{\sqrt{2\pi}}\left[\exp\left(-\frac{l^2}{2\sigma^2}\right) - 1\right] + l - l\Phi\left(-\frac{l}{\sigma}\right) \quad (4.222)$$

$$\int_0^l \Phi\left(\frac{-x}{\sigma}\right) dx = \frac{\sigma}{\sqrt{2\pi}}\left[1 - \exp\left(-\frac{l^2}{2\sigma^2}\right)\right] + l - l\Phi\left(\frac{l}{\sigma}\right) \quad (4.223)$$

it follows,

$$\iiiint \mathbb{C}[\varepsilon(x, y), \varepsilon(x', y')] dx' dy' dx dy'$$

$$= \pi l_c^2 \sigma_\varepsilon^2 \left\{\frac{l_c}{\sqrt{\pi}}\left[\exp\left(-\frac{l_x^2}{l_c^2}\right) - 1\right] - l_x + 2l_x \Phi\left(\frac{\sqrt{2}l_x}{l_c}\right)\right\} \cdot \left\{\frac{l_c}{\sqrt{\pi}}\left[\exp\left(-\frac{l_y^2}{l_c^2}\right) - 1\right] - l_y + 2l_y \Phi\left(\frac{\sqrt{2}l_y}{l_c}\right)\right\}$$

(4.224)

As a quick verification, if in Eq. (4.219), the effective length l_c is large enough (i.e., the correlation between $\varepsilon(x, y)$ and $\varepsilon(x', y')$ is close to 1 even for a relative large distance), then $\exp\left(-\frac{l_y^2}{l_c^2}\right) \approx 1 - \frac{l_y^2}{l_c^2}$ and $\Phi\left(\frac{\sqrt{2}l_y}{l_c}\right) \approx \frac{1}{2}$, and

$$\iiiint \mathbb{C}[\varepsilon(x, y), \varepsilon(x', y')] dx' dy' dx dy' = \pi l_c^2 \sigma_\varepsilon^2 \left(-\frac{l_c}{\sqrt{\pi}}\frac{l_x^2}{l_c^2}\right) \cdot \left(\frac{-l_c}{\sqrt{\pi}}\frac{l_y^2}{l_c^2}\right) = A^2 \sigma_\varepsilon^2 \quad (4.225)$$

which is consistent with previous result.

Furthermore, Eq. (4.224) finally yields

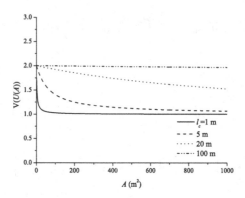

Fig. 4.23 Dependence of $\mathbb{V}[U(A)]$ on A

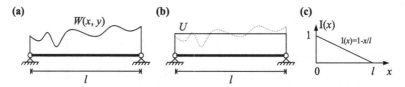

Fig. 4.24 A simply supported beam subjected to random load $W(x)$

$$\mathbb{V}[U(A)] = \sigma_{\overline{W}}^2$$
$$+ \frac{\pi l_c^2 \sigma_\varepsilon^2}{A^2} \left\{ \frac{l_c}{\sqrt{\pi}} \left[\exp\left(-\frac{l_x^2}{l_c^2}\right) - 1 \right] - l_x + 2l_x \Phi\left(\frac{\sqrt{2}l_x}{l_c}\right) \right\}$$
$$\cdot \left\{ \frac{l_c}{\sqrt{\pi}} \left[\exp\left(-\frac{l_y^2}{l_c^2}\right) - 1 \right] - l_y + 2l_y \Phi\left(\frac{\sqrt{2}l_y}{l_c}\right) \right\} \tag{4.226}$$

For illustration purpose, Fig. 4.23 shows the dependence of $\mathbb{V}[U(A)]$ on A for different values of l_c, assuming that $\sigma_{\overline{W}} = \sigma_\varepsilon = 1$ and $l_x = l_y = \sqrt{A}$. It can be seen that for all l_c, $\mathbb{V}[U(A)]$ varies within $[\sigma_{\overline{W}}^2, \sigma_{\overline{W}}^2 + \sigma_\varepsilon^2]$ as expected. A greater l_c leads to a slower decay rate of $\mathbb{V}[U(A)]$ with respect to A.

Example 4.34

Consider a simply supported beam as shown in Fig. 4.24a, which is subjected to a static continuous load $W(x)$ along the beam length.

(1) Derive the EUDL (c.f. Fig. 4.24b) in term of the shear force at the left-hand support.
(2) Let $W(x) = \overline{W} + \varepsilon(x)$, where $\overline{W}$ has a mean value of 1 kN/m and a COV of 0.5, and $\varepsilon(x)$ has a zero-mean, a standard deviation of 0.3 kN/m and a correlation structure of $\rho(\varepsilon(x_1), \varepsilon(x_2)) = \exp(-|x_1 - x_2|)$. Compute the mean value and standard deviation of EUDL. Assume $l = 2$ m.

Solution

(1) Let $I(x)$ be the influence function of the shear force at the left-hand support, $x \in [0, L]$, which is derived as

$$I(x) = 1 - \frac{x}{l}, \quad 0 \le x \le l \tag{4.227}$$

as shown in Fig. 4.24c.

Since the EUDL, denoted by U, produces the same shear force at the left end as $W(x)$, one has

$$\int_0^l W(x)I(x)dx = \int_0^l U \cdot I(x)dx = U \int_0^l I(x)dx \tag{4.228}$$

which further gives

$$U = \frac{2}{l} \int_0^l W(x)\left(1 - \frac{x}{l}\right)dx \tag{4.229}$$

(2) Similar to Eq. (4.215), the mean value of U is estimated by

$$\mathbb{E}(U) = \mathbb{E}\left[\frac{2}{l}\int_0^l W(x)\left(1 - \frac{x}{l}\right)dx\right] = \frac{2}{l}\int_0^l \mathbb{E}[W(x)]\left(1 - \frac{x}{l}\right)dx$$

$$= \frac{2\mathbb{E}(\overline{W})}{l}\int_0^l \left(1 - \frac{x}{l}\right)dx = \mathbb{E}(\overline{W}) = 1 \text{ kN/m} \tag{4.230}$$

Next, by referring to Eq. (4.216), the variance of U, $\mathbb{V}(U)$, is calculated as

$$\mathbb{V}(U) = \mathbb{V}\left[\frac{2}{l}\int_0^l W(x)\left(1 - \frac{x}{l}\right)dx\right] = \frac{4}{l^2}\mathbb{V}\left[\int_0^l [\overline{W} + \varepsilon(x)]\left(1 - \frac{x}{l}\right)dx\right]$$

$$= \frac{4}{l^2}\mathbb{V}\left[\int_0^l \overline{W}\left(1 - \frac{x}{l}\right)dx\right] + \frac{4}{l^2}\mathbb{V}\left[\int_0^l \varepsilon(x)\left(1 - \frac{x}{l}\right)dx\right]$$

$$= \frac{4}{l^2}\mathbb{V}\left[\overline{W}\int_0^l \left(1 - \frac{x}{l}\right)dx\right] + \frac{4}{l^2}\int_0^l\int_0^l \mathbb{C}\left[\varepsilon(x_1)\left(1 - \frac{x_1}{l}\right), \varepsilon(x_2)\left(1 - \frac{x_2}{l}\right)\right]dx_1 dx_2 \tag{4.231}$$

Note that

$$\mathbb{C}\left[\varepsilon(x_1)\left(1 - \frac{x_1}{l}\right), \varepsilon(x_2)\left(1 - \frac{x_2}{l}\right)\right]$$

$$= \rho\left[\varepsilon(x_1)\left(1 - \frac{x_1}{l}\right), \varepsilon(x_2)\left(1 - \frac{x_2}{l}\right)\right] \cdot \sqrt{\mathbb{V}\left[\varepsilon(x_1)\left(1 - \frac{x_1}{l}\right)\right] \cdot \mathbb{V}\left[\varepsilon(x_2)\left(1 - \frac{x_2}{l}\right)\right]}$$

$$= \rho[\varepsilon(x_1), \varepsilon(x_2)] \cdot \left(1 - \frac{x_1}{l}\right)\left(1 - \frac{x_2}{l}\right)\sqrt{\mathbb{V}[\varepsilon(x_1)] \cdot \mathbb{V}[\varepsilon(x_2)]}$$

$$= \exp(-|x_1 - x_2|) \cdot \left(1 - \frac{x_1}{l}\right)\left(1 - \frac{x_2}{l}\right)\mathbb{V}(\varepsilon(x)) \tag{4.232}$$

with which

$$\frac{4}{l^2} \int_0^l \int_0^l \mathbb{C}\left[\varepsilon(x_1)\left(1 - \frac{x_1}{l}\right), \varepsilon(x_2)\left(1 - \frac{x_2}{l}\right)\right] dx_1 dx_2$$

$$= \frac{4\mathbb{V}(\varepsilon(x))}{l^2} \int_0^l \int_0^l \exp(-|x_1 - x_2|) \cdot \left(1 - \frac{x_1}{l}\right)\left(1 - \frac{x_2}{l}\right) dx_1 dx_2$$

$$= \frac{4\mathbb{V}(\varepsilon(x))}{l^2} \int_0^l \left(1 - \frac{x_1}{l}\right)\left\{\int_0^{x_1} \exp(-(x_1 - x_2)) \cdot \left(1 - \frac{x_2}{l}\right) dx_2 + \int_{x_1}^l \exp(-(x_2 - x_1)) \cdot \left(1 - \frac{x_2}{l}\right) dx_2\right\} dx_1$$

$$= \frac{4\mathbb{V}(\varepsilon(x))}{l^2} \int_0^l \left(1 - \frac{x_1}{l}\right)\left[1 - \exp(-x_1) - \frac{x_1 + \exp(-x_1) - 1}{l} + \frac{\exp(x_1 - l) + l - x_1 - 1}{l}\right] dx_1$$

$$= \frac{4\mathbb{V}(\varepsilon(x))}{l^2} \cdot \frac{2l^3 - 3l^2 + 6 - 6(l+1)\exp(-l)}{3l^2}$$

$$\tag{4.233}$$

Thus, with $l = 2$ m, $\mathbb{V}(\overline{W}) = 0.25$ (kN/m)^2, and $\mathbb{V}(\varepsilon(x)) = 0.09$ (kN/m)^2, one has

$$\mathbb{V}(U) = \mathbb{V}(\overline{W}) + \frac{4\mathbb{V}(\varepsilon(x))}{l^2} \cdot \frac{2l^3 - 3l^2 + 6 - 6(l+1)\exp(-l)}{3l^2} = 0.25 + 0.09 \times 0.63 = 0.3067 \text{ (kN/m)}^2$$

$$\tag{4.234}$$

yielding a standard deviation of 0.554 kN/m of the EUDL.

4.4.3 Load Combination

As mentioned earlier, structures are in most cases subjected to loads that vary in time. We have mentioned in Sect. 3.4 that a stochastic process $X(t)$ can be used to describe the time-variation of a process with uncertainties (e.g., a time-varying load process herein). Further details regarding the modelling of a stochastic load process on the temporal scale can be found in Chap. 5.

When a structure is subjected to only one time-varying load, the structural reliability can be estimated by considering the maximum load within a reference period of interest. For example, if a structure is subjected to live load $L(t)$ plus a dead load D, then for a service period of $[0, T]$, the maximum live load $L_{\max} = \max_{t \in [0,T]} L(x)$, combined with the dead load, is to be incorporated in reliability assessment. However, in the presence of the simultaneous actions of n (two or more) time-varying loads $L_i(t)$, $i = 1, 2, \ldots, n$, i.e.,

$$L_{\text{total}}(t) = L_1(t) + L_2(t) + \cdots + L_n(t) \tag{4.235}$$

the combination of the n individual maximum loads would result in an unlikely event of the simultaneous occurrence of the n extreme loads and correspondingly a load combination that is too conservative. Alternatively, the Turkstra's rule [29] provides a reasonable estimate of the maximum load, max L_{total}, by considering the lifetime maximum value of one load and the instantaneous values of the remaining loads, although it may be unconservative when the probability of the joint occurrence of more than one maximum load is non-negligible. With the Turkstra's rule, max L_{total} is approximated by

$$\max L_{\text{total}} = \max_i \left[\max_{t \in [0,T]} L_i(t) + \sum_{j \neq i} L_j(t) \right] \tag{4.236}$$

As an example of applying Eq. (4.236), we consider a structure exposed to dead load D, live load L and wind load W within a service period of $[0, T]$. Let $L_{\max}$, $W_{\max}$ denote the maximum live load and wind load within $[0, T]$, and let L_{apt}, W_{apt} denote the arbitrary-point-in-time live load and wind load respectively. The calculation of structural reliability would consider one of the following two load combinations [9] instead of simply $D + L_{\max} + W_{\max}$: (1) $D + L_{\max} + W_{\text{apt}}$; (2) $D + W_{\max} + L_{\text{apt}}$.

Example 4.35

Consider the joint actions of two independent load processes $L_1(t)$ and $L_2(t)$ on a yearly scale. Within a reference period of 100 years, $L_1(i)$ (i.e., L_1 in the ith year, $i = 1, 2, \ldots, 100$) follows an Extreme Type I distribution with a mean value of 1 and a standard deviation of 0.4; $L_2(i)$ also follows an Extreme Type I distribution having a mean value of 0.5 and a standard deviation of 0.3. Assume that $L_1(i)$ is independent of $L_1(j)$, and $L_2(i)$ is independent of $L_2(j)$ if $i \neq j$. Define $L_{1+2,\max} = \max_{i=1}^{100}(L_1(i) + L_2(i))$, $L_{1,\max} = \max_{i=1}^{100} L_1(i)$, and $L_{2,\max} = \max_{i=1}^{100} L_2(i)$. Compare the PDFs of $L_{1+2,\max}$, $L_{1,\max} + L_2(i)$ and $L_{1,\max} + L_{2,\max}$. Comments on the results.

Solution

First, note that the PDF of $L_1(i) + L_2(i)$ can be calculated by the convolution of the PDFs of $L_1(i)$ and $L_2(i)$ (c.f. Example 1.45), that is

$$f_{L_1+L_2}(x) = \int_{-\infty}^{\infty} f_{L_1}(x - \tau) f_{L_2}(\tau) d\tau \tag{4.237}$$

which further gives the CDF of $L_1(i) + L_2(i)$, $F_{L_1+L_2}(x) = \int_{-\infty}^{x} f_{L_1+L_2}(\tau) d\tau$. Thus, the PDF of $L_{1+2,\max}$ is obtained as

$$f_{L_{1+2,\max}}(x) = 100 F_{L_1+L_2}^{99}(x) f_{L_1+L_2}(x) \tag{4.238}$$

Next, the PDFs of $L_{1,\max}$ and $L_{2,\max}$ are respectively

$$f_{L_{1,\max}}(x) = 100 F_{L_1}^{99}(x) f_{L_1}(x), \quad f_{L_{2,\max}}(x) = 100 F_{L_2}^{99}(x) f_{L_2}(x) \tag{4.239}$$

with which the PDFs of $L_{1,\max} + L_2(i)$ and $L_{1,\max} + L_{2,\max}$ can be derived similar to Eq. (4.237).

The PDFs of $L_{1+2,\max}$, $L_{1,\max} + L_2(i)$ and $L_{1,\max} + L_{2,\max}$ are plotted in Fig. 4.25, from which it can be seen that $L_{1,\max} + L_{2,\max}$ gives a too much conservative estimate of the maximum load, while the PDF of $L_{1,\max} + L_2(i)$ is closer to that of $L_{1+2,\max}$. In this example, however, $L_{1,\max} + L_2(i)$ leads to a slightly unconservative estimate of the maximum load.

Fig. 4.25 PDFs of
$L_{1+2,\max}$, $L_{1,\max} + L_2$ and
$L_{1,\max} + L_{2,\max}$

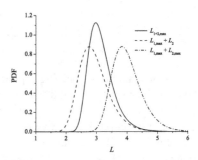

4.4.4 Reliability Calibration of Existing Design Criteria

We have before discussed the computation method for structural reliability (or the reliability index) of both structural members and systems. These methods can be used to calibrate the reliability level of existing design criteria. This topic will be discussed in this section. As an *inverse problem*, given a target failure probability (or reliability index), it is also important to develop a practical design criterion to yield, approximately, the desired reliability target. This will be later discussed in Sect. 4.4.5.

Most currently-enforced standards and codes for structural design have been based on the concept of probability-based design. In terms of safety check, the basic idea is to examine the probability of "Required strength $\leq$ Design strength", by noting that both the required strength and the design strength may be random variables with uncertainties arising from both structural properties and the external loads. With this regard, two methods have been widely used, namely load and resistance factor design (LRFD) and allowable stress design (ASD). When one considers the structural serviceability level, however, the requirement would be "Deformation due to service loads $\leq$ Deformation limit". It is noticed that the subsequent discussions on the safety check may not be applicable to serviceability check.

The LRFD gives a design criterion taking a form of

$$\varphi R_{\mathrm{n}} \geq \sum_{i}^{m} \gamma_i Q_{\mathrm{n}m} \tag{4.240}$$

where R_{n} and $Q_{\mathrm{n}i}$ are the nominal resistance and the ith nominal load effect, respectively, φ is the resistance factor, and γ_i is the load factor for Q_i. Typically, φ has a value of less than unity to reflect the variability and importance of member strength. Similarly, the determination of each load factor depends on the statistics and importance (e.g., principle or companion) of the load effect.

The ASD governs structural design with

$$\sigma \leq \frac{\sigma_{\mathrm{u}}}{\mathrm{FS}} \tag{4.241}$$

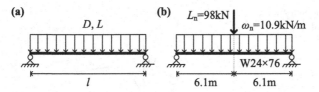

Fig. 4.26 Two illustrative steel beams

where σ is the stress induced by external loads, σ_u is the limit stress, and FS is the factor of safety. Engineering practice during the past decades have revealed the advantage of LRFD over ASD [10], because that the LRFD takes into consideration the limit state (with, potentially, the occurrence of plastic hinge) of a structural member and thus may account for the redundancy and load redistribution subsequent to the initial yielding, while ASD deals with the elastic design with a focus on the first yielding. Additionally, more factors are involved in LRFD compared with ASD (one factor of FS), yielding a more uniform reliability level for different load conditions.

In the presence of an existing design criterion, one may employ the reliability assessment methods (c.f. Sect. 4.2) to check the safety level of a structure. To demonstrate this point, we consider the reliability of a simply supported steel beam in flexure [14], as shown in Fig. 4.26a.

Considering the joint action of live load L and dead load D only, the LRFD gives

$$\varphi R_n = \gamma_D D_n + \gamma_L L_n \tag{4.242}$$

where φ is the resistance factor, $R_n = F_{yn} \cdot Z_n$ is the nominal resistance (cross-sectional plastic moment), Z_n is the nominal plastic modulus, F_{yn} is the nominal ultimate stress, D_n and L_n are the nominal dead load and live load respectively, γ_D and γ_L are the dead load and live load factors.

Now we examine the reliability level of a structure designed according to Eq. (4.242). The limit state function is as follows,

$$G = R - D - L = B \cdot R_n - D - L \tag{4.243}$$

where B is the bias factor of resistance. Clearly, totally three random variables are involved in Eq. (4.243), namely B, D and L. Substituting Eq. (4.242) into Eq. (4.243), one has

$$G = B \frac{\gamma_D D_n + \gamma_L L_n}{\varphi} - D - L$$

$$\Rightarrow G' = \frac{G}{D_n} = B \cdot \frac{\gamma_D + \gamma_L \frac{L_n}{D_n}}{\varphi} - \underbrace{\frac{D}{D_n}}_{X_1} - \underbrace{\frac{L}{L_n}}_{X_2} \cdot \frac{L_n}{D_n} \tag{4.244}$$

Fig. 4.27 Reliability index
as a function of $\frac{L_n}{D_n}$

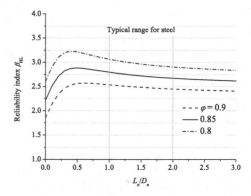

If the probability distribution of the three variables, B, X_1, X_2, are known, one can use the FORM (c.f. Sect. 4.2.2) to calculate the reliability index β_{HL}. It would be expected from Eq. (4.244) that the reliability index depends on the resistance/load factors (φ, γ_D and γ_L) as well as $\frac{L_n}{D_n}$.

For illustration purpose, we let $\gamma_D = 1.2$ and $\gamma_L = 1.6$, as adopted from ASCE-7 [1]. Furthermore, X_1 is assumed to have a normal distribution with a mean value of 1.05 and a COV of 0.1; X_2 is assumed to follow an Extreme Type I distribution with a mean value of 1 and a COV of 0.25. Also, B has a lognormal distribution with a mean value of 1.07 and a COV of 0.13 (c.f. Example 4.33). Based on this configuration, Fig. 4.27 presents the dependence of the reliability index on the variation of $\frac{L_n}{D_n}$ for different values of φ. It can be seen that a greater value of φ leads to a smaller reliability index as expected. For all cases, the reliability index increases slightly with $\frac{L_n}{D_n}$ at the early stage and then decreases as $\frac{L_n}{D_n}$ becomes larger. Specifically, within a range of $\frac{L_n}{D_n} \in [1, 2]$, which is typical for steel beams (for RC beams, however, the typical range would be [0.5, 1.5]) [9], the reliability index varies insignificantly, indicating that the LRFD design criterion in Eq. (4.242) yields a uniform reliability level for different load cases.

Example 4.36

Consider a simply supported steel girder subjected to the combination of live load L and dead load D, as in Fig. 4.26b [11]. The nominal dead load is constantly distributed with an intensity of 10.9 kN/m and the nominal live load is concentrated at the mid-span with a magnitude of 98 kN. The girder, with a cross section of W24 × 76, was designed for flexure according to $0.9F_{yn}Z_n = 1.2M_D + 1.6M_L$, where F_{yn} is the nominal yield stress, Z_n is the plastic section modulus, $M_\bullet$ is the moment due to load $\bullet$ ($\bullet = D, L$). The probabilistic information of the random variables can be found in Table 4.7. Calculate the probability of failure (collapse) within a reference period of 50 years using FORM.

Solution

First, the ultimate moment (after the occurrence of plastic hinge) at the mid-span due to D and M are calculated as $M_D = \frac{l^2}{8}\omega$, and $M_L = \frac{l}{4}L$ respectively, where l is the beam span (12.2 m in this example). The nominal plastic section modulus of W24 × 76 is 200

Table 4.7 Statistics of the variables in Fig. 4.26b

Variable	Mean	COV	Distribution type
Dead load ω	$1.05\omega_n$	0.1	Normal
50-year maximum live load L	$1.00L_n$	0.25	Extreme Type I
Yield stress F_y	$1.05F_{yn}$	0.11	Lognormal
Section modulus Z	$1.00Z_n$	0.06	Normal
Flexure model bias B	1.02	0.05	Normal

in^3, or equivalently 3.277×10^{-3} m^3. With this, the girder has a nominal yield stress of 245 MPa according to the design criterion. The limit state function is

$$G = BF_yZ - M_D - M_L = BF_yZ - \frac{l^2}{8}\omega - \frac{l}{4}L \qquad (4.245)$$

With the probability information of the variables in Table 4.7, applying FORM gives a reliability index of 2.49, which leads to a failure probability of 0.0063. This safety level corresponds to a risk category I failure that is not sudden and does not lead to wide-spread progression of damage, as presented in Table 4.1.

Example 4.37

Reconsider the steel beam as shown in Fig. 4.26b. Assume that the beam is designed with ASD, $0.66F_{yn}S_n = D_n + L_n$, where S_n is the nominal elastic section modulus. The beam has, as before, a cross section of W24 × 76 and its yielding stress is determined according to the design criterion. Suppose that both $\frac{S}{S_n}$ and $\frac{Z}{Z_n}$ have a common PDF. Calculate the probability of failure (collapse) within a reference period of 50 years using FORM.

Solution

Note that the nominal elastic section modulus of W24 × 76 is 176 in^3, or 2.884×10^{-3} m^3. With this, the girder has a nominal yield stress of 380 MPa according to the design criterion. The limit state is the same as that in Eq. (4.245). Using the FORM gives a reliability index of 4.2, leading to a failure probability of 1.33×10^{-5}.

Comparing Examples 4.36 and 4.37, it can be seen that, ensuring the reliability index above the minimum of approximately 2.5 with respect to a reference period of 50 years, on which the LRFD Specifications is based [12], the LRFD permits approximately a 55% reduction in beam yielding stress compared with ASD.

Example 4.38

Reconsider Example 4.36. If we would like to extend the life time of the steel beam by 50 years from the present and have some updated statistical information of the basic variables,

Fig. 4.28 Reliability index
of formwork shores as a
function of $\frac{L_n}{D_n}$

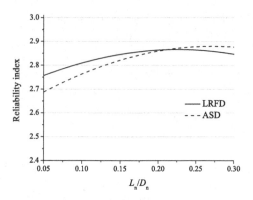

the safety check for the subsequent 50 years should be conducted under the governance
of a new criterion instead of that for a new structure [11]. Now, we assume that the yield
stress has an updated mean value of 318 MPa and a COV of 0.07. The dead load is revised
to have a mean value of $1.16D_n$ and a COV of 0.05. No deterioration is observed (so no
change to the cross section size) and no further information on the live load is available.
Estimate the new reliability index with the updated statistical information.

Solution

We substitute the updated information to Eq. (4.245), with which applying FORM gives
a reliability index of 3.4. This indicates that the reliability is improved with the updated
information. As a byproduct, a longer length of the beam would be permitted if setting the
minimum value of the reliability index as 2.5. Analytical solution shows that the length
could be increased by approximately 18%.

Example 4.39

We consider the reliability of formwork shores during concrete replacement. It is obtained
by synthesizing both in situ load survey and the data reported in the literature that the
dead load is a normal variable with a mean-to-nominal value of 1 and a COV of 0.25;
the live load has a mean-to-nominal ratio of 0.9 and a COV of 0.5. The resistance has a
mean-to-nominal ratio of 1.05 and a COV of 0.15. The LRFD design criterion is suggested
[38] as $0.9R_n = 1.7D_n + 1.5L_n$. If using ASD, however, the design would be governed
by $0.54R_n = D_n + L_n$.

Calculate the reliability index with different values of $\frac{L_n}{D_n} \in [0.05, 0.3]$.

Solution

For the two design criteria (LRFD and ASD), the dependence of the reliability index on
$\frac{L_n}{D_n}$ is presented in Fig. 4.28. It can be seen that both design methods yield a reliability
index of approximately 2.8 for the considered range of $\frac{L_n}{D_n}$. The LRFD method give a
more unified reliability level compared with ASD.

4.4.5 Reliability-Based Limit State Design

It is often the case in practice to develop a design criterion that is based on some different assumptions from existing standards and codes. For example, if we are concerned about structural reliability within a service period of 40 years instead of 50 years as specified in the standards, or if we have some probability information from in situ survey on the structural resistance and/or loads, a new design criterion would be essentially required. In other words, given the description of the basic random variables such as the environmental conditions, structural geometry, construction materials, performance objectives and the target reliability index β (or equivalently, the limit failure probability $\mathbb{P}_f$), the task is to design structural member sizes, strengths, connection properties and other details under the guidance of a new LRFD design criterion.

In the presence of m load effects $Q_1, Q_2, \ldots, Q_m$, we consider a design criterion of

$$\varphi R_n \geq \sum_i^m \gamma_i Q_{ni} \tag{4.246}$$

and correspondingly, a limit state function as follows,

$$
\begin{aligned}
G &= R - Q_1 - Q_2 - \cdots - Q_m = B R_n - Q_1 - Q_2 - \cdots - Q_m \\
&= B \frac{\sum_i^m \gamma_i Q_{ni}}{\varphi} - Q_1 - Q_2 - \cdots - Q_m \\
\Rightarrow G' &= \frac{G}{Q_{n1}} = B \cdot \frac{\sum_i^m \gamma_i \frac{Q_{ni}}{Q_{n,1}}}{\varphi} - \underbrace{\frac{Q_1}{Q_{n1}}}_{X_1} - \underbrace{\frac{Q_2}{Q_{n2}} \cdot \frac{Q_{n2}}{Q_{n1}}}_{X_2} - \cdots - \underbrace{\frac{Q_m}{Q_{nm}} \cdot \frac{Q_{nm}}{Q_{n1}}}_{X_m}
\end{aligned}
\tag{4.247}
$$

where R is resistance, and B is the bias factor of resistance, and φ, γ_i are the resistance/load factors. Technically, if the probability information on the random variable $B, X_1, X_2, \ldots, X_m$ is accessible, one can use FORM (c.f. Sect. 4.2.2) to find the reliability index β_{HL} and the checking point $(R^*, Q_1^*, \ldots, Q_m^*)$. With this, one has

$$\varphi = \frac{R^*}{R_n}, \quad \gamma_i = \frac{Q_i^*}{Q_{ni}}, \quad i = 1, 2, \ldots, m \tag{4.248}$$

Now, given a target reliability index of β_{target}, our objective is to determine the set of factors $\varphi, \gamma_1, \gamma_2, \ldots, \gamma_m$. This is, however, not straightforward because one cannot find the desired factors explicitly with a given β_{target}. Nonetheless, an iteration-based approach can be used to accomplish this task.

Defining two vectors $\mathbf{c} = (\varphi, \gamma_1, \gamma_2, \ldots, \gamma_m)$ and $\mathbf{Z}^* = (R^*, Q_1^*, \ldots, Q_m^*)$, the iteration-based procedure is as follows,

Fig. 4.29 Illustration of optimizing the resistance and load factors

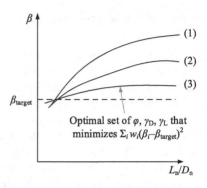

(1) Preliminary: set the initial realization of $\mathbf{c}$ (e.g., $\mathbf{c}^{(0)} = (1, 1, 1)$).
(2) For the jth iteration ($j = 1, 2, \ldots$), find the checking point $\mathbf{Z}^{*(j)}$ using FORM with $\mathbf{c}^{(j-1)}$, and then calculate $\mathbf{c}^{(j)}$ with $\mathbf{Z}^{*(j)}$ according to Eq. (4.248).
(3) The procedure terminates if such a j is found that $|\mathbf{Z}^{*(j)} - \mathbf{Z}^{*(j-1)}| \le \varepsilon$, where ε is a predefined threshold (e.g., 0.001).

Finally, $\mathbf{c}^{(j)}$ stabilizes at the desired vector $\mathbf{c}$.

Note that the resistance/load factors in Eq. (4.248) may vary for different load cases (e.g., different coefficients of X_i in Eq. (4.247)). Thus, finally, we need to determine such a set of $\varphi, \gamma_1, \gamma_2, \ldots, \gamma_m$ which minimizes $\sum_i w_i (\beta_i - \beta_{target})^2$ over the design space (for example, the range of $\frac{L_n}{D_n}$ of interest if considering the combination of live load L and dead load D only), where β_i and w_i are the reliability index and the weight of the ith load case, as illustrated in Fig. 4.29. The expression $\sum_i w_i (\beta_i - \beta_{target})^2$ herein is to measure the *closeness* between the overall reliability index and the target β_{target}.

It is finally noticed that in the presence of different optimal sets of resistance/load factors (they may account for different structure types or load combinations, for example), one can select one set of load factors first that maximizes the closeness between the reliability index over the considered design space and the target reliability. Subsequently, once the load factors are determined, the selection of resistance factor φ is done governed by the target reliability index.

Example 4.40

We consider a steel beam subjected to the joint action of live load L and dead load D, as shown in Fig. 4.26a. The limit state function is $G = R - D - L$. We assume that the statistics of R, D, L are the same as those in Fig. 4.27. The target reliability index is set as 2.5.

(1) Calculate $\varphi, \gamma_L, \gamma_D$ for different values of $\frac{L_n}{D_n} \in [0.5, 3]$.
(2) Determine a design criterion for steel beams with a typical range of $\frac{L_n}{D_n} \in [1, 2]$, assuming that each w_i is identical.

Fig. 4.30 The dependence of resistance and load factors on $\frac{L_n}{D_n}$

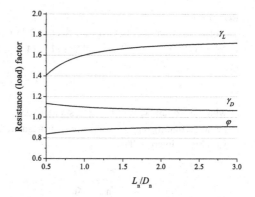

Fig. 4.31 An example of optimizing the resistance and load factors

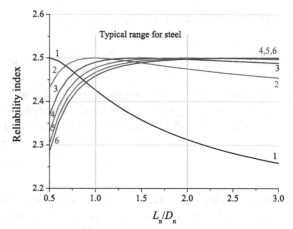

Solution

(1) For each value of $\frac{L_n}{D_n}$, we employ an iteration-based approach to find the resistance/load factors. The results are shown in Fig. 4.30. It can be seen that φ takes a value less than unity. γ_D decreases with $\frac{L_n}{D_n}$ while γ_L increases as the relative importance of live load is enhanced.

(2) To choose an appropriate set of φ, γ_L, γ_D, we consider six candidates from Fig. 4.30 corresponding to $\frac{L_n}{D_n} = 0.5, 1.0, 1.5, 2.0, 2.5, 3.0$ respectively (marked as cases 1–6). With each of the candidates, the dependence of reliability index on $\frac{L_n}{D_n}$ is presented in Fig. 4.31. It can be seen that for a typical range of $\frac{L_n}{D_n} \in [1, 2]$ for steel beam, case 3 gives the "best approximation", with which the design criterion is $0.9 R_n = 1.1 D_n + 1.7 L_n$.

Remark

Note that if we use FOSM (c.f. Sect. 4.2.1) to determine $\mathbf{c} = (\varphi, \gamma_1, \gamma_2, \ldots, \gamma_m)$ and $\mathbf{X}^* = (R^*, Q_1^*, \ldots, Q_m^*)$, similar to Eq. (4.59), it follows,

$$X_i^* = \mu_i - \beta \alpha_i \sigma_i = \mu_i (1 - \beta \alpha_i v_i) \tag{4.249}$$

where μ_i, σ_i and v_i are the mean, standard deviation and COV of X_i respectively. It should be noted that α_i may take a positive or negative value (c.f. Eq. (4.57)), depending on the variable of interest. Thus,

$$\varphi = \frac{R^*}{R_n} = \frac{\mu_R}{R_n}(1 - \beta \alpha_R v_R) \approx \frac{\mu_R}{R_n} \exp(-\beta \alpha_R v_R) \tag{4.250}$$

Similarly, the load factor γ_i for load Q_i is

$$\gamma_i = \frac{Q_i^*}{Q_{ni}} = \frac{\mu_{Q_i}}{Q_{ni}}(1 - \beta \alpha_{Q_i} v_{Q_i}) \tag{4.251}$$

Equations (4.250) and (4.251) provide a reasonable approximation for the resistance/load factors, and have been specified in some standards for structural design such as ASCE-7 [4], where it reads that the load factor γ for load effect Q can be approximated by

$$\gamma = \left(\frac{\mu_Q}{Q_n}\right) \cdot \left(1 + \alpha_Q \beta v_Q\right) \tag{4.252}$$

where μ_Q is the mean value of load, Q_n is the nominal load effect, v_Q is the COV of Q, β is the reliability index, and α_Q is a sensitivity coefficient that has a value of 0.8 when Q is a principle action and 0.4 when Q is a companion action.

Similarly, the resistance factor φ for resistance R can be approximately calculated according to

$$\varphi = \left(\frac{\mu_R}{R_n}\right) \cdot \exp\left(-\alpha_R \beta v_R\right) \tag{4.253}$$

where μ_R is the mean value of R, R_n is the nominal resistance, v_R is the COV of R, β is the reliability index, and α_R is a sensitivity coefficient that has a value of approximately 0.7.

For example, with a reliability target of $\beta_{\text{target}} = 3.0$, if a principle load Q_1 has a mean-to-nominal value of 1.0 and a COV of 0.25 (c.f. the live load in Fig. 4.27), then $\gamma_{Q_1} = 1.0 \cdot (1 + 0.8 \times 3 \times 0.25) = 1.6$. Similarly, if a companion action Q_2 has a mean-to-nominal value of 0.3 and a COV of 0.6, then the load factor $\gamma_{Q_2} = 0.3 \cdot (1 + 0.4 \times 3 \times 0.60) = 0.52$. In ASCE-7 [4], the load combination for dead load D, live load L and (companion) snow load S is recommended as $1.2D + 1.6L + 0.5S$.

4.4.6 Cost Analysis-Based Design Optimization

We have discussed before the structural design from a perspective of structural safety and serviceability. In practice, the economy is another important indicator that should be taken into account for structural design. For engineered structures such as buildings, bridges and other infrastructures, provided that the safety level is satisfied (beyond an expected baseline), the minimization of economic cost (or an optimized structural design with minimum cost) would also be desired [19].

The total cost of a structure, C_T, is expressed as follows,

$$C_T = C_I + C_F + C_M + C_S + \cdots \tag{4.254}$$

$$\mathbb{E}(C_T) = \mathbb{E}(C_I) + \mathbb{E}(C_F) + \mathbb{E}(C_M) + \mathbb{E}(C_S) + \cdots \tag{4.255}$$

where C_I is the initial cost (design, construction, materials and labor, fixed cost, etc), C_F is the cost due to structural failure, C_M is the maintenance and repair cost, and C_S is the salvage cost. Some additional items may also be included in the calculation of C_T depending on the considered case. In the following discussion, we consider C_I and C_F only since the two items contribute to the majority of C_T for most cases, with which $C_T = C_I + C_F$ and $\mathbb{E}(C_T) = \mathbb{E}(C_I) + \mathbb{E}(C_F)$.

We further model the initial cost as a linear function of $\frac{r_n}{s_n}$ [16], where r_n is the nominal resistance, and s_n is the nominal load effect, i.e.,

$$\mathbb{E}(C_I) = c_0 + k \left(\frac{r_n}{s_n} \right) \tag{4.256}$$

where both c_0 and k are two constants. Furthermore, the cost due to failure is described by

$$\mathbb{E}(C_F) = \mathbb{P}_f \cdot c_F \tag{4.257}$$

where $\mathbb{P}_f$ is the failure probability as a function of $\frac{r_n}{s_n}$, and $c_F = \gamma c_0$, with γ being a constant reflecting the relative magnitude between the initial cost and failure cost. Thus, the total cost is equivalently expressed as follows,

$$\mathbb{E}(C_T) = \mathbb{E}(C_I) + \mathbb{E}(C_F) = c_0 + k \left(\frac{r_n}{s_n} \right) + \gamma c_0 \mathbb{P}_f \tag{4.258}$$

which is a function of $\frac{r_n}{s_n}$, as shown in Fig. 4.32a.

Illustratively, we consider a structure in the presence of resistance R and a single load effect S. The variation of R is typically negligible compared with that of S in most cases (c.f. Table 4.6), with which it is assumed that R deterministically equals r_n. Next, we consider the case where S follows an Extreme Type I distribution (c.f. Sect. 2.2.8). Since r_n is, for a well-designed structure, at the upper tail of the distribution of S (c.f. Fig. 4.32b), it follows,

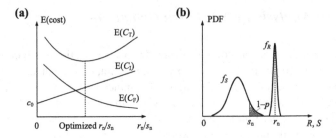

Fig. 4.32 Dependence of total cost on nominal resistance and nominal load

$$\mathbb{P}_f = 1 - \mathbb{P}(S \le R) \approx 1 - F_S(r_n) = 1 - \exp\left[-\exp\left(-\frac{r_n - \mu}{\sigma}\right)\right] \approx \exp\left(-\frac{r_n - \mu}{\sigma}\right) \tag{4.259}$$

where F_S is the CDF of S. Thus, Eq. (4.258) becomes

$$\mathbb{E}(C_T) = c_0 + k\left(\frac{r_n}{s_n}\right) + \gamma c_0 \cdot \exp\left(-\frac{r_n - \mu}{\sigma}\right) \tag{4.260}$$

Conceptually, we can minimizes $\mathbb{E}(C_T)$ in Eq. (4.260) to determined the value of r_n. Motivated by this, we have [19]

$$\frac{d\mathbb{E}(C_T)}{dr_n} = 0 \Rightarrow \frac{k}{s_n} - \frac{\gamma c_0}{\sigma} \exp\left(-\frac{r_n - \mu}{\sigma}\right) = 0$$
$$\Rightarrow r_{n,opt} = \mu - \sigma \ln\left(\frac{k\sigma}{\gamma c_0 s_n}\right) \tag{4.261}$$

Correspondingly, the optimized failure probability is

$$\mathbb{P}_{f,opt} = \frac{k\sigma}{s_n \gamma c_0} \tag{4.262}$$

Example 4.41

We assume that the load effect S has an Extreme Type I distribution with a mean value of 1 and a COV of 0.4, and the nominal load s_n is at the 99.9% percentile of S. Let $c_F = 10c_0$ and $k = 0.1c_0$. Using Eqs. (4.261) and (4.262), calculate the optimized nominal resistance and failure probability that minimizes the total cost.

Solution

The CDF of S is

$$F_S(x) = \exp\left(-\exp\left(-\frac{x - \mu}{\sigma}\right)\right) \tag{4.263}$$

where μ and σ are the location and scale parameters respectively. With Eqs. (2.234) and (2.237), these two parameters are determined by

$$\sigma = \frac{\sqrt{6}}{\pi} \times 0.4 = 0.3119, \quad \mu = 1 - 0.5772\sigma = 0.8200 \tag{4.264}$$

Furthermore, note that

$$F_S(s_n) = \exp\left(-\exp\left(-\frac{s_n - \mu}{\sigma}\right)\right) = 0.999 \Rightarrow s_n = \mu - \sigma \ln(-\ln 0.999) = 2.974 \tag{4.265}$$

Now, since $\frac{k}{c_0} = 0.1$ and $\gamma = 10$, with Eqs. (4.261) and (4.262),

$$r_{n,opt} = \mu - \sigma \ln\left(\frac{k\sigma}{\gamma c_0 s_n}\right) = 0.8200 - 0.3119 \ln\left(\frac{0.1 \times 0.3119}{10 \times 2.974}\right) = 2.96$$

$$\mathbb{P}_{f,opt} = \frac{k\sigma}{s_n \gamma c_0} = \frac{0.1 \times 0.3119}{2.974 \times 10} = 0.001 \tag{4.266}$$

The optimized failure probability corresponds to a reliability index of 3.1. It is also noticed that the optimized r_n is approximately identical to the nominal load s_n.

Example 4.42

In Eq. (4.260), if the load effect S is modelled by an Extreme Type II distribution, (1) derive the optimized r_n and the corresponding failure probability; (2) recalculate the optimized nominal resistance and failure probability in Example 4.41.

Solution

(1) Since S has an Extreme Type II distribution, its CDF $F_S(x)$ is as follows,

$$F_S(x) = \exp\left[-\left(\frac{x}{\varepsilon_2}\right)^{-k_2}\right] \tag{4.267}$$

where ε_2 is scale parameter and k_2 is shape parameter. As before, r_n typically lies at the upper tail of F_S, with which the failure probability is approximated by

$$\mathbb{P}_f = 1 - F_S(r_n) = 1 - \exp\left[-\left(\frac{r_n}{\varepsilon_2}\right)^{-k_2}\right] \approx \left(\frac{r_n}{\varepsilon_2}\right)^{-k_2} \tag{4.268}$$

Furthermore, the mean value of total cost is

$$\mathbb{E}(C_T) = \mathbb{E}(C_I) + \mathbb{E}(C_F) = c_0 + k\left(\frac{r_n}{s_n}\right) + \gamma c_0 \cdot \left(\frac{r_n}{\varepsilon_2}\right)^{-k_2} \tag{4.269}$$

To minimize $\mathbb{E}(C_T)$, we let $\frac{d\mathbb{E}(C_T)}{dr_n} = 0$, which gives

$$\frac{d\mathbb{E}(C_T)}{dr_n} = 0 \Rightarrow \frac{k}{s_n} - \gamma c_0 k_2 \varepsilon_2^{k_2} r_n^{-k_2-1} = 0$$

$$\Rightarrow r_{n,opt} = \left(\frac{\gamma c_0 k_2 \varepsilon_2^{k_2} s_n}{k}\right)^{\frac{1}{k_2+1}} \tag{4.270}$$

Correspondingly, the optimized failure probability is

$$\mathbb{P}_{f,opt} = \left(\frac{r_{n,opt}}{\varepsilon_2}\right)^{-k_2} = \varepsilon_2^{k_2}\left(\frac{\gamma c_0 k_2 \varepsilon_2^{k_2} s_n}{k}\right)^{\frac{-k_2}{k_2+1}} \tag{4.271}$$

(2) We first compute ε_2 and k_2. According to Eq. (2.242),

$$\frac{\sqrt{\Gamma\left(1-\frac{2}{k_2}\right) - \Gamma^2\left(1-\frac{1}{k_2}\right)}}{\Gamma\left(1-\frac{1}{k_2}\right)} = 0.4 \Rightarrow k_2 = 4.17 \tag{4.272}$$

Furthermore, $\varepsilon_2 = \dfrac{1}{\Gamma\left(1-\frac{1}{k_2}\right)} = 0.82$. Next, the nominal load effect s_n is determined by

$$F_S(s_n) = \exp\left[-\left(\frac{s_n}{\varepsilon_2}\right)^{-k_2}\right] = 0.999 \Rightarrow s_n = 4.320 \tag{4.273}$$

Since $\frac{k}{c_0} = 0.1$ and $\gamma = 10$, with Eqs. (4.270) and (4.271),

$$r_{n,opt} = \left(\frac{\gamma c_0 k_2 \varepsilon_2^{k_2} s_n}{k}\right)^{\frac{1}{k_2+1}} = 3.65, \quad \mathbb{P}_{f,opt} = \varepsilon_2^{k_2}\left(\frac{\gamma c_0 k_2 \varepsilon_2^{k_2} s_n}{k}\right)^{\frac{-k_2}{k_2+1}} = 0.002 \tag{4.274}$$

It is noticed that in Eqs. (4.261) and (4.262), the load effect S and the failure probability $\mathbb{P}_f$ depend on the service period of interest, say, t years. We in the following discuss the optimal annual failure probability. Assume that structural failure occurs independently on a yearly scale, and that the annual failure probability is small enough and constant over time, denoted by $\mathbb{P}_{f,annual}$. With this,

$$1 - \mathbb{P}_f = (1 - \mathbb{P}_{f,annual})^t \Rightarrow \mathbb{P}_{f,annual} = 1 - (1 - \mathbb{P}_f)^{\frac{1}{t}} \approx \frac{1}{t}\mathbb{P}_f \tag{4.275}$$

Let r be the discount rate, which is assumed to be time-invariant. The present value of C_I for a reference period of t years is calculated as follows,

$$\mathbb{E}(C_T) = \mathbb{E}(C_I) + \sum_{i=1}^{t} C_{F,i} \mathbb{P}_{f,annual} = \mathbb{E}(C_I) + \sum_{i=1}^{t} \frac{c_F}{(1+r)^i} \cdot \mathbb{P}_{f,annual}$$

$$\approx \mathbb{E}(C_I) + c_F \mathbb{P}_{f,annual} \int_0^t \frac{1}{(1+r)^\tau} d\tau \tag{4.276}$$

$$= \mathbb{E}(C_I) + \frac{c_F \mathbb{P}_{f,annual}}{\ln(1+r)} \left[1 - \exp\left(-t \ln(1+r)\right)\right]$$

As before, we let $\mathbb{E}(C_I) = c_0 + k\left(\frac{r_n}{s_n}\right)$, and $c_F = \gamma c_0$. If we model the load effect S with an Extreme Type I distribution as in Eq. (4.259), according to Eq. (4.275), $\mathbb{P}_{f,annual} = \frac{1}{t} \exp\left(-\frac{r_n - \mu}{\sigma}\right)$. Thus,

$$\mathbb{E}(C_T) = c_0 + k\left(\frac{r_n}{s_n}\right) + \frac{\gamma c_0 \exp\left(-\frac{r_n - \mu}{\sigma}\right)}{t \ln(1+r)} \left[1 - \exp\left(-t \ln(1+r)\right)\right] \tag{4.277}$$

To find such a nominal resistance r_n that minimizes $\mathbb{E}(C_T)$, we let $\frac{d\mathbb{E}(C_T)}{dr_n} = 0$, with which

$$\frac{k}{s_n} - \frac{\gamma c_0 \left[1 - \exp\left(-t \ln(1+r)\right)\right]}{\sigma t \ln(1+r)} \exp\left(-\frac{r_n - \mu}{\sigma}\right) = 0 \tag{4.278}$$

yielding the optimized r_n as follows,

$$r_{n,opt} = \mu - \sigma \ln\left[\frac{kt \ln(1+r)\sigma}{s_n \gamma c_0 \left[1 - \exp\left(-t \ln(1+r)\right)\right]}\right] \tag{4.279}$$

Correspondingly, the optimized annual failure probability is

$$\mathbb{P}_{f,annual,opt} = \frac{1}{t} \exp\left(-\frac{r_n - \mu}{\sigma}\right) = \frac{k\sigma \ln(1+r)}{s_n \gamma c_0 \left[1 - \exp\left(-t \ln(1+r)\right)\right]} \tag{4.280}$$

Specifically, in Eq. (4.280), if $r \to 0$,

$$\lim_{r \to 0} \mathbb{P}_{f,annual,opt} = \lim_{r \to 0} \frac{kr\sigma}{s_n \gamma c_0 \cdot rt} = \frac{k\sigma}{s_n \gamma c_0 t} \tag{4.281}$$

With this, $\mathbb{P}_f = t \mathbb{P}_{f,annual} = \frac{k\sigma}{s_n \gamma c_0}$, which is consistent with Eq. (4.262).

Furthermore, if in Eq. (4.280) $t \to \infty$, then $\exp\left(-t \ln(1+r)\right) \to 0$, and

$$\lim_{t \to \infty} \mathbb{P}_{f,annual,opt} = \frac{k\sigma \ln(1+r)}{s_n \gamma c_0} \tag{4.282}$$

which is independent of the time period of interest (i.e., t years).

4.5 Reliability Assessment with Imprecise Information

When the distribution function of a random variable is unknown but only some partial information such as the mean value and variance, the estimate of structural failure probability incorporating the variable may be fairly sensitive to the choice of the distribution type. This has been discussed in Example 4.5. Unfortunately, it is often the case that the identification of a variable's distribution function is difficult or even impossible due to limited information/data. Rather, only incomplete information such as the first- and the second- order moments (mean and variance) of the variable can be reasonably estimated. In such a case, the incompletely-informed random variable can be quantified by a *family* of candidate probability distributions rather than a single known distribution function. A practical way to represent an imprecise probability is to use a probability bounding approach by considering the lower and upper bounds of the imprecise probability functions. Under this context, approaches of interval estimate of reliability have been used to deal with reliability problems with imprecise probabilistic information [3, 5, 13, 21, 27, 32].

4.5.1 Probability-Box

Recall Eq. (4.117), which, repeated in the following, gives the general formulation of structural reliability problem.

$$\mathbb{P}_f = \mathbb{P}(G(\mathbf{X}) < 0) = \int \cdots \int_{G(\mathbf{X}) < 0} f_{\mathbf{X}}(\mathbf{x}) d\mathbf{x} \tag{4.117}$$

One can use the Monte Carlo simulation method (c.f. Sect. 4.2.3) to estimate the failure probability with $\mathbb{P}_f \approx \widetilde{P} = \frac{1}{m} \sum_{i=1}^{m} \mathbb{I}[G(\mathbf{x}_i) < 0]$ (c.f. Eq. (4.118)), where $\mathbf{x}_i$ can be generated with its probability distribution. However, if the distribution function of $\mathbf{X}$ (or some elements of $\mathbf{X}$) cannot be determined uniquely, a family of all possible distribution functions of the imprecise variable(s) should be considered instead.

For an incompletely-informed random variable X, a *probability-box* (or *p-box* for short) describes a family of distribution functions by specifying the lower and upper bounds of the CDF, i.e.,

$$\underline{F}_X(x) \leq F_X(x) \leq \overline{F}_X(x), \quad x \in (-\infty, \infty) \tag{4.283}$$

where $F_X(x)$ is the (unknown) CDF of X, $\underline{F}_X$ and $\overline{F}_X$ are the lower and upper bounds of F_X respectively.

If only the mean and standard deviation of X are known, denoted by μ_X and σ_X respectively, and the distribution type is unknown, the Chebyshev's inequality gives a lower and an upper bound of F_X as follows [23]

$$\underline{F}_X(x) = \begin{cases} 0, & x \le \mu_X + \sigma_X \\ 1 - \dfrac{\sigma_X^2}{(x - \mu_X)^2}, & x \ge \mu_X + \sigma_X \end{cases} \tag{4.284a}$$

$$\overline{F}_X(x) = \begin{cases} \dfrac{\sigma_X^2}{(x - \mu_X)^2}, & x \le \mu_X - \sigma_X \\ 1, & x \ge \mu_X - \sigma_X \end{cases} \tag{4.284b}$$

Furthermore, if it is additionally known that X varies with $[\underline{x}, \overline{x}]$, a tighter bounds of F_X is as follows [13]

$$\underline{F}_X(x) = \begin{cases} 0, & x \le \mu_X + \sigma_X^2/(\mu_X - \overline{x}) \\ 1 - [b(1 + a) - c - b^2]/a, & \mu_X + \sigma_X^2/(\mu_X - \overline{x}) < x < \mu_X + \sigma_X^2/(\mu_X - \underline{x}) \\ 1/[1 + \sigma_X^2/(x - \mu_X)^2], & \mu_X + \sigma_X^2/(\mu_X - \underline{x}) \le x < \overline{x} \\ 1, & x \ge \overline{x} \end{cases}$$

$$\tag{4.285a}$$

$$\overline{F}_X(x) = \begin{cases} 0, & x \le \underline{x} \\ 1/[1 + (x - \mu_X)^2/\sigma_X^2], & \underline{x} \le x < \mu_X + \sigma_X^2/(\mu_X - \overline{x}) \\ 1 - (b^2 - ab + c)/(1 - a), & \mu_X + \sigma_X^2/(\mu_X - \overline{x}) < x < \mu_X + \sigma_X^2/(\mu_X - \underline{x}) \\ 1, & x \ge \mu_X + \sigma_X^2/(\mu_X - \underline{x}) \end{cases}$$

$$\tag{4.285b}$$

where $a = (x - \underline{x})/(\overline{x} - \underline{x})$, $b = (\mu_X - \underline{x})/(\overline{x} - \underline{x})$, and $c = \sigma_X^2/(\overline{x} - \underline{x})^2$.

Another approach to determine the lower and upper bounds of F_X is to use a linear programming-based approach [32], as will be discussed in Sect. 4.5.3.1.

4.5.2 Interval Monte Carlo Simulation

If the random vector $\mathbf{X}$ (or some of its elements) in Eq. (4.117) is incompletely-informed, the probability of failure cannot be uniquely determined but is expected to vary within an interval $[\underline{\mathbb{P}}_f, \overline{\mathbb{P}}_f]$. The *interval Monte Carlo simulation* method can be used to find the two bounds $\underline{\mathbb{P}}_f$ and $\overline{\mathbb{P}}_f$ [37]. The basis is that

$$\underline{\mathbb{P}}_f = \min \left\{ \frac{1}{m} \sum_{i=1}^{m} \mathbb{I}[G(\mathbf{x}_i) < 0], \text{ for all possible } F_X \right\} \tag{4.286}$$

and

$$\overline{\mathbb{P}}_f = \max \left\{ \frac{1}{m} \sum_{i=1}^{m} \mathbb{I}[G(\mathbf{x}_i) < 0], \text{ for all possible } F_X \right\} \tag{4.287}$$

To implement the interval Monte Carlo simulation, in the presence of the CDF envelope of an imprecise variable X (c.f. Eq. (4.283)), we generate in each simulation run two samples from the lower and upper bounds of F_X respectively according to

$$\underline{x}_i = \overline{F}_X^{-1}(u_i), \quad \overline{x}_i = \underline{F}_X^{-1}(u_i), \quad i = 1, 2, \ldots, m \qquad (4.288)$$

where u_i is a sample of uniform distribution over $[0, 1]$ associated with the ith replication. As such, the interval $[\underline{x}_i, \overline{x}_i]$ covers all possible simulated numbers from the family of distributions contained in the p-box for a given u_i.

Now, we let $\min G(\mathbf{x}_i)$ and $\max G(\mathbf{x}_i)$ denote the minima and maxima of the limit state function $G(\mathbf{x})$ respectively, with which

$$\mathbb{I}[\max G(\mathbf{x}_i) < 0] \leq \mathbb{I}[G(\mathbf{x}_i) < 0] \leq \mathbb{I}[\min G(\mathbf{x}_i) < 0] \qquad (4.289)$$

which further gives

$$\frac{1}{m}\sum_{i=1}^{m}\mathbb{I}[\max G(\mathbf{x}_i) < 0] \leq \frac{1}{m}\sum_{i=1}^{m}\mathbb{I}[G(\mathbf{x}_i) < 0] \leq \frac{1}{m}\sum_{i=1}^{m}\mathbb{I}[\min G(\mathbf{x}_i) < 0]$$
$$(4.290)$$

Thus, the lower and upper bounds of $\mathbb{P}_f$, $\underline{\mathbb{P}}_f$ and $\overline{\mathbb{P}}_f$, are obtained as follows,

$$\underline{\mathbb{P}}_f = \frac{1}{m}\sum_{i=1}^{m}\mathbb{I}[\max G(\mathbf{x}_i) < 0], \quad \overline{\mathbb{P}}_f = \frac{1}{m}\sum_{i=1}^{m}\mathbb{I}[\min G(\mathbf{x}_i) < 0] \qquad (4.291)$$

4.5.3 Linear Programming-Based Approach

In this section, we discuss a linear programming-based approach to determine the CDF envelope of an imprecise random variables, and to estimate the lower and upper bounds of structural reliability (or failure probability) in the presence of incompletely-informed random variables [32].

4.5.3.1 Reliability Problems with One Imprecise Random Variable

Consider a reliability analysis problem involving random variables $[Z, \mathbf{Y}]$, in which Z is a random variable with an imprecise distribution function, and $\mathbf{Y} = [Y_1, Y_2, \ldots]$ is the remaining random vector with a known joint distribution function. It is assumed that Z and $\mathbf{Y}$ are statistically independent. The failure probability is given by

$$\mathbb{P}_f = \int_{G(\mathbf{Y}, Z)<0} f_Z(z) f_{\mathbf{Y}}(\mathbf{y}) dz d\mathbf{y} = \int f_Z(z) \xi_Z(z) dz \qquad (4.292)$$

in which $f_Z(z)$ and $f_{\mathbf{Y}}(\mathbf{y})$ are the (joint) PDFs of $\mathbf{Y}$ and Z, respectively; $\xi_Z(z)$ is the conditional failure probability on $Z = z$, i.e.,

$$\xi_Z(z) = \mathbb{P}(G(\mathbf{Y}, Z = z) < 0) = \int_{G(\mathbf{Y}, Z=z)<0} f_{\mathbf{Y}}(\mathbf{y}) d\mathbf{y} \tag{4.293}$$

The conditional failure probability $\xi_Z(z)$ for a given value of $Z = z$ can be obtained analytically through the integration in Eq. (4.293) or numerically using simulation-based methods.

To facilitate further derivation, Z is normalized into $[0, 1]$ by introducing a reduced random variable $X = \frac{Z - Z_{\min}}{Z_{\max} - Z_{\min}}$, where $Z_{\max}$ and $Z_{\min}$ are the maxima and minima of Z, respectively. With this, Eq. (4.292) becomes

$$\mathbb{P}_f = \int_0^1 f_X(x)\xi(x) dx \tag{4.294}$$

where $f_X(x)$ is the PDF of X, and $\xi(x) = \xi_Z((Z_{\max} - Z_{\min})x + Z_{\min})$.

We first consider the case where the mean value and standard deviation of Z are known, denoted by μ_Z and σ_Z respectively. The maxima and minima of Z can be reasonably approximated by $\mu_Z \pm k\sigma_Z$, in which k is sufficiently large (e.g., $k = 5$). With this, the mean and standard deviation of the reduced variable X are

$$\mu_X = \frac{\mu_Z - \min Z}{\max Z - \min Z}, \quad \sigma_X = \frac{\sigma_Z}{\max Z - \min Z} \tag{4.295}$$

Note that in Eq. (4.294), the distribution type of X is unknown, and thus the values of $f_X(x)$ for each x cannot be uniquely determined. We now discretize the domain of X (i.e., $[0, 1]$) into n identical sections, $[x_0 = 0, x_1], [x_1, x_2], \ldots, [x_{n-1}, x_n = 1]$, where n is sufficiently large such that $\left| f_X(x) - f_X\left(\frac{x_{i-1} + x_i}{2}\right) \right|$ is negligible for $\forall i = 1, 2, \ldots, n$ and $\forall x \in [x_{i-1}, x_i]$. The sequence $f_X\left(\frac{x_{i-1} + x_i}{2}\right), \forall i = 1, 2, \ldots, n$ is denoted by $\{f_1, f_2, \ldots, f_n\}$ for the purpose of simplicity. With this, Eq. (4.294) can be approximated by

$$\mathbb{P}_f = \int_0^1 \xi(x) f_X(x) dx = \lim_{n \to \infty} \sum_{i=1}^n \xi\left(\frac{i - 0.5}{n}\right) \frac{1}{n} \cdot f_i \tag{4.296}$$

Note that the mean value and variance of X, as well as the basic characteristics of a distribution function simultaneously give

$$\begin{cases} \sum_{i=1}^n f_i \cdot \frac{1}{n} = 1 \\ \sum_{i=1}^n f_i \cdot \frac{1}{n} \cdot \frac{i}{n} = \mu_X \\ \sum_{i=1}^n f_i \cdot \frac{1}{n} \left(\frac{i}{n}\right)^2 = \mu_X^2 + \sigma_X^2 \\ 0 \le f_i \le n, \forall i = 1, 2, \ldots, n \end{cases} \tag{4.297}$$

Thus, Eqs. (4.296) and (4.297) indicate that the bound estimate of $\mathbb{P}_f$ can be converted into a classic linear programming problem, i.e., Eq. (4.296) is the objec-

Fig. 4.33 Example of a
simple linear
programming-based problem

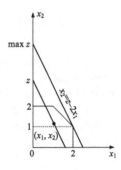

tive function to be optimized, $\mathbf{f} = \{f_1, f_2, \ldots, f_n\}$ is the vector of variables to be determined, and Eq. (4.297) represents the constraints.

Furthermore, one can also apply Eqs. (4.296) and (4.297) to determine the CDF envelope for X (as mentioned earlier) by letting, for an arbitrary value of τ

$$\xi(x) = \mathbb{I}(\tau \geq x) = \begin{cases} 1, & x \leq \tau \\ 0, & \text{otherwise.} \end{cases} \tag{4.298}$$

with which

$$\int_0^1 \xi(x) f_X(x) \mathrm{d}x = \int_0^\tau f_X(x) \mathrm{d}x = F_X(\tau) \tag{4.299}$$

Thus, by solving the linear programming problem defined in Eqs. (4.296) and (4.297), the best-possible bounds for $F_X(\tau)$ can be obtained.

If it is known, additionally, that X varies within the range $[\underline{x}, \overline{x}]$, where $0 \leq \underline{x} \leq \overline{x} \leq 1$, we can introduce a new variable $X' = \frac{X - \underline{x}}{\overline{x} - \underline{x}}$ to enable the applicability of Eq. (4.296). Furthermore, if the mean value of X is an interval estimate of $[\underline{\mu}_X, \overline{\mu}_X]$ rather than a point estimate, then the second constraint equation in Eq. (4.296), $\sum_{i=1}^n f_i \cdot \frac{1}{n} \cdot \frac{i}{n} = \mu_X$, is modified as

$$\begin{cases} \sum_{i=1}^n f_i \cdot \frac{-1}{n} \cdot \frac{i}{n} \leq -\underline{\mu}_X \\ \sum_{i=1}^n f_i \cdot \frac{1}{n} \cdot \frac{i}{n} \leq \overline{\mu}_X \end{cases} \tag{4.300}$$

A similar modification can be made to the third constraint equation in Eq. (4.296) if the standard deviation of X is known to have a predefined range.

Remark

A linear programming problem takes a standard form of

$$\min \mathbf{c}^T \mathbf{x}, \quad \text{subjected to} \quad \mathbf{A} \mathbf{x} \preceq \mathbf{b} \quad \text{and} \quad \mathbf{x} \succeq \mathbf{0} \tag{4.301}$$

where $\mathbf{x}$ is a variable vector to be determined, $\mathbf{b}$ and $\mathbf{c}$ are two known vectors, $\mathbf{A}$ is a coefficient matrix. The operator $\preceq$ (or $\succeq$) in Eq. (4.301) means that each element in the left-hand vector is no more (or less) than the corresponding element in the right-hand vector. The constraints $\mathbf{Ax} \preceq \mathbf{b}$ and $\mathbf{x} \succeq \mathbf{0}$ simultaneously define a convex poly-tope in which the objective function, $\mathbf{c}^T\mathbf{x}$, is to be optimized [2, 6].

An extended form of Eq. (4.301) takes the form of

$$\min \ \mathbf{c}^T\mathbf{x}, \ \text{subjected to} \ \mathbf{A}_{eq}\mathbf{x} = \mathbf{b}_{eq}, \ \mathbf{A}_{in}\mathbf{x} \preceq \mathbf{b}_{in} \ \text{and} \ \mathbf{x} \succeq \mathbf{0} \quad (4.302)$$

where $\mathbf{b}_{eq}$ and $\mathbf{b}_{in}$ are two known vectors, and $\mathbf{A}_{eq}$ and $\mathbf{A}_{in}$ are two coefficient matrices. By noting the fact that "$\mathbf{A}_{eq}\mathbf{x} = \mathbf{b}_{eq}$" is equivalent to "$\mathbf{A}_{eq}\mathbf{x} \preceq \mathbf{b}_{eq} \bigcap -\mathbf{A}_{eq}\mathbf{x} \preceq -\mathbf{b}_{eq}$", Eq. (4.302) can be transferred into Eq. (4.301) assigning $\mathbf{A} = \begin{bmatrix} \mathbf{A}_{eq} \\ -\mathbf{A}_{eq} \\ \mathbf{A}_{in} \end{bmatrix}$ and $\mathbf{b} = \begin{bmatrix} \mathbf{b}_{eq} \\ -\mathbf{b}_{eq} \\ \mathbf{b}_{in} \end{bmatrix}$ in Eq. (4.301).

For illustration purpose, we consider a simple example. With two variables x_1 and x_2 satisfying $0 \le x_1, x_2 \le 2$ and $x_1 + x_2 \le 3$, the objective is to maximize $z = 2x_1 + x_2$. The desired (x_1, x_2) can be found by letting, in Eq. (4.301),

$$\mathbf{x} = \begin{bmatrix} x_1 \\ x_2 \end{bmatrix}, \quad \mathbf{A} = \begin{bmatrix} 1 & 0 \\ -1 & 0 \\ 0 & 1 \\ 0 & -1 \\ 1 & 1 \end{bmatrix}, \quad \mathbf{b} = \begin{bmatrix} 2 \\ 0 \\ 2 \\ 0 \\ 3 \end{bmatrix}, \quad \mathbf{c} = \begin{bmatrix} -2 \\ -1 \end{bmatrix} \quad (4.303)$$

with which $(x_1, x_2) = (2, 1)$ and $\max z = 5$. Figure 4.33 presents the visualization of maximizing z subjected to the constraints on (x_1, x_2).

Example 4.43

We consider a rigid-plastic portal frame as shown in Fig. 2.1a, which is subjected to a horizontal load H and a vertical load V. The structure may fail due to one of the following three limit states,

$$G_1(\mathbf{X}) = M_1 + 2M_3 + 2M_4 - H - V$$
$$G_2(\mathbf{X}) = M_2 + 2M_3 + M_4 - V \quad (4.304)$$
$$G_3(\mathbf{X}) = M_1 + M_2 + M_4 - H$$

in which M_1 through M_4 are the plastic moment capacities at the joints. The random variables considered in this example, $\{M_1, M_2, M_3, M_4, V, H\}$, are assumed to be statistically independent with each other, and their probabilistic information is presented in Table 4.8. Limited statistical information on H is available so that its distribution type is unknown. With this, consider the following three representative cases: (1) H has a mean value of 1.9 and a standard deviation of 0.45; (2) H has a mean value of 1.9 and a standard deviation of 0.45, and is strictly defined within [1.0, 3.0]; (3) H has a mean value within [1.87, 1.93] and a standard deviation of 0.45.

Table 4.8 Statistics of the random variables in Example 4.43

Variable	Distribution type	Mean	Standard deviation
M_1, M_2, M_3, M_4	Normal	1.0	0.3
V	Normal	1.5	0.3

(1) For case (1), compute and compare the CDF envelopes with Eq. (4.284) and linear-programming-based approach.

(2) For case (2), compute and compare the CDF envelopes with Eq. (4.285) and linear-programming-based approach.

(3) For the three cases, find the lower and upper bounds of failure probability.

Solution

(1) The p-boxes for case (1) obtained from linear programming-based method (c.f. Eqs. (4.296) and (4.299)) and the Chebyshev's inequality (Eq. (4.284)) are presented in Fig. 4.34a. It can be seen that the p-box from the Chebyshev's inequality is significantly wider than that from linear programming.

(2) The CDF bounds for case (2) are given in Fig. 4.34b, obtained from linear programming and Eq. (4.285) respectively, showing that both methods give a consistent estimate of the p-box.

(3) We first obtain the conditional failure probability in Eq. (4.293) by fitting as follows,

$$\xi_H(h) = \Phi(0.0007h^6 - 0.0067h^5 + 0.0036h^4 + 0.133h^3 - 0.2856h^2 + 1.2389h - 3.7204) \tag{4.305}$$

The lower and upper bounds of failure probability are presented in Table 4.9 for the three cases. For comparison purpose, the second column gives the failure probability bounds computed by the interval Monte Carlo simulation, where the p-box of H is first obtained with Eqs. (4.284) and (4.285) and then the interval Monte Carlo method is used. This method is referred to as IMC1. The third column of Table 4.9 was also computed using the interval Monte Carlo method with the probability-box for H being constructed using linear programming-based method first. This method is referred to as IMC2. The fourth column of Table 4.9 presents the results computed by the linear programming method directly. This method is referred to as *direct optimization*. It can be seen that the failure probability intervals obtained with the direct optimization method are significantly narrower than those based on interval Monte Carlo method with p-boxes.

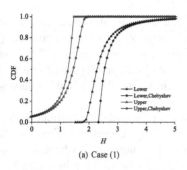
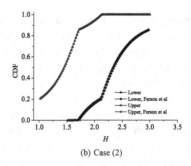

(a) Case (1)   (b) Case (2)

Fig. 4.34 CDF bounds of H for cases (1) and (2)

Table 4.9 Lower and upper bounds of failure probability for Example 4.43

Case no.	IMC1	IMC2	Direct optimization
(1)	[0.0090, 0.3678]	[0.0184, 0.2593]	[0.0597, 0.1057]
(2)	[0.0223, 0.2490]	[0.0223, 0.2490]	[0.0831, 0.1106]
(3)	/	[0.0097, 0.4233]	[0.0523, 0.1918]

4.5.3.2 Problems with Multiple Imprecise Random Variables

In this section, we discusses the reliability problem involving multiple imprecise random variables. Suppose that the reliability problem involves a mixture of imprecise random variables and conventional random variables, $[\mathbf{Z}, \mathbf{Y}]$, in which $\mathbf{Z} = \{Z_1, Z_2, \ldots, Z_k\}$ is the vector of k imprecise random variables with unknown distribution functions, while $\mathbf{Y}$ is the conventional random vector with known distribution function. Similar to Eq. (4.292), the probability of failure is given by

$$\mathbb{P}_f = \int_{G(\mathbf{Y},\mathbf{Z}) \leq 0} f_{\mathbf{Z}}(\mathbf{z}) f_{\mathbf{Y}}(\mathbf{y}) d\mathbf{z} d\mathbf{y} \qquad (4.306)$$

where $f_{\mathbf{Z}}(\mathbf{z})$ is the joint distribution of $\mathbf{Z}$. We assume that each element in $\mathbf{Z}$, Z_1 through Z_k, is statistically independent, with which Eq. (4.306) becomes

$$\mathbb{P}_f = \int \cdots \int \xi_{\mathbf{Z}}(\mathbf{z}) f_{\mathbf{Y}}(\mathbf{y}) d\mathbf{y} \prod_{i=1}^{k} f_{Z_i}(z_i) d\mathbf{z} \qquad (4.307)$$

where $\xi_{\mathbf{Z}}(\mathbf{z})$ is the conditional failure probability on $\mathbf{Z} = \mathbf{z}$, i.e.,

$$\xi_{\mathbf{Z}}(\mathbf{z}) = \mathbb{P}(G(\mathbf{Y}, \mathbf{Z} = \mathbf{z}) < 0) = \int_{G(\mathbf{Y},\mathbf{Z}=\mathbf{z}) < 0} f_{\mathbf{Y}}(\mathbf{y}) d\mathbf{y} \qquad (4.308)$$

In an attempt to find the lower and upper bounds of $\mathbb{P}_f$, the objective is to find the optimized distribution function of each element in $\mathbf{Z}$, Z_i, so as to maximize (or minimize) $\mathbb{P}_f$ in Eq. (4.306). To begin with, consider the case where $k = 2$ (i.e., two imprecise random variables are involved). The PDFs of Z_1 and Z_2 are written as f_{Z_1} and f_{Z_2}, respectively. The failure probability $\mathbb{P}_f$ in Eq. (4.306) becomes a function of f_{Z_1} and f_{Z_2}, denoted by $\mathbb{P}_f = h(f_{Z_1}, f_{Z_2})$.

Consider the lower bound of $\mathbb{P}_f$. Note that a set of candidate distribution types exists for both f_{Z_1} and f_{Z_2}, denoted by Ω_{Z_1} and Ω_{Z_2}, respectively. First, an arbitrary distribution is assigned for Z_1 and Z_2 (e.g., a normal distribution), whose PDFs are $_1 f_{Z_1} \in \Omega_{Z_1}$ and $_1 f_{Z_2} \in \Omega_{Z_2}$. Next, we find $_2 f_{Z_2} \in \Omega_{Z_2}$ which minimizes $h(_1 f_{Z_1}, f_{Z_2})$ for $\forall f_{Z_2} \in \Omega_{Z_2}$, followed by determining $_2 f_{Z_1} \in \Omega_{Z_1}$ which minimizes $h(f_{Z_1}, _2 f_{Z_2})$ for $\forall f_{Z_1} \in \Omega_{Z_1}$. The approach to find $_2 f_{Z_2}$ and $_2 f_{Z_1}$ has been discussed in Sect. 4.5.3.1. As such, it is easy to see that

$$h(_2 f_{Z_1}, _2 f_{Z_2}) \leq h(_1 f_{Z_1}, _2 f_{Z_2}) \leq h(_1 f_{Z_1}, _1 f_{Z_2}). \tag{4.309}$$

This fact implies that the pair $(_2 f_{Z_1}, _2 f_{Z_2})$ leads to a reduced $\mathbb{P}_f$ compared with the pair $(_1 f_{Z_1}, _1 f_{Z_2})$. Similarly, one can further find the subsequent sequences $(_3 f_{Z_1}, _3 f_{Z_2})$ through $(_n f_{Z_1}, _n f_{Z_2})$, in which n is a sufficiently large number of iteration. By noting that $h(f_{Z_1}, f_{Z_2})$ is bounded, $h(_n f_{Z_1}, _n f_{Z_2})$ converges to the lower bound of $\mathbb{P}_f$ as n is large enough [36]. Further, the upper bound of the failure probability can also be found using a similar procedure.

Now consider the more generalized case where $k > 2$. The failure probability in Eq. (4.306) is rewritten as follows,

$$\mathbb{P}_f = h(f_{Z_1}, f_{Z_2}, \dots, f_{Z_k}) \tag{4.310}$$

where f_{Z_i} is the PDF of Z_i for $i = 1, 2, \dots, k$. Let Ω_{Z_i} denote the set of all the possible candidate distribution functions of element Z_i. In terms of the lower bound of $\mathbb{P}_f$, an iteration-based approach is as follows,

(1) Preliminary: assign an arbitrary distribution for each element in $\mathbf{Z}$, i.e., $_1 f_{Z_1}$ through $_1 f_{Z_k}$, and calculate $h_1 = h(_1 f_{Z_1}, _1 f_{Z_2}, \dots, _1 f_{Z_k})$.
(2) Find $_j f_{Z_i} = f_{Z_i} \in \Omega_{Z_i}$ which minimizes $h(_j f_{Z_1}, _j f_{Z_2}, \dots, _j f_{Z_{i-1}}, f_{Z_i}, \dots, _{j-1} f_{Z_{i+1}}, \dots, _{j-1} f_{Z_k})$ for $i = 1, 2, \dots, k$ and $j = 2$, and calculate $h_j = h(_j f_{Z_1}, _j f_{Z_2}, \dots, _j f_{Z_k})$.
(3) For each j, if $|h_j - h_{j-1}|$ is smaller than the predefined error limit (say, 10^{-5}), then h_j is found to be the lower bound of $\mathbb{P}_f$; otherwise, return to step (2) with j replaced by $j + 1$.

It can be seen that for each $j = 1, 2, \dots, h_j \leq h_{j-1}$. This observation is guaranteed by the fact that

$$h(_j f_{Z_1}, _j f_{Z_2}, \dots, _j f_{Z_k}) \leq h(_j f_{Z_1}, _j f_{Z_2}, \dots, _{j-1} f_{Z_k})$$
$$\leq h(_j f_{Z_1}, _j f_{Z_2}, \dots, _{j-1} f_{Z_{k-1}}, _{j-1} f_{Z_k}) \leq \cdots \leq h(_{j-1} f_{Z_1}, _{j-1} f_{Z_2}, \dots, _{j-1} f_{Z_k}).$$
$$\tag{4.311}$$

Fig. 4.35 Schematic representation of an oscillation system

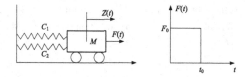

with which the sequence $\{h_j\}$ converges to the lower bound of $\mathbb{P}_f$ as j is sufficiently large [36].

Finally, for the upper bound of the probability of failure, a similar procedure can be used, with the operation "minimize" replaced by "maximize".

Example 4.44

Consider a non-linear single degree of freedom system without damping [17], as shown in Fig. 4.35. The limit state function is defined by the case where the maximum displacement response exceeds the limit, i.e.,

$$G = 3D - |Z_{max}| = 3D - \left| \frac{2F_0}{M\Omega_0^2} \sin\left(\frac{\Omega_0^2 t_0}{2}\right) \right| \tag{4.312}$$

where Z_{max} is the maximum displacement response of the system, $\Omega_0 = \sqrt{(C_1 + C_2)/M}$, and D is the displacement at which one of the two springs yields. The system is deemed to "fail" if $G(\mathbf{X}) < 0$ and "survive" otherwise. The probabilistic information regarding the six random variables in Eq. (4.312) is summarized in Table 4.10. It is assumed that the variables C_1 and C_2 are statistically independent and imprecise with their distribution types unknown.

Find the interval for system failure probability [32].

Solution
First, the fragility curve of the system with respect to C_1 and C_2 is fitted through numerical simulation as follows,

$$\xi_{C_1,C_2}(c_1, c_2) = 0.072\Phi(-0.016c^6 + 0.138c^5 - 0.348c^4 + 0.182c^3 + 0.202c^2 + 1.919c - 3.656) \tag{4.313}$$

where $c = 3 - c_1 - c_2$.

Setting an error threshold of 10^{-4}, the bounds of failure probability are obtained with five iterations, yielding an interval of failure probability of $[0.0171, 0.0311]$. The details are shown in Table 4.11. For comparison purpose, the interval of failure probability is found to be $[0.0001, 0.0655]$ if using IMC1, and $[0.0020, 0.0579]$ for IMC2.

Table 4.10 Statistics of the random variables in Fig. 4.35

Variable	Distribution type	Mean	Standard deviation
M	Normal	1	0.05
D	Normal	0.5	0.05
F_0	Normal	1	0.2
t_0	Normal	1	0.2
C_1	Unknown	1	0.6
C_2	Unknown	0.5	0.3

Table 4.11 Bounds of failure probability from the proposed iteration-based approach

Iteration no.	Operation	Lower bound	Upper bound
1	$_1 f_{C_1}, _1 f_{C_2} \sim$ normal distribution	0.0250	0.0250
2	$_1 f_{C_1}$ fixed, $_2 f_{C_2}$ optimized	0.0245	0.0260
3	$_2 f_{C_2}$ fixed, $_2 f_{C_1}$ optimized	0.0171	0.0310
4	$_2 f_{C_1}$ fixed, $_3 f_{C_2}$ optimized	0.0171	0.0311
5	$_3 f_{C_2}$ fixed, $_3 f_{C_1}$ optimized	0.0171	0.0311

Problems

4.1 In the presence of the limit state function of $Z = R - S$, where both R and S are normal variables, let θ be the central factor of safety, ν_R, ν_S the COVs of R and S respectively. Show that conditional on ν_R, ν_S, the reliability index β is an increasing function of θ.

4.2 The annual maximum peak ground acceleration (PGA) is modeled by a Fréchet distribution with scale parameter ε and shape parameter k. Let A_n be the nominal PGA with an exceeding probability of α in T years. Derive A_n.

4.3 If the reliability index is 2.5 for a reference period of 50 years, what is the corresponding annual failure probability?

4.4 The structural resistance R follows a Gamma distribution with shape parameter $a > 0$ and scale parameter $b > 0$; the load effect, S, which is independent of R, is an exponential random variable with mean $\frac{1}{\lambda}$. Let $Y = \frac{R}{S}$ be the overall factor of safety, and $F_Y(y)$ the CDF of Y. Show that

$$F_Y(y) = \frac{1}{\left(1 + \frac{b\lambda}{y}\right)^a}, \quad y \geq 0 \tag{4.314}$$

4.5 The structural resistance R follows a lognormal distribution with mean 3 and standard deviation 0.75; the load effect S follows a Weibull distribution with mean 1 and standard deviation 0.5.

Fig. 4.36 A 2-span beam subjected to concentrated load F

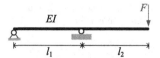

(1) What is the failure probability if R and S are statistically independent?
(2) Modeling the correlation between R and S with Gumbel–Hougaard copula, what is the failure probability if the Kendall's tau for R and S is 0.5?

4.6 Consider a limit state function of $G = R - S$, with which the structural failure is deemed to occur when $G < 0$. The resistance R has a mean value of 3 and a standard deviation of 0.5, while the load effect S has a mean value of 1 and a standard deviation of 0.4.

(1) Find the checking point in the normalized space and the corresponding one in the original space.
(2) If the nominal resistance and load effect are as in Eq. (4.12), compute the resistance factor φ_R and the load factor γ_S.

4.7 Tail sensitivity problem. Consider a limit state function of $G = R - S$. Suppose that the resistance, R, is lognormally distributed with a mean value of 3 and a COV of 0.2. The load effect S has a mean value of 1 and a COV of 0.3. Compute the failure probability for the following two cases: (1) S is a normal variable; (2) S follows an Extreme Type I distribution.

4.8 For a structure subjected to live load L and dead load D, the limit state function is $G = R - L - D$, where R is the structural resistance. If $\mu_R = 3, \mu_D = 1, \mu_L = 0.6, \sigma_R = 0.45, \sigma_D = 0.1, \sigma_L = 0.3$, where $\mu_\bullet$ and $\sigma_\bullet$ are the mean and standard deviation of random variable $\bullet$ respectively, determine the resistance and load factors (use the nominal resistance and nominal loads as defined in Eq. (4.12)).

4.9 Consider a 2-span beam as shown in Fig. 4.36. The maximum deflection for the left span (with a length of l_1) is $u_{\max} = \frac{F l_1^2 l_2}{9\sqrt{3}EI}$, where EI is the bending rigidity. If F has a mean of 10 kN and a COV of 0.4, EI has a mean of 500 kN·m^2 and a COV of 0.2, compute the Hasofer–Lind reliability index. Assume that $l_1 = l_2 = 5$ m. The beam fails if $u_{\max} > l_1/75$.

4.10 Consider a simply supported beam with a length of L. Suppose that the moment resistance at the mid-span, M_{mid}, has a mean of 200 kN·m and a COV of 0.15. The beam is subjected to uniform load W along the length, which has a mean of 10 kN/m and a COV of 0.2. The length L has a mean of 8 m and a COV of 0.125. Suppose that each variable (M_{mid}, L, W) is independent mutually. Focusing on the moment at the mid-span of the beam, calculate the failure probability using Eq. (4.69).

4.11 Recall Problem 4.10. Recalculate the reliability index β_{HL} using Eq. (4.82) or (4.90). Set $\varepsilon = 0.01$. Compare the result with that in Problem 4.10.

Fig. 4.37 A beam subjected
to uniform load W

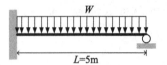

$L=5\text{m}$

4.12 In Problem 4.10, if we additionally know that M_{mid} is a lognormal variable, L is a Gamma variable, and W follows an Extreme Type I distribution, recalculate the probability of failure using FORM.

4.13 In Problem 4.12, use Monte Carlo simulation method to find the failure probability and the reliability index β.

4.14 Consider a simply supported beam with a length of $l = 1$ and a square cross section (the side length is a). A vertical concentrated load F, which is a normal variable with a mean of 0.3 and a standard deviation of 0.2, is applied at the middle of the beam. The normal yield stress is a normal variable with a mean of 1 and a COV of 0.15, while the shear yield stress is a normal variable with a mean of 1 and a COV of 0.1. If the target reliability index is 2.5, what is the minimum value for a?

4.15 Consider a five-meter beam as in Fig. 4.37, which is subjected to uniformly distributed load W. Suppose that W is an Extreme Type I variable with a mean value of 45 kN/m and a COV of 0.3, the yield stress of the beam material, F_y, is a lognormal variable with mean 300 MPa and COV 0.15, the section modulus Z_x is deterministically 800 cm³. What is the probability of failure of the beam by plastic collapse?

4.16 Consider a limit state function of $G = R - S$. Suppose that the resistance, R, is a lognormal variable with a mean value of 3 and a COV of 0.2, and the load effect S is a Rayleigh variable with a mean value of 1. Calculate the probability of failure using Eq. (4.132).

4.17 Reconsider Problem 4.16. If we use a simulation-based approach to estimate the failure probability, show that the use of importance sampling can increase the computational efficiency.

4.18 Consider a structure subjected to the combined effect of live load L and dead load D. The structure was designed using the *allowable stress design* according to $\frac{R_n}{FS} \geq D_n + L_n$, where FS is the factor of safety, R_n, D_n, L_n are the nominal resistance, nominal dead load and nominal live load respectively. Suppose that R, D, L are normal variables with $\mu_R = 1.10R_n, \sigma_R = 0.15\mu_R, \mu_D = 1.05D_n, \sigma_D = 0.05\mu_D, \mu_L = 0.9L_n, \sigma_L = 0.3\mu_L$, where $\mu_\bullet$ and $\sigma_\bullet$ are the mean and standard deviation of $\bullet = R, D, L$ respectively. Assume that $FS = \frac{5}{3}$. When $\frac{L_n}{D_n}$ varies within a range of $[0.5, 4]$, determine the reliability index β for different values of $\frac{L_n}{D_n}$.

Fig. 4.38 A structure
subjected to wind load

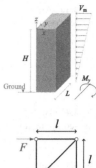

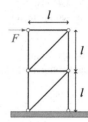

Fig. 4.39 An eight-bar truss

4.19 Consider a structure subjected to wind load, as shown in Fig. 4.38. The height and width of the surface normal to wind load are $H = 8$ m and $L = 3$ m respectively. The wind speed varies linearly with the height, and equals zero at the ground level. The wind pressure p is calculated by $p = \frac{1}{2}\rho V^2 C_{\text{fig}}$, where ρ is the air density (1.2 kg/m^3), V is the wind speed, C_{fig} is a constant that reflects the structural shape and aerodynamics. Take C_{fig} as 0.45 in the following.

(1) When the design wind speed at height H, V_{m}, is 40 m/s, with a resistance factor of 0.85, what is the design resisting moment of the structure at the bottom?
(2) If the wind speed at height H has a mean of 40 m/s and a COV of 0.3, and the structural resisting moment at the bottom is a lognormal variable with a COV of 0.15, determine the mean of the resisting moment with a target reliability index of 2.5.
(3) If we additionally know in (2) that the wind speed at height H is a Weibull variable, recalculate the mean of the resisting moment using FORM.

4.20 Consider an eight-bar truss as shown in Fig. 4.39, which is subjected to a horizontal load F. Suppose that F follows an Extreme Type I distribution with a mean value of 1 and a standard deviation of 0.3. Each bar has a statistically independent and identically lognormally distributed resistance of axial force (either compression or tension), with a mean value of 3 and a standard deviation of 0.5. Compute the failure probability of the truss.

4.21 In Problem 4.20, if the resistances of any two bars are correlated with a linear correlation coefficient of 0.8, what is the failure probability of the truss?

4.22 In Problem 4.20, if the occurrence of the horizontal load F is a Poisson process with rate $\lambda = 0.2$/year, and the magnitude of each load, conditional on occurrence, follows an Extreme Type I distribution with a mean value of 1 and a standard deviation of 0.3, what is the failure probability of the truss for a reference period of 10 years?

Fig. 4.40 An n-component parallel system

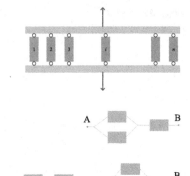

Fig. 4.41 A three-component system

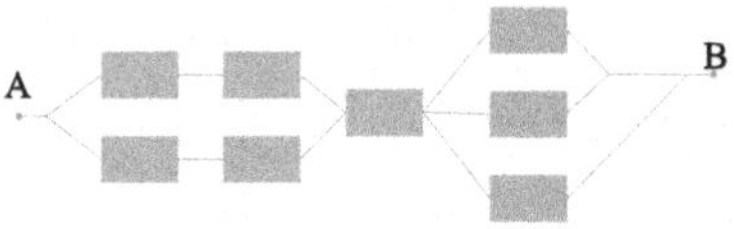

Fig. 4.42 An eight-component system

4.23 Consider an n-component parallel system as shown in Fig. 4.40. Suppose that each component has an identical failure probability of 0.2. The performance of component i is only dependent on that of component $i - 1$ for $i = 2, 3, \ldots, n$. Provided that component $i - 1$ fails, the probability of failure for component i is $1 - \frac{1}{n}$. What is the system failure probability as $n \to \infty$?

4.24 Reconsider Example 4.29. If the capacities of each rope, R_1, R_2 and R_3, are mutually correlated with an identical linear correlation coefficient of 0.6 for any two capacities,

(1) Compute the system failure probability.
(2) Compare the result from (1) and the bounds obtained in Example 4.29.

4.25 Consider a 5-out-of-8 system with a failure probability of $p = 0.01$ for all components. What is the system failure probability?

4.26 Consider a three-component system as shown in Fig. 4.41. Each box represents a component. The system is deemed as survival if points A and B are connected by any path. We introduce a Bernoulli random variable for each of the three components, denoted by B_1, B_2, B_3, which returns 1 if the component fails and 0 otherwise. If $\mathbb{P}(B_i = 1) = 0.01$ for $i = 1, 2, 3$, and the correlation coefficient of B_i and B_j is 0.5 for $i, j \in \{1, 2, 3\}$, what is the system failure probability?

4.27 Consider an eight-component system as shown in Fig. 4.42. Each box represents a component. The system is deemed as survival if points A and B are connected by any path. If each component behaves independently and has an identical failure probability of $p = 0.01$, what is the system failure probability?

4.28 In Example 4.28, if V and H are correlated with a linear correlation coefficient of 0.7, recompute the system failure probability.

4.29 Consider a 5-out-of-8 system, where each component has an identically log-normally distributed resistance with mean 2 and COV 0.3. If all the components are subjected to a deterministic load effect of 1, what is the system failure probability if any two of the component resistances are correlated with a linear correlation coefficient of 0.7?

4.30 In Example 4.34, let $W(x) = \overline{W} + \varepsilon(x)$, where $\overline{W}$ has a mean value of 1 kN/m and a COV of 0.5, and $\varepsilon(x)$ has a zero-mean, a standard deviation of 0.3 kN/m and a correlation structure of $\rho(\varepsilon(x_1), \varepsilon(x_2)) = \exp(-(x_1 - x_2)^2)$. Compute the mean value and standard deviation of EUDL. Assume $l = 2$ m. Compare the result with that in Example 4.34 and comment on the difference.

4.31 Reconsider Example 4.41. If the load effect S is associated with a reference period of 40 years, taking into account the impact of the discount rate, $r = 5\%$, what is the optimal annual failure probability?

4.32 Reconsider Example 4.42. If the load effect S is associated with a reference period of 40 years, taking into account the impact of the discount rate, $r = 5\%$, what is the optimal annual failure probability?

4.33 We consider the reliability of a scaffolding system during concrete replacement [34]. Both the live load L and the dead load D are present. The live load follows an Extreme Type I distribution with a mean-to-nominal ratio of 0.85 and a COV of 0.6, the dead load is normally distributed having a mean-to-nominal value of 1.05 and a COV of 0.3, and the scaffolding resistance is lognormally distributed with a mean-to-nominal value of 1.10 and a COV of 0.15. The target reliability index is set as 3.0.

(1) Using a LRFD-based safety check as follows, $\varphi R_n = \gamma_D D_n + \gamma_L L_n$, calculate φ, γ_L, γ_D for different values of $\frac{L_n}{D_n} \in [0.1, 0.3]$.
(2) Determine a LRFD-based design criterion for the scaffolding system with a typical range of $\frac{L_n}{D_n} \in [0.1, 0.3]$.

4.34 Consider the serviceability of a lining structure subjected to water seepage, as discussed in Problem 2.32. The lining structure has a surface of 10 m × 10 m and a thickness of 0.15 m. Assume that the hydraulic conductivity K and the water pressure p are mutually independent. At any single location of the surface, the hydraulic conductivity K is lognormally distributed having a mean value of 5.67×10^{-13} m/s and a COV of 0.3, and the water pressure is a Gamma variable with a mean value of 0.06 MPa and a COV of 0.2. The water pressure p is constant for the whole surface, while the hydraulic conductivity is a random filed. The K's at any two locations with a distance of x (in meters) have a correlation coefficient of $\exp(-x)$. At time t, the structural performance is deemed as "satisfactory" if the water seepage depth does not exceed the thickness for at least 99% area of the surface. What are the probabilities of satisfactory structural performance at the end of 20 and 50 years respectively?

4.35 Recall Example 4.43. We additionally consider the following case: H has a mean value within [1.87, 1.93], a standard deviation of 0.45, and is strictly defined within [1.0, 3.0].

(1) Compute the CDF envelope of H with linear-programming-based approach.
(2) Find the lower and upper bounds of failure probability.

4.36 In Example 4.44, if we additionally know that $C_1 \in [0.5, 1.5]$, recompute the interval for system failure probability.

References

1. ACI Committee 318-14 (2014) Building code requirements for structural concrete and commentary. American Concrete Institute, Farmington Hills, MI
2. Alevras D, Padberg MW (2001) Linear optimization and extensions: problems and solutions. Springer Science & Business Media, New York
3. Alvarez DA, Uribe F, Hurtado JE (2018) Estimation of the lower and upper bounds on the probability of failure using subset simulation and random set theory. Mech Syst Signal Process 100:782–801. https://doi.org/10.1016/j.ymssp.2017.07.040
4. ASCE/SEI 7-10 (2011) Minimum design loads for buildings and other structures. American Society of Civil Engineers
5. Baudrit C, Dubois D, Perrot N (2008) Representing parametric probabilistic models tainted with imprecision. Fuzzy Sets Syst 159(15):1913–1928. https://doi.org/10.1016/j.fss.2008.02.013
6. de Berg M, Cheong O, van Kreveld M, Overmars M (2008) Computational geometry: algorithms and applications, 3rd edn. Springer, Berlin. https://doi.org/10.1007/978-3-540-77974-2
7. Bjarnadottir S, Li Y, Stewart MG (2011) A probabilistic-based framework for impact and adaptation assessment of climate change on hurricane damage risks and costs. Struct Saf 33(3):173–185. https://doi.org/10.1016/j.strusafe.2011.02.003
8. Cui L, Xie N (2005) On a generalized k-out-of-n system and its reliability. Int J Syst Sci 36(5):267–274. https://doi.org/10.1080/00207720500062470
9. Ellingwood B, Galambos TV, MacGregor JG, Cornell CA (1980) Development of a probability based load criterion for American National Standard A58: building code requirements for minimum design loads in buildings and other structures. US Department of Commerce, National Bureau of Standards
10. Ellingwood BR (1994) Probability-based codified design: past accomplishments and future challenges. Struct Saf 13(3):159–176. https://doi.org/10.1016/0167-4730(94)90024-8
11. Ellingwood BR (1996) Reliability-based condition assessment and LRFD for existing structures. Struct Saf 18(2–3):67–80. https://doi.org/10.1016/0167-4730(96)00006-9
12. Ellingwood BR (2000) LRFD: implementing structural reliability in professional practice. Eng Struct 22(2):106–115. https://doi.org/10.1016/S0141-0296(98)00099-6
13. Ferson S, Kreinovich V, Ginzburg L, Myers DS, Sentz K (2003) Constructing probability boxes and Dempster-Shafer structures. Technical report. SAND2002–4015, Sandia National Laboratories
14. Galambos TV, Ellingwood B, MacGregor JG, Cornell CA (1982) Probability based load criteria: assessment of current design practice. J Struct Div 108(5):959–977
15. Griffiths D, Huang J, Fenton GA (2011) Probabilistic infinite slope analysis. Comput Geotech 38(4):577–584. https://doi.org/10.1016/j.compgeo.2011.03.006
16. Hanai M (1975) Assessment of load factors by means of structural reliability theory. Trans Archit Inst Jpn 231:13–20

17. Huang X, Chen J, Zhu H (2016) Assessing small failure probabilities by AK–SS: an active learning method combining Kriging and subset simulation. Struct Saf 59:86–95. https://doi. org/10.1016/j.strusafe.2015.12.003
18. Kamalja KK (2017) Reliability computing method for generalized k-out-of-n system. J Comput Appl Math 323:111–122. https://doi.org/10.1016/j.cam.2017.02.045
19. Kanda J, Ellingwood B (1991) Formulation of load factors based on optimum reliability. Struct Saf 9(3):197–210. https://doi.org/10.1016/0167-4730(91)90043-9
20. Kuo W, Zuo MJ (2003) Optimal reliability modeling: principles and applications. Wiley, New York
21. Limbourg P, De Rocquigny E (2010) Uncertainty analysis using evidence theory–confronting level-1 and level-2 approaches with data availability and computational constraints. Reliab Eng Syst Saf 95(5):550–564. https://doi.org/10.1016/j.ress.2010.01.005
22. Nowak AS, Collins KR (2012) Reliability of structures, 2nd edn. CRC Press, Boca Raton
23. Oberguggenberger M, Fellin W (2008) Reliability bounds through random sets: non-parametric methods and geotechnical applications. Comput Struct 86(10):1093–1101. https://doi.org/10. 1016/j.compstruc.2007.05.040
24. Rackwitz R, Fiessler B (1976) Note on discrete safety checking when using non-normal stochastic models for basic variables. Loads project working session. MIT, Cambridge
25. Shepherd DK (2008) k-out-of-n systems. Encyclopedia of statistics in quality and reliability. Wiley, New York. https://doi.org/10.1002/9780470061572.eqr342
26. Song S, Lu Z, Qiao H (2009) Subset simulation for structural reliability sensitivity analysis. Reliab Eng Syst Saf 94(2):658–665. https://doi.org/10.1016/j.ress.2008.07.006
27. Tang WH, Ang A (2007) Probability concepts in engineering: emphasis on applications to civil and environmental engineering. Wiley, Hoboken
28. Tang XS, Li DQ, Rong G, Phoon KK, Zhou CB (2013) Impact of copula selection on geotechnical reliability under incomplete probability information. Comput Geotech 49:264–278. https:// doi.org/10.1016/j.compgeo.2012.12.002
29. Turkstra CJ (1972) Theory of structural design decisions. Solid Mechanics Study No. 2, University of Waterloo, Waterloo, Ontario, Canada
30. Vickery P, Skerlj P, Twisdale L (2000) Simulation of hurricane risk in the US using empirical track model. J Struct Eng 126(10):1222–1237. https://doi.org/10.1061/(ASCE)0733-9445(2000)126:10(1222)
31. Wang C (2020) Reliability-based design of lining structures for underground space against water seepage. Undergr Space. https://doi.org/10.1016/j.undsp.2020.03.004
32. Wang C, Zhang H, Beer M (2018) Computing tight bounds of structural reliability under imprecise probabilistic information. Comput Struct 208:92–104. https://doi.org/10.1016/j. compstruc.2018.07.003
33. Wang C, Zhang H, Li Q (2019) Moment-based evaluation of structural reliability. Reliab Eng Syst Saf 181:38–45. https://doi.org/10.1016/j.ress.2018.09.006
34. Wang C, Zhang H, Rasmussen KJ, Reynolds J, Yan S (2020) System reliability-based limit state design of support scaffolding systems. Eng Struct 216:110677. https://doi.org/10.1016/j. engstruct.2020.110677
35. Xue G, Dai H, Zhang H, Wang W (2017) A new unbiased metamodel method for efficient reliability analysis. Struct Saf 67:1–10. https://doi.org/10.1016/j.strusafe.2017.03.005
36. Yeh J (2006) Real analysis: theory of measure and integration, 2nd edn. World Scientific Publishing Company, Singapore
37. Zhang H, Mullen RL, Muhanna RL (2010) Interval Monte Carlo methods for structural reliability. Struct Saf 32(3):183–190. https://doi.org/10.1016/j.strusafe.2010.01.001
38. Zhang H, Reynolds J, Rasmussen KJ, Ellingwood BR (2015) Reliability-based load requirements for formwork shores during concrete placement. J Struct Eng 142(1):04015094. https:// doi.org/10.1061/(ASCE)ST.1943-541X.0001362

Chapter 5
Time-Dependent Reliability Assessment

Abstract This chapter discusses the approaches for structural time-dependent reliability assessment. The significant difference between the time-dependent reliability and the classical reliability (c.f. Chap. 4) is the involvement of the time-variant characteristics in the analysis, where the variation of both the structural resistance and the external loads on the temporal scale should be reasonably modelled. This chapter starts from the motivation of time-dependent reliability assessment, followed by the modelling techniques of the resistance deterioration and the external load processes. Both the discrete and continuous load processes are discussed. Subsequently, the time-dependent reliability assessment approaches in the presence of both the discrete and the continuous loads are addressed. The comparison between the reliabilities associated with the two types of load processes is also presented.

5.1 Why Time-Dependent Reliability?

The failure of important engineered structures within a community, triggered by either environmental or anthropogenic extreme events, may lead to substantial economic losses and disruption to the society. Motivated by the increasing awareness from the public and asset owners regarding the safety of structures and infrastructure systems, the research community and engineers have been seeking for advanced implementations of risk mitigation and construction practice. Many factors such as the environmental conditions, severe load intensity, may pose a significant threat to the structural safety. The exact impacts of these factors are, unfortunately, often difficult to predict in a deterministic way, and thus should be assessed using a probability-based method taking into account the uncertainties associated with both the structural performance and the external load effects [4, 6, 9, 15, 20, 31, 32, 38, 43, 44].

With this regard, structural reliability assessment provides a useful tool of evaluating and managing structural safety and serviceability level, which is informative of the structural ability of withstanding future extreme events within its service life under a probability-based framework. As discussed in Sect. 4.1, the structural reliability, $\mathbb{L}$, can be quantitatively measured by the probability that the load effect (S) does not exceed the structural resistance (R), i.e.,

© The Author(s), under exclusive license to Springer Nature Switzerland AG 2021   263
C. Wang, *Structural Reliability and Time-Dependent Reliability*,
Springer Series in Reliability Engineering,
https://doi.org/10.1007/978-3-030-62505-4_5

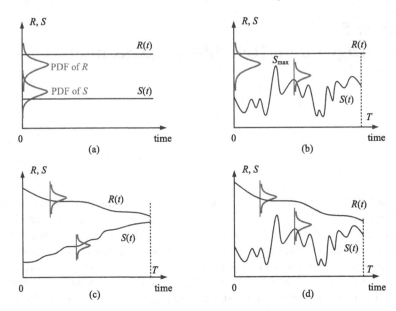

Fig. 5.1 Time-variation of both the resistance and load processes

$$\mathbb{L} = \mathbb{P}(R > S) \tag{5.1}$$

Note that both R and S in Eq. (5.1) are time-invariant, and thus the reliability $\mathbb{L}$ is independent of the considered reference period, as schematically shown in Fig. 5.1a. That is, for a service period of interest, say, $[0, T]$, both the resistance and the load effect can be reasonably represented by a single random variable. However, realistic structures are often subjected to severe operating or environmental conditions during their service life, and thus may suffer from the deterioration of structural strengthen or stiffness due to these factors. Moreover, the external load intensity and/or frequency may also change with time [17, 34, 35, 37]. Examples 5.1–5.5 in the following present some simple illustrations of the time-variant resistance and load processes for a realistic structure. As such, a reasonable approach for structural reliability assessment should take into account the time-variant characteristics of both the resistance and the external load processes. This is especially the case when a relatively long reference period (say, 50 years) is considered and the variation of both R and S with time is non-negligible.

Example 5.1

Consider a simply supported RC beam that is subjected to marine environment, as shown in Fig. 5.2. The chloride ingression into the concrete may initiate the corrosion of steel bar when the chloride concentration at the steel surface reaches a predefined threshold, and further result in a reduction of structural resistance. The relationship between the steel cross section loss and time t can be described by the Faraday's laws of electrolysis [24], which takes a form of

Fig. 5.2 Illustration of the moment bearing capacity of a simply supported RC beam

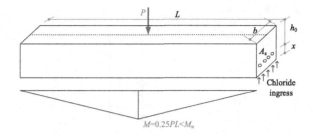

$$\Delta A_s = k\psi_0 \cdot i_{cor} \cdot (t - t_i), \quad t \geq t_i \tag{5.2}$$

where A_s is the steel cross section area (mm^2), ψ_0 is the bar diameter (mm), k is a constant, i_{cor} is the corrosion rate (μA/cm^2), and t_i is the time of corrosion initiation, which equals the time at which the chloride concentration at the steel bar surface reaches a threshold c_{cr} (c.f. Examples 2.26 and 3.27). We assume that the reduction of moment bearing capacity is dominated by the section loss of steel bars (note that some other factors may also have a significant impact on the reduction of the resistance, e.g., [46]). With this, the time-variant resistance (moment bearing capacity) of the beam, $M_u(t)$, is obtained as follows,

$$M_u(t) = f_y A_s \left(h_0 - \frac{f_y A_s}{2 f_c b} \right) = f_y (A_0 - \Delta A_s) \left[h_0 - \frac{f_y (A_0 - \Delta A_s)}{2 f_c b} \right] \tag{5.3}$$

where b is the width of the cross section, f_c is the concrete compressive strength, f_y is the steel tensile strength, A_0 is the initial cross section area of the steel bars, and h_0 is the effective depth of the section.

Assume that in Eq. (5.2), $k = 0.0366$, and i_{cor} follows a lognormal distribution with a mean value of 0.67 μA/cm^2 and a COV of 0.58. The time of corrosion initiation follows a normal distribution with a mean value of 30 years and a COV of 0.2. Totally four reinforced bars, with a diameter of 18 mm, are evenly placed at a depth of x to the surface. The compressive strength of the concrete is 20 MPa, and the tensile strength of the steel bars is 335 MPa. In terms of the geometry of the beam section, $b = 250$ mm, $h_0 = 460$ mm and $x = 40$ mm.

(1) Use a simulation-based approach to generate sample trajectories of the time-variant moment bearing capacity of the beam, and plot them.

(2) Assume that the length of the beam is $L = 3$ m. A point load of $P = 16$ kN is applied at the midpoint of the beam. Find the probability of failure of the beam at 50 years and 100 years respectively. The failure is defined as the case where the moment at the mid-span due to P exceeds the ultimate moment bearing capacity.

Solution

(1) The initial cross section area of the four steel bars is $A_0 = 4 \times \pi \times \left(\frac{18}{2} \right)^2 = 1018$ mm^2. Before the initiation of corrosion, the resistance (ultimate moment-bearing capacity) is calculated according to Eq. (5.3) as follows,

$$M_u(0) = f_y A_0 \left(h_0 - \frac{f_y A_0}{2 f_c b} \right) = 335 \times 1018 \times \left(460 - \frac{335 \times 1018}{2 \times 20 \times 250} \right) \times 10^{-6} \, \text{kN} \cdot \text{m} = 14.52 \, \text{kN} \cdot \text{m} \tag{5.4}$$

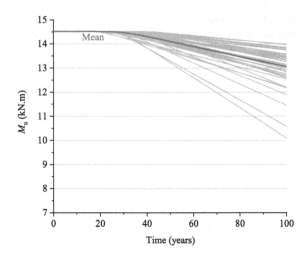

Fig. 5.3 Time-variant moment bearing capacity of the beam in Fig. 5.2

When employing a simulation-based approach to generate a sample trajectory of M_u with time, the procedure is as follows for each simulation run.

(a) Generate two samples for i_{cor} and t_i, denoted by i_0 and t_{i0} respectively.
(b) For a given time t, calculate $\Delta A_s(t)$ according to $\Delta A_s(t) = 4 \times 18 \times 0.0366 \times i_0 \cdot \max(0, t - t_{i0})$.
(c) Calculate $M_u(t)$ with $M_u(t) = f_y(A_0 - \Delta A_s(t))\left[h_0 - \frac{f_y(A_0 - \Delta A_s(t))}{2 f_c b}\right]$.

The sampled curves of the time-variant $M_u(t)$ are presented in Fig. 5.3, from which it can be seen that the moment bearing capacity degrades by approximately 10% on average over a reference period of 100 years.

(2) Let M_P denote the moment at the mid-span posed by the point load P. Since the beam is simply supported, $M_P = \frac{1}{4}PL = \frac{1}{4} \times 16 \times 3 = 12\,\text{kN} \cdot \text{m}$. The probability of failure at time t is calculated as the probability of $M_u(t) < M_P$.

Given $t = 50$ or 100 years, one can use Monte Carlo simulation to approximate the failure probability (c.f. Sect. 3.5.1 for more information). Here, 10^7 samples are used for each reference period. Results show that at the end of 50 years, the failure probability is 9.84×10^{-4}; at the end of 100 years, the probability of failure is 0.146. The comparison between the two probabilities of failure clearly demonstrates the impact of resistance deterioration on the structural safety level.

Example 5.2

Consider the deterioration scenario of a wharf pile, as shown in Fig. 5.4a. The cross section of the pile may be reduced due to some environmental factors. Consider an ideal case in Fig. 5.4b, where the circular cross section has an initial radius of 10 cm and the annual loss of the radius follows a Gamma distribution with a mean value of 0.05 cm and a COV of 0.2. Assume that the compressive strength of the pile is 40 MPa and the degraded cross section is always circular.

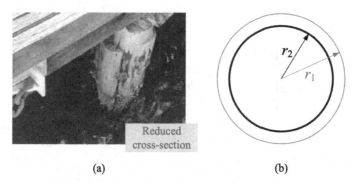

Fig. 5.4 Deterioration of a wharf pile

Use a simulation-based approach to find the time-variant ultimate axial force of the pile for a reference period of 50 years.

Solution

The ultimate axial force of the pile at year t is

$$P(t) = \sigma \cdot [\pi r^2(t)], \quad t = 1, 2, \ldots 50 \tag{5.5}$$

where $\sigma = 40$ MPa is the compressive strength of the pile, and $r(t)$ is the radius of the cross section at year t. In order to sample a trajectory of $P(t)$, a simulation-based method is used. For each simulation run,

(a) Simulate a sequence of samples of the annual radius loss, denoted by $\Delta r_1, \Delta r_2, \ldots$ Δr_{50} respectively.
(b) The radius at the tth year is $r(t) = r_0 - \sum_{i=1}^{t} \Delta r_i$, where r_0 is the initial radius.
(c) The ultimate axial force at year t is calculated according to Eq. (5.5).

Some sampled trajectories of $P(t)$ are shown in Fig. 5.5. On average, the ultimate axial force degrades by approximately 44% within a service period of 50 years.

Example 5.3

Consider a rigid column subjected to vertical load P and horizontal load F, as shown in Fig. 5.6a. The bottom of the column is connected with the ground by a rotational spring, whose moment-rotation relationship is shown in Fig. 5.6b. The length of the column is L.

(1) Find the maximum force F_{max} that the column-spring system can resist;

(2) Assume that the yielding moment of the spring degrades to half of the original case (i.e., $\frac{M_0}{2}$, as shown in Fig. 5.6c), recalculate the maximum horizontal force that the system can resist.

Solution

(1) As the column-spring system is subjected to the joint action of F and P, the equilibrium of the system gives $FL + P\Delta = M$, where Δ is the horizontal displacement at the top of the

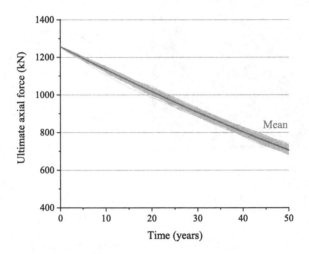

Fig. 5.5 Time-variant ultimate axial force $P(t)$ in Fig. 5.4b

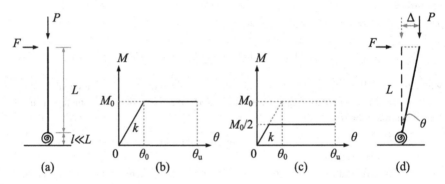

Fig. 5.6 Illustration of a column-spring system

column, and M is the moment at the bottom of the column (i.e., the moment that the spring resists, c.f. Fig. 5.6d). During the elastic stage of the spring,

$$FL + PL\theta = M = \frac{M_0}{\theta_0}\theta, \quad 0 \leq \theta \leq \theta_0$$
$$\Rightarrow F = \frac{\frac{M_0}{\theta_0}\theta - PL\theta}{L} = \left(\frac{M_0}{\theta_0 L} - P\right)\theta \Rightarrow F_{\max} = \frac{M_0}{L} - P\theta_0 \tag{5.6}$$

Subsequently, for the plastic stage,

$$FL + PL\theta = M_0, \quad \theta_0 \leq \theta \leq \theta_u$$
$$\Rightarrow F = \frac{M_0}{L} - P\theta \Rightarrow F_{\max} = \frac{M_0}{L} - P\theta_0 \tag{5.7}$$

Thus, the maximum horizontal load for the column-spring system is $F_{max} = \frac{M_0}{L} - P\theta_0$.

(2) For the "degraded" moment-rotation relationship as in Fig. 5.6c, the yielding moment becomes $\frac{M_0}{2}$ and the corresponding yielding rotational angle becomes $\frac{\theta_0}{2}$. Thus, the maximum horizontal load is obtained as $F'_{max} = \frac{\frac{M_0}{2}}{L} - P\frac{\theta_0}{2} = \frac{1}{2}F_{max}$. Comparing the result from (1), it can be seen that the maximum horizontal force that the system can resist degrades by 50% due to the reduction of the yielding moment of the spring.

Example 5.4

For an in-service bridge, the traffic volume may increase with time due to the development of local economics and social functionalities. For instance, Pan et al. [37] analyzed the in situ traffic data of a highway bridge in China covering a period of 1997–2007, and found that the daily traffic volume increased from 14.8 to 61.4 thousand over the 11 years. This change will accordingly increase the intensity of the "extreme" traffic loads. Illustratively, we assume that the weight of each vehicle follows a normal distribution with a mean value of (normalized) 1 and a standard deviation of 0.2.

(1) Find and illustrate the PDFs of the daily maximum vehicle weight for years 1997 and 2007, respectively.

(2) Calculate the mean value of the daily maximum vehicle weight for years 1997 and 2007, respectively.

Solution

(1) Let W_n denote the daily maximum vehicle weight with n vehicles for that day. With this, the CDF of each vehicle is $F_{W_1}(x) = \Phi\left(\frac{x-1}{0.2}\right)$. Furthermore, one has

$$F_{W_n}(x) = F_{W_1}^n(x) = \left[\Phi\left(\frac{x-1}{0.2}\right)\right]^n, \quad f_{W_n}(x) = n\left[\Phi\left(\frac{x-1}{0.2}\right)\right]^{n-1} \cdot \frac{1}{0.2}\phi\left(\frac{x-1}{0.2}\right) \quad (5.8)$$

where $F_{W_n}(x)$ and $f_{W_n}(x)$ are the CDF and PDF of W_n respectively. For the years 1997 and 2007, substituting $n = 14800$ and $n = 61400$ into Eq. (5.8) gives the PDFs of the daily maximum vehicle weight. The results are presented in Fig. 5.7, from which we can see that the distribution of the maximum daily vehicle weight shifts to the right with a greater number of vehicles.

(2) Based on Eq. (5.8), the mean value of the daily maximum vehicle weight is calculated as follows,

$$\mathbb{E}(W_n) = \int_0^\infty x f_{W_n}(x)dx = \int_0^\infty x \cdot n\left[\Phi\left(\frac{x-1}{0.2}\right)\right]^{n-1}\frac{1}{0.2}\phi\left(\frac{x-1}{0.2}\right)dx \quad (5.9)$$

Substituting $n = 14800$ and $n = 61400$ into Eq. (5.9) yields

$$\mathbb{E}(W_{14800}) = 1.79, \quad \mathbb{E}(W_{61400}) = 1.86 \quad (5.10)$$

By noting that $\frac{\mathbb{E}(W_{61400})-\mathbb{E}(W_{14800})}{\mathbb{E}(W_{14800})} = 4\%$, the maximum daily vehicle weight increased by 4% during a reference period of 1997–2007.

Fig. 5.7 The PDFs of the
normalized daily vehicle
weight

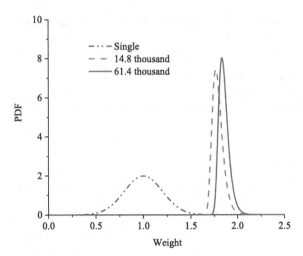

In Example 5.4, we have assumed a normal distribution for the weight of each vehicle. Recall
that in Example 4.5, we discussed the *tail sensitivity problem*. If we assign a lognormal
distribution for the vehicle weight in Example 5.4 instead of a normal distribution, then
$\mathbb{E}(W_{14800}) = 2.15$ and $\mathbb{E}(W_{61400}) = 2.29$. With this, $\frac{\mathbb{E}(W_{61400}) - \mathbb{E}(W_{14800})}{\mathbb{E}(W_{14800})} = 6.7\%$, implying
that the maximum daily vehicle weight increased by 6.7% during 1997–2007.

Example 5.5
Reconsider Example 4.2. For a location at Miami-Dade County, Florida, historical records
show that the annual maximum wind speed (3-s gust) follows a Weibull distribution with a
mean value of 24.3 m/s and a COV of 0.584. We assume that the mean value of the future
wind speed will increase by 0.5% annually due to the potential impact of climate change,
while the COV remains unchanged. Let the nominal wind speed v_n be that with an exceeding
probability of 7% in the future 50 years. Compute v_n.
Solution
Similar to Example 4.2, the CDF of annual maximum wind speed V (m/s) is given as follows,

$$F_V(v) = 1 - \exp\left[-\left(\frac{v}{u}\right)^\alpha\right], \quad v \ge 0 \tag{5.11}$$

where $u = 27.301$ and $\alpha = 1.769$ are the scale and shape parameters, respectively.
Now, considering the annual increment of 0.5% in the mean value of V, the CDF of the
maximum wind speed over 50 years, V_{50}, is determined by

$$F_{V_{50}}(v) = \prod_{i=1}^{50} \left\{1 - \exp\left[-\left(\frac{v}{u\left(1 + \frac{i}{100}\right)}\right)^\alpha\right]\right\}, \quad v \ge 0 \tag{5.12}$$

Fig. 5.8 The CDFs of V_{50} considering/not considering the increasing mean of V

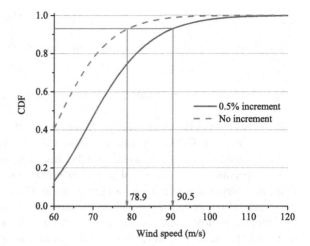

Since the nominal wind speed v_n has an exceeding probability of 7%, one has

$$F_{V_{50}}(v_n) = \prod_{i=1}^{50}\left\{1 - \exp\left[-\left(\frac{v_n}{u\left(1 + \frac{i}{100}\right)}\right)^\alpha\right]\right\} = 1 - 7\% = 0.93 \quad (5.13)$$

In Eq. (5.13), a numerical approach is used to approximate the value of v_n. By referring to Fig. 5.8, v_n is found as 90.5 m/s. Comparing the result in Example 4.2, the nominal wind speed increases by 14.6% over a reference period of 50 years.

For a reference period of $[0, T]$, the structural reliability, denoted by $\mathbb{L}(0, T)$, is the probability that the load effect S does not exceed the resistance R for an arbitrary time point within the considered interval. Mathematically, it follows,

$$\mathbb{L}(0, T) = \mathbb{P}\{R(t) > S(t), \forall t \in [0, T]\} \quad (5.14)$$

where $R(t)$ and $S(t)$ are the resistance and load effect at time t, respectively. The reliability $\mathbb{L}(0, T)$ in Eq. (5.14) is called *time-dependent reliability* as it is dependent on the duration of the considered time period, T.

The comparison between Eqs. (5.1) and (5.14) implies that the time-dependent reliability is featured by the involvement of the time-variant characteristics of the quantities. In Eq. (5.14), the symbol $\forall$ (for all) indicates that, to achieve a survival state of a structure within the time period $[0, T]$, for an arbitrary point of time, the corresponding load effect shall not exceed the resistance at that time. The failure probability, $\mathbb{P}_f(0, T)$, is the complementary part of $\mathbb{L}(0, T)$ (c.f. Eq. (4.1)) and is expressed as follows,

$$\mathbb{P}_f(0, T) = 1 - \mathbb{L}(0, T) = \mathbb{P}\{R(t) < S(t), \exists t \in [0, T]\} \qquad (5.15)$$

Let T_f denote the time of structural failure, and F_{T_f} the CDF of T_f. By definition, it follows,

$$F_{T_f}(\tau) = \mathbb{P}(T_f \leq \tau) = \mathbb{P}(\text{failure within } [0, \tau]) = \mathbb{P}_f(0, T) = 1 - \mathbb{L}(0, \tau) \quad (5.16)$$

It can be seen from Eq. (5.16) that the failure probability within a reference period of $[0, \tau]$ equals the CDF of the time to failure valued at τ.

Recall that in Sect. 4.1, we mentioned the structural resistance may take different forms depending on the structural target performance of interest. This also applies to the time-dependent reliability problem as in Eq. (5.14). For example, if we define the structural failure as the case where the cumulative structural damage exceeds a predefined threshold, then in Eq. (5.14), we can treat $S(t)$ as the cumulative damage up to time t, and $R(t) = $ the predefined threshold (either time-variant or constant). This type of failure will further be discussed in Sect. 5.4.5.2.

Note that Eq. (5.14) presents a general definition of the time-dependent reliability problem, where the involvement of the time domain makes the solution of Eq. (5.14) more difficult than that of a classic reliability problem (c.f. Eqs. (4.1) and (5.1)). The basic ideas to simplify/solve Eq. (5.14) will be discussed in the following.

To begin with, we consider a simple case of Eq. (5.14) where the resistance does not vary with time while the load process is time-variant, as shown in Fig. 5.1b. With this, the reliability for $[0, T]$ can be estimated by

$$\mathbb{L}(0, T) = \mathbb{P}(R > S_{\max}) \qquad (5.17)$$

where $S_{\max} = \max\{S(t)|t \in [0, T]\}$. As such, the time-dependent reliability problem is converted into a classic one since both R and $S_{\max}$ are random variables.

Example 5.6

For a building subjected to wind hazards located at Miami-Dade County, suppose that the future wind scenario is as in Example 5.5. The resistance (wind-resisting capacity) of the building follows a lognnormal distribution with a mean value of 100 m/s and a COV of 0.2. Compute the probability of failure over subsequent reference periods up to 50 years.

Solution

Let R denote the resistance of the building, and S_i the maximum wind speed for the ith year, $i = 1, 2, \ldots 50$. For a service period of $[0, n]$ (years), the probability of failure is, according to Eq. (5.15), calculated as follows,

$$\mathbb{P}_f(0, n) = 1 - \mathbb{L}(0, n) = 1 - \mathbb{P}(R > S_1 \cap R > S_2 \cap \cdots \cap R > S_n), \quad n = 1, 2, \ldots 50$$
$$(5.18)$$

Using the law of total probability, it follows,

Fig. 5.9 The probability of failure for reference periods up to 50 years

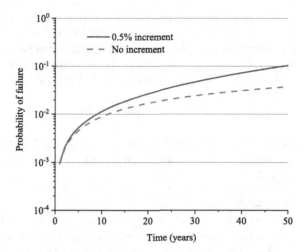

$$\mathbb{P}_f = 1 - \int_0^\infty \mathbb{P}(r > S_1 \cap r > S_2 \cap \cdots \cap r > S_n) \cdot f_R(r)\mathrm{d}r$$

$$= 1 - \int_0^\infty \mathbb{P}[r > \max(S_1, S_2, \ldots S_n)] \cdot f_R(r)\mathrm{d}r \tag{5.19}$$

$$= 1 - \int_0^\infty F_{S_{\max,n}}(r) \cdot f_R(r)\mathrm{d}r, \quad n = 1, 2, \ldots 50$$

where $f_R(r)$ is the PDF of R, $S_{\max,n} = \max(S_1, S_2, \ldots S_n)$, and $F_{S_{\max,n}}(r)$ is the CDF of $S_{\max,n}$. Similar to Eq. (5.12), one has

$$F_{S_{\max,n}}(r) = \prod_{i=1}^{n} \left\{ 1 - \exp\left[-\left(\frac{r}{u\left(1 + \frac{i}{100}\right)} \right)^\alpha \right] \right\}, \quad r \geq 0 \tag{5.20}$$

where $u = 27.301$ and $\alpha = 1.769$. Thus,

$$\mathbb{P}_f(0, n) = 1 - \int_0^\infty \prod_{i=1}^{n} \left\{ 1 - \exp\left[-\left(\frac{r}{u\left(1 + \frac{i}{100}\right)} \right)^\alpha \right] \right\} \cdot f_R(r)\mathrm{d}r \tag{5.21}$$

Equation (5.21) can be solved numerically. The result is presented in Fig. 5.9. Clearly, a longer reference period will lead to a greater failure probability, due to the accumulation of the wind hazards. This observation also demonstrates the dependence of the time-dependent reliability on the duration of the reference period of interest. For comparison purpose, the failure probability associated with a stationary wind process (i.e., not considering the annual increment in the wind speed) is also presented in Fig. 5.9. Evidently, the annual increase in the wind speed results in a greater failure probability.

If the resistance in Eq. (5.17) also varies with time, then the reliability within a service period of $[0, T]$ can not be transformed into a classical problem by simply considering two variables that are representative of both $R(t)$ and $S(t)$. For instance, if one compares the maximum of the load effect and the minimum of resistance to evaluate the reliability, denoted by $\mathbb{L}'(0, T)$, it follows,

$$\mathbb{L}'(0, T) = \mathbb{P}(R_{min} > S_{max}) \tag{5.22}$$

where $S_{max} = \max\{S(t)\}$ and $R_{min} = \min\{R(t)\}$ for $t \in [0, T]$. It is demonstrated in Example 5.7 that the structural reliability may be significantly underestimated by $\mathbb{L}'(0, T)$.

Example 5.7

Suppose that $R(t) = X_R \cdot \left(4 - \frac{t}{50}\right)$, and $S(t) = X_S \cdot \left(2 - \frac{t}{50}\right)$, where the time t is in years, X_R and X_S are two statistically independent and identically distributed variables, following a lognormal distribution with a mean value of 1 and a standard deviation of 0.2.
(1) Calculate the probability of structural failure within a service period of 50 years using Eq. (5.14).
(2) Repeat (1) with Eq. (5.22), and comment on the results.
Solution
(1) According to Eq. (5.14), the failure probability is given by

$$\mathbb{P}_f(0, 50) = 1 - \mathbb{P}(4X_R > 2X_S \cap 3X_R > X_S) = 1 - \mathbb{P}\left(\frac{X_R}{X_S} > \frac{1}{2}\right) = 0.0067 \tag{5.23}$$

(2) The failure probability in Eq. (5.22), which is estimated by comparing the maximum of $S(t)$ and the minimum of $R(t)$, denoted by $\mathbb{P}'_f(0, 50)$, is

$$\mathbb{P}'_f(0, 50) = 1 - \mathbb{P}(3X_R > 2X_S) = 1 - \mathbb{P}\left(\frac{X_R}{X_S} > \frac{2}{3}\right) = 0.0738 \approx 11\mathbb{P}_f(0, 50) \tag{5.24}$$

Thus, the failure probability is significantly overestimated by simply considering the maximum of the load effect and the minimum of the resistance.

Clearly, the time-variant characteristics of both $R(t)$ and $S(t)$ should be reasonably addressed in time-dependent reliability analyses. In an attempt to perform the reliability assessment on the temporal scale (c.f. Eq. (5.15)), a straightforward idea is to use a generalization of Eq. (5.1) in the time domain taking the form of

$$\mathbb{L}_{in}(t) = \mathbb{P}[R(t) > S(t)] \tag{5.25}$$

which gives an estimate of the instantaneous reliability at time t. In Eq. (5.25), the subscription "$_{in}$" is to emphasize that the reliability is an instantaneous estimate at a specific time point. Given that the PDFs of $R(t)$ and $S(t)$ are $f_{R(t)}$ and $f_{S(t)}$, and that the CDFs of $R(t)$ and $S(t)$ are $F_{R(t)}$ and $F_{S(t)}$ respectively, similar to Eq. (4.2), the reliability $\mathbb{L}_{in}(t)$ can be estimated as follows,

$$\mathbb{L}_{in}(t) = \int_0^\infty [1 - F_{R(t)}(s)] f_{S(t)}(s) ds = \int_0^\infty F_{S(t)}(r) f_{R(t)}(r) dr \qquad (5.26)$$

It is noticed that Eqs. (5.25) and (5.26) work only if the following two requirements are satisfied simultaneously: (1) $S(t)$ is non-decreasing; and (2) $R(t)$ is non-increasing. Conditional on these two requirements, for a reference period of $[0, T]$, the reliability $\mathbb{L}(0, T)$ can be simply measured by $\mathbb{L}_{in}(T)$ in Eq. (5.25), as illustrated in Fig. 5.1c. An example is given as follows to demonstrate the applicability of Eq. (5.25).

Example 5.8

Recall Example 5.1, where the RC beam is subjected to the chloride ingress-induced corrosion of steel bars. We consider a limit state that the chloride concentration at the steel surface $C(x, t)$ reaches a predefined threshold c_{cr} (i.e., the initiation of steel corrosion). With this, we treat the threshold c_{cr} as the structural resistance, which is time-invariant (non-increasing), and the chloride concentration $C(x, t)$ as the time-variant load effect $S(t)$ at time t (non-decreasing with time). The beam is deemed as "failure" (failure to satisfy the durability requirements) once $C(x, t) > c_{cr}$.

Given that the time of corrosion initiation follows a normal distribution with a mean value of 30 years and a COV of 0.2, compute the probability of failure for a reference period of 50 years (note that the definition of structural failure in this example differs from that in Example 5.1).

Solution

Since the "resistance", $R(t) \equiv c_{cr}$, is non-increasing, and the "load effect" $S(t)$ is non-decreasing on the temporal scale, Eq. (5.25) applies. Thus, the failure probability with in a reference period of $[0, T]$ is evaluated by

$$\mathbb{P}_f(0, T) = 1 - L(0, T) = 1 - \mathbb{L}_{in}(T) = 1 - \mathbb{P}[C(x, T) \leq c_{cr}] \qquad (5.27)$$

Let $F_{t_i}(t)$ denote the CDF of the time of corrosion initiation, t_i. By definition,

$$F_{t_i}(t) = \mathbb{P}(t_i \leq t) = \mathbb{P}(\text{Failure occurs within } [0, t]) = \mathbb{P}[C(x, t) > c_{cr}] \qquad (5.28)$$

Thus,

$$\mathbb{P}_f(0, T) = 1 - F_{t_i}(T) \qquad (5.29)$$

with which

$$\mathbb{P}_f(0, 50) = 1 - \Phi\left(\frac{50 - 30}{30 \times 0.2}\right) = 4.29 \times 10^{-4} \qquad (5.30)$$

That is, the probability of failure for a reference period of 50 years is 4.29×10^{-4}, if the limit state is that the corrosion at the steel surface initiates. Compared with the result in Question (1) of Example 5.1, it can be seen that different definitions of structural failure (limit state) may lead to different failure probabilities.

Recall that Eq. (5.25) is based on the monotonicity requirements on both $R(t)$ and $S(t)$. While for typical engineered structures, $R(t)$ is non-increasing without maintenance or repair measures, the monotonicity of the load process $S(t)$ as in Fig. 5.1c is often not the case in practical engineering. Nonetheless, Eq. (5.25) does provide a snapshot for the instantaneous risk that the structure suffers from.

More generally, the time-variant load $S(t)$ can be modelled as a stochastic process with random fluctuation, as illustrated in Fig. 5.1d. Correspondingly, the time-dependent reliability within a reference period of $[0, T]$ should be estimated according to Eq. (5.14) with advanced methods, where the analysis, for most cases, cannot be simply transformed into a classical reliability problem.

The resistance process is by nature a continuous process; the modelling techniques will be discussed in Sect. 5.2. The load process, however, can be typically classified into two types: continuous and discrete, as schematically illustrated in Fig. 5.10 [66]. Intuitively, the two categories of load process can be converted into each other mutually when used for structural reliability assessment. For instance, for a continuous load process $S(t)$ as in Fig. 5.10a, only those extreme loads with a severe intensity (say, exceeding a predefined threshold) will pose a significant threat to structural safety; these significant loads can be modelled as a discrete process, as shown in Fig. 5.10b. Inversely, a discrete load process can be treated as a continuous load process $S(t) = 0$ superimposed by the discrete extreme loads S_i for $i = 1, 2, \ldots n$, as in Fig. 5.10b. However, the methods for time-dependent reliability assessment considering the two types of load processes may be different, providing different insights into the evaluation of structural safety level.

Fig. 5.10 Comparison of two types of load process: continuous and discrete

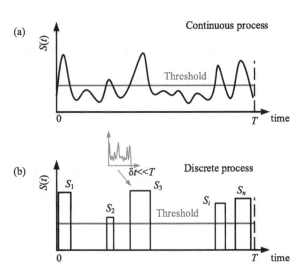

In this Chapter, both the discrete and continuous load processes will be discussed (c.f. Sect. 5.3). For the case of a discrete load process, we assume that the load intensity varies negligibly during the interval in which it occurs without dynamic response, and the duration of each load δt is small enough compared with the whole service period of interest (i.e., $\delta t \ll T$, where T is the duration of the considered reference period). The time-dependent reliability assessment will be discussed in Sect. 5.4 considering a discrete load process. The reliability analysis in the presence of a continuous load process will be further discussed in Sect. 5.5.

5.2 Resistance Deterioration Process

The deterioration process of structural resistance (e.g., strength or stiffness) may include multifarious mechanisms due to the complexity of the environmental factors [9, 57], and thus is difficult to predict based on the mechanism analysis only. For instance, a bridge pier in a marine environment may be exposed to the joint action of chloride ingress-induced corrosion [22, 24] and random ship collision [41, 45]. Some existing studies have used the data from experiments in laboratories to capture the deterioration characteristics [3, 12, 27, 30, 67]. However, the extrapolation of experimental data is often questionable when used for reliability analysis of realistic structures [9], due to the fact that the in situ environmental conditions may differ significantly from those in the laboratory. Moreover, the data fitting-based models may suffer from biased estimate in the presence of limited size of samples. In an attempt to avoid the drawbacks of purely fitting-based deterioration models, some researches have employed the stochastic processes to reflect the main characteristics of deterioration, where the unknown parameters can be calibrated with the help of observed data.

Mathematically, for an aging structure, the time-variation of structural resistance at time t, $R(t)$, can be described as follows,

$$R(t) = R_0 \cdot G(t) \tag{5.31}$$

where R_0 is the initial resistance (a random variable), and $G(t)$ is the deterioration function, which is a monotonically non-increasing stochastic process without maintenance or rehabilitation measures. Mori and Ellingwood [32] used the following expression to model the deterioration function,

$$g(t) = 1 - a \cdot t^\eta \tag{5.32}$$

where a is a parameter that accounts for the deterioration rate, and η is a parameter that reflects the dominant deterioration type. For example, for the deterioration mechanisms of corrosion, sulfate attack or diffusion-controlled aging, respectively, η takes a value of 1, 2 or 0.5 [32]. In Eq. (5.32), $g(t)$ is used instead of $G(t)$ to emphasize that the deterioration process herein is a deterministic one.

Example 5.9

Recall Example 5.1. Show that a linear relationship could be employed to model the time-dependent behaviour of $M_u(t)$ based on Eqs. (5.2) and (5.3).

Solution

With Eq. (5.3), for typical cases where the variation of $\frac{f_y \Delta A_s}{2 f_c b}$ with time is negligible compared with $h_0 - \frac{f_y A_0}{2 f_c b}$, one has

$$
\begin{aligned}
M_u(t) &= f_y A_s \left(h_0 - \frac{f_y A_s}{2 f_c b} \right) = f_y (A_0 - \Delta A_s) \left[h_0 - \frac{f_y (A_0 - \Delta A_s)}{2 f_c b} \right] \\
&\approx f_y (A_0 - \Delta A_s) \left(h_0 - \frac{f_y A_0}{2 f_c b} \right) \\
&= f_y [A_0 - k \psi_0 \cdot i_{cor} \cdot (t - t_i)] \left(h_0 - \frac{f_y A_0}{2 f_c b} \right)
\end{aligned}
\tag{5.33}
$$

which indicates that M_u decreases linearly with time after the corrosion initiation, which is consistent with that reported in [32].

In order to incorporate the uncertainty associated with the deterioration process, one may use the following equations to model the time-variation of a deterioration process $G(t)$ (or an indicative parameter of deterioration such as the loss of steel cross section area, depth of penetration, and others) [9]

$$
\begin{aligned}
G(t) &= \alpha(t - t_i)^\beta \epsilon_1(t); & t \geq t_i \\
\text{or } G(t) &= \alpha(t - T_i)^\beta + \epsilon_2(t); & t \geq t_i
\end{aligned}
\tag{5.34}
$$

where t_i is the time of deterioration (e.g., corrosion) initiation, α and β are two parameters that reflect the deterioration rate and type respectively, $\epsilon_1(t)$ and $\epsilon_2(t)$ are two random error terms, which can be modelled customarily as random variables (lognormal for ϵ_1 and normal for ϵ_2) or stochastic processes [28]. However, the monotonicity (non-increasing characteristics) of the deterioration process cannot be guaranteed by the formulas in Eq. (5.34), since ϵ_1 may be smaller than 1, or ϵ_2 may be negative.

To guarantee the monotonicity of the deterioration process, some researchers have also used the Gamma process to describe the resistance deterioration of aging structures [7, 16, 25, 39, 40, 57], which yields a simple and efficient expression for the deterioration process. For a reference period of $[0, T]$, the deterioration function $G(t), t \in [0, T]$, is derived through dividing the interval $[0, t]$ into n identical sections, i.e., $[t_0 = 0, t_1], [t_1, t_2], \dots [t_{n-1}, t_n = t]$. If the increment of deterioration within the ith interval is denoted by D_i, which is a Gamma random variable, it follows,

$$
G(t) = G(t_n) = 1 - \sum_{i=1}^{n} D_i
\tag{5.35}
$$

If each D_i has an identical scale parameter, then $\sum D_i$ also follows a Gamma distribution with the same scale parameter of each D_i (c.f. Sect. 2.2.7). This property indeed ensures that the deterioration function $G(t)$ at an arbitrary time could be described by a Gamma distribution and thus is beneficial for practical use. Moreover, the model in Eq. (5.35) describes a monotonically non-increasing process and accounts for the auto-correlation in the deterioration process (i.e., $G(t_i)$ and $G(t_j)$ are correlated for two times t_i and t_j). In fact, according to Eq. (5.35), the correlation coefficient between $G(t_i)$ and $G(t_j)$ for times t_i and $t_j (t_j > t_i)$ is determined by

$$\rho(G(t_i), G(t_j)) = \frac{\mathbb{C}\left(G(t_i), G(t_i) - \sum_{k=i+1}^{j} D_k\right)}{\sqrt{\mathbb{V}(G(t_i))\mathbb{V}(G(t_j))}} = \sqrt{\frac{\mathbb{V}(G(t_j))}{\mathbb{V}(G(t_i))}} \qquad (5.36)$$

The importance of reasonably modelling the auto-correlation in the deterioration process for structural reliability assessment is demonstrated in Example 5.10.

Example 5.10

For an aging structure, suppose that the resistances at times t_1 and t_2 ($t_2 > t_1$) are R_1 and R_2, both of which are Gamma distributed and normalized with respective to the nominal load effect S_n. R_1 has a shape parameter of 7.5 and a scale parameter of $\frac{1}{3}$; R_2 has a shape parameter of 6 and a scale parameter of $\frac{1}{3}$. It is known that totally two load events occur, denoted by S_1 and S_2, at times t_1 and t_2 respectively. The load intensity of both S_1 and S_2 has an Extreme Type I distribution with a mean value of S_n and a COV of 0.3. Compute the probability of structural failure for the following four cases. Assume that S_1 and S_2 are mutually independent.
(1) R_1 and R_2 are fully correlated.
(2) R_1 and R_2 are statistically independent.
(3) $R_1 = R_2 + \widetilde{R}$, where $\widetilde{R}$ is a Gamma variable with a shape parameter of 1.5 and a scale parameter of $\frac{1}{3}$. Assume that $\widetilde{R}$ is independent of R_2.
(4) R_1 and R_2 are correlated, and the correlation coefficient is the same as in (3).
Solution
In this example, the probability of failure, $\mathbb{P}_f$, is calculated as follows,

$$\mathbb{P}_f = 1 - \mathbb{L} = 1 - \mathbb{P}(R_1 > S_1 \cap R_2 > S_2) \qquad (5.37)$$

(1) Since R_1 and R_2 are fully correlated, one has

$$\mathbb{P}_f = 1 - \mathbb{P}\left[F_{R_1}(R_1) > F_{R_1}(S_1) \cap F_{R_2}(R_2) > F_{R_2}(S_2)\right]$$

$$= 1 - \mathbb{P}\left[U > F_{R_1}(S_1) \cap U > F_{R_2}(S_2)\right] = 1 - \int_0^1 \mathbb{P}\left[u > F_{R_1}(S_1) \cap u > F_{R_2}(S_2)\right] du$$

$$= 1 - \int_0^1 \mathbb{P}\left[S_1 < F_{R_1}^{-1}(u) \cap S_2 < F_{R_2}^{-1}(u)\right] du$$

$$= 1 - \int_0^1 F_{S_1}[F_{R_1}^{-1}(u)] \cdot F_{S_2}[F_{R_2}^{-1}(u)] du = 0.1411$$

$$(5.38)$$

where U is a uniform variable within $[0, 1]$, F_{R_1} and F_{R_2} are the CDFs of R_1 and R_2, F_{S_1} and F_{S_2} are the CDFs of S_1 and S_2, respectively.

(2) The deterioration process is modelled as statistically independent. Then

$$
\begin{aligned}
\mathbb{P}_f &= 1 - \mathbb{P}(R_1 > S_1 \cap R_2 > S_2) = 1 - \mathbb{P}(R_1 > S_1) \cdot \mathbb{P}(R_2 > S_2) \\
&= 1 - \left[\int_0^\infty F_{S_1}(r) f_{R_1}(r) dr \right] \cdot \left[\int_0^\infty F_{S_2}(r) f_{R_2}(r) dr \right] = 0.1605
\end{aligned}
\tag{5.39}
$$

(3) The correlation between R_1 and R_2, as well as the monotonicity (i.e., $R_1 \geq R_2$), is guaranteed by $R_1 = R_2 + \widetilde{R}$. With this, it follows,

$$
\begin{aligned}
\mathbb{P}_f &= 1 - \mathbb{P}(R_1 > S_1 \cap R_2 > S_2) = 1 - \mathbb{P}(R_2 + \widetilde{R} > S_1 \cap R_2 > S_2) \\
&= 1 - \int_0^\infty \int_0^\infty F_{S_1}(r_2 + \widetilde{r}) F_{S_2}(r_2) f_{R_2}(r_2) f_{\widetilde{R}}(\widetilde{r}) dr_2 d\widetilde{r} = 0.1442
\end{aligned}
\tag{5.40}
$$

Since $R_1 = R_2 + \widetilde{R}$, the correlation coefficient between R_1 and R_2 is uniquely determined by

$$
\rho(R_1, R_2) = \frac{\mathbb{C}(R_2 + \widetilde{R}, R_2)}{\sqrt{\mathbb{V}(R_1)\mathbb{V}(R_2)}} = \sqrt{\frac{\mathbb{V}(R_2)}{\mathbb{V}(R_1)}} = 0.8944
\tag{5.41}
$$

(4) The correlation between R_1 and R_2 is considered, with a linear correlation coefficient of ρ. To begin with, we transfer R_1 and R_2 into two standard normal variables Y_1 and Y_2 respectively with $F_{R_1}(R_1) = \Phi(Y_1)$ and $F_{R_2}(R_2) = \Phi(Y_2)$. The correlation coefficient between Y_1 and Y_2, ρ', is determined according to Eq. (3.37) as in Sect. 3.3. The numerical solution gives $\rho' = \rho(1.002 - 0.012c_1 + 0.125c_1^2 + 0.022\rho + 0.001\rho^2 - 0.077\rho c_1 - 0.012c_2 + 0.125c_2^2 - 0.077\rho c_2 + 0.014c_1c_2)$ with a maximum error of 4% [29], where c_1 and c_2 are the COVs of R_1 and R_2. With this, we introduce a normal variable $\widetilde{Y}$ which is independent of Y_1 and satisfies $Y_2 = \rho'Y_1 + \widetilde{Y}$. The mean value and standard deviation of $\widetilde{Y}$ are accordingly obtained as 0 and $\sqrt{1 - \rho'^2}$ respectively. Further, one has

$$
\begin{aligned}
\mathbb{P}_f &= 1 - \mathbb{P}(R_1 > S_1 \cap R_2 > S_2) = 1 - \mathbb{P}\left[F_{R_1}(R_1) > F_{R_1}(S_1) \cap F_{R_2}(R_2) > F_{R_2}(S_2)\right] \\
&= 1 - \mathbb{P}\left[\Phi(Y_1) > F_{R_1}(S_1) \cap \Phi(Y_2) > F_{R_2}(S_2)\right] \\
&= 1 - \mathbb{P}\left[Y_1 > \Phi^{-1}[F_{R_1}(S_1)] \cap \rho'Y_1 + \widetilde{Y} > \Phi^{-1}[F_{R_2}(S_2)]\right] \\
&= 1 - \int_{-\infty}^\infty \int_{-\infty}^\infty F_{S_1}\left[F_{R_1}^{-1}(\Phi(y_1))\right] \cdot F_{S_2}\left[F_{R_2}^{-1}(\Phi(\rho'y_1 + \widetilde{y}))\right] f_{Y_1}(y_1) f_{\widetilde{Y}}(\widetilde{y}) dy_1 d\widetilde{y} \\
&= -0.0008\rho^3 - 0.0064\rho^2 - 0.0120\rho + 0.1605
\end{aligned}
\tag{5.42}
$$

In Eq. (5.42), assigning $\rho = 1$ and 0 respectively gives a failure probability of 0.1413 and 0.1605, consistent with Cases 1 (fully correlated deterioration) and 2 (independent deterioration). Moreover, if $\rho = 0.8944$ in Eq. (5.42), then $\mathbb{P}_f$ equals 0.1440, which agrees well with that given by Case 3. This comparison suggests that for the purpose of structural reliability assessment, the deterioration process is sufficiently described by considering the probability information of $G(t)$ at each time and the autocorrelation between different times.

Example 5.11

Recall the deterioration function in Eq. (5.35). We use the following expression for Eq. (5.35) [20],

$$G(t_n) = 1 - \sum_{i=1}^{n} \hat{d}(t_i) \cdot \epsilon(t_i) \tag{5.43}$$

in which $\hat{d}(t_i)$ is the average degradation between t_{i-1} and t_i, taking a form of $\hat{d}(t_i) = k \cdot t^m \cdot |t_i - t_{i-1}|$, $\epsilon(t)$ is a sequence of independent Gamma random variables, with a time-dependent shape factor of $\alpha(t) = \frac{\hat{d}(t)}{\xi}$ and a scale factor of $\beta(t) = \frac{\xi}{\hat{d}(t)}$. As n is large enough,

(1) Derive the mean value and standard deviation of $G(t)$ in Eq. (5.43) in terms of m, k and ξ.

(2) Find the correlation coefficient of $G(t_k)$ and $G(t_j)$ for times t_k and t_j $(k \geq j)$.

Solution

(1) Since $\epsilon(t)$ has a shape factor of $\alpha(t)$ and a scale factor of $\beta(t)$, the mean value and variance of $\epsilon(t)$ are respectively,

$$\mathbb{E}(\epsilon(t)) = 1, \quad \mathbb{V}(\epsilon(t)) = \frac{\hat{d}(t)}{\xi} \cdot \frac{\xi^2}{\hat{d}^2(t)} = \frac{\xi}{\hat{d}(t)} \tag{5.44}$$

Thus,

$$\mathbb{E}(\hat{d}(t) \cdot \epsilon(t)) = \hat{d}(t), \quad \mathbb{V}(\hat{d}(t) \cdot \epsilon(t)) = \hat{d}^2(t)\frac{\xi}{\hat{d}(t)} = \xi\hat{d}(t) \tag{5.45}$$

With this, the mean and the standard deviation of the degradation function at time $t = t_n$, $G(t)$ are determined as follows,

$$\mathbb{E}(G(t)) = 1 - \sum_{i=1}^{n} \mathbb{E}(\hat{d}(t_i) \cdot \epsilon(t_i)) = 1 - \sum_{i=1}^{n} \hat{d}(t_i) = 1 - \int_{0}^{t} k\tau^m d\tau = 1 - \frac{k}{m+1}t^{m+1}$$

$$\mathbb{V}(G(t)) = \sum_{i=1}^{n} \mathbb{V}(\hat{d}(t_i) \cdot \epsilon(t_i)) = \xi \sum_{i=1}^{n} \hat{d}(t_i) = \xi \int_{0}^{t} k\tau^m d\tau = \xi\frac{k}{m+1}t^{m+1}$$

$$\tag{5.46}$$

yielding $\sigma_{G(t)} = \sqrt{\frac{k\xi}{m+1}t^{m+1}}$.

(2) Since $\epsilon(t_k)$ and $\epsilon(t_j)$ are independent, the auto-covariance of $G(t)$ at two different time instants, t_k and t_j $(k \geq j)$, is

$$\mathbb{C}(G(t_j), G(t_k)) = \mathbb{C}\left(G(t_j), G(t_j) - \sum_{i=j+1}^{k} \hat{d}(t_i) \cdot \epsilon(t_i)\right) = \mathbb{V}(G(t_j)) \tag{5.47}$$

Thus, the correlation coefficient of $G(t_k)$ and $G(t_j)$, $\rho_{k,j}$, is determined by

$$\rho_{k,j} = \frac{\sigma_{G(t_j)}}{\sigma_{G(t_k)}} = \sqrt{\frac{t_j^{m+1}}{t_k^{m+1}}} \tag{5.48}$$

The disadvantage of the Gamma process-based deterioration model is that (1) the dependence of the variance of $G(t)$ on its mean value is time-invariant; and (2) the auto-correlation between the deterioration functions at two different times is uniquely determined once the mean values of $G(t)$ at the two times are known. The former is explained by the fact that each increment D_i has an identical scale parameter (say, b), and thus $\mathbb{V}\left(\sum_{k=1}^{i} D_k\right) = b\mathbb{E}\left(\sum_{k=1}^{i} D_k\right)$. The latter is explained by applying some simple algebra to Eq. (5.36), which gives

$$
\rho(G(t_i), G(t_j)) = \sqrt{\frac{\mathbb{V}(1 - G(t_j))}{\mathbb{V}(1 - G(t_i))}} = \sqrt{\frac{\mathbb{V}\left(\sum_{k=1}^{j} D_k\right)}{\mathbb{V}\left(\sum_{k=1}^{i} D_k\right)}} \sqrt{\frac{\mathbb{E}\left(\sum_{k=1}^{j} D_k\right)}{\mathbb{E}\left(\sum_{k=1}^{i} D_k\right)}}
$$

$$(5.49)$$

The aforementioned deterioration models in Eqs. (5.32), (5.34) and (5.35) have considered the degrading process as independent of the load process. This is reasonable when the deterioration process is dominated by the environmental factor-driven mechanisms such as the chloride ingress-induced corrosion. However, in many practical cases, the external load process will also have an impact on the deterioration process. One simple example is that a greater load intensity is more likely to result in a severer physical damage to the structure (e.g., cracks on the concrete surface) and subsequently an acceleration to the deterioration process [36]. Also, the accumulated effects of repeated significant loads will lead to a more fragile structure subjected to the load actions [2, 18, 19], indicating again the potential dependence of the deterioration process on the applied load process. Naturally, a model is needed to consider the deterioration process as a combination of gradual and shock deteriorations, and to model the mutual correlations between gradual deterioration, shock deterioration and the magnitude of load effect [62].

For a reference period of $[0, T]$, suppose that a load sequence $\{\hat{S}_{t_i}\}$ $(i = 1, 2, \ldots n)$ occurs at times $t_1, t_2, \ldots t_n$. Modelling the deterioration process as a combination of gradual and shock deteriorations, the resistance at time t_n, $R(t_n)$, is obtained as follows [62]

$$
R(t_n^-) = R_0 - \sum_{i=1}^{n-1} \Delta R_{t_i}^S - \sum_{i=1}^{n} \Delta R_{t_i}^G, \quad R(t_n^+) = R(t_n^-) - \Delta R_{t_n}^S \qquad (5.50)
$$

where t_n^- and t_n^+ are the time instants immediately before and after t_n, $\Delta R_{t_i}^S$ is the shock deterioration at time t_i, and $\Delta R_{t_i}^G$ is the gradual deterioration during time interval $[t_{i-1}, t_i]$ for $i = 1, 2, \ldots n$. Further, with Eq. (5.31), the deterioration function at time t_n, $G(t_n)$, is given by

$$
G(t_n^-) = 1 - \sum_{i=1}^{n-1} \Delta Q_{t_i} - \sum_{i=1}^{n} \Delta P_{t_i}, \quad G(t_n^+) = 1 - \sum_{i=1}^{n} \Delta Q_{t_i} - \sum_{i=1}^{n} \Delta P_{t_i} \quad (5.51)
$$

Fig. 5.11 The mutual correlation between S_{t_i}, $\Delta P_{t_{i+1}}$ and ΔQ_{t_i} in the modelling of resistance deterioration

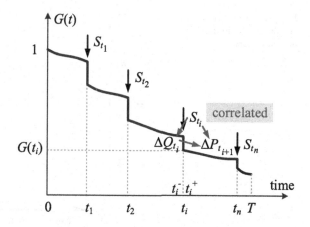

where $\Delta Q_{t_i} = \Delta R_{t_i}^S/R_0$ and $\Delta P_{t_i} = \Delta R_{t_i}^G/R_0$ for $i = 1, 2, \ldots n$. The generalized form of Eq. (5.51) for time τ is given as follows,

$$G(\tau^-) = 1 - \left(\sum_{i=1}^{n_1(\tau)} \Delta Q_{t_i} + \sum_{i=1}^{n_1(\tau)+1} \Delta P_{t_i} \right), \quad G(\tau^+) = 1 - \left(\sum_{i=1}^{n_2(\tau)} \Delta Q_{t_i} + \sum_{i=1}^{n_2(\tau)+1} \Delta P_{t_i} \right)$$
(5.52)

where $n_1(\tau) = \max\{j : t_j < \tau\}$, and $n_2(\tau) = \max\{j : t_j \leq \tau\}$.

Due to the fact that the physical relationship between the deteriorations (shock and gradual) and the magnitude of shock event is often challenging to determine in many cases, which practically contains significant uncertainties and nonlinear mechanisms, a positive correlation coefficient can be used to reflect the dependence of the deterioration process on the shock intensity [62], i.e., both $\Delta P_{t_{i+1}}$ and ΔQ_{t_i} in Eq. (5.52) are correlated with the load effect S_{t_i} for each $i = 1, 2, \ldots n$, where $t_{n+1} = T$, and $S_{t_i} = \hat{S}_{t_i}/R_0$ is the normalized load effect by the initial resistance. Also, $\Delta P_{t_{i+1}}$ and ΔQ_{t_i} are positively correlated, as stated before. With this, $\Delta P_{t_{i+1}}$, ΔQ_{t_i} and S_{t_i} are mutually correlated with each other, as illustrated in Fig. 5.11.

The Gaussian copula function (c.f. Sect. 2.3.3) can be used to construct the joint CDF of the correlated $\Delta P_{t_{i+1}}$, ΔQ_{t_i} and S_{t_i}. Recall that, mathematically, for mutually correlated random variables X, Y and Z with marginal CDFs of $F_X(x)$, $F_Y(y)$ and $F_Z(z)$, respectively, the joint CDF, $F_{X,Y,Z}(x, y, z)$, is given as follows with the help of the Gaussian copula function,

$$F_{X,Y,Z}(x, y, z) = \Phi_G \left[\Phi^{-1}(F_X(x)), \Phi^{-1}(F_Y(y)), \Phi^{-1}(F_Z(z)) \right] \quad (5.53)$$

where Φ_G is the joint CDF of a multivariate standard normal distribution. It is noticed that with Eq. (5.53), the dependency of the resistance deterioration (either $\Delta P_{t_{i+1}}$ or ΔQ_{t_i}) on the load intensity (S_{t_i}) can be reflected by the conditional distribution on S_{t_i}. Example 5.12 in the following demonstrates the conditional distribution of ΔQ_{t_i} on S_{t_i}.

Fig. 5.12 The PDF of ΔQ_{t_i} and the unconditional PDFs on $s = 0.2$ and 0.4 respectively

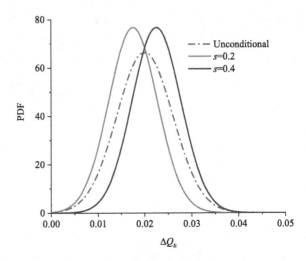

Example 5.12

Suppose that in Fig. 5.11, S_{t_i} follows a normal distribution with a mean value of 0.3 and a COV of 0.4; ΔQ_{t_i} also follows a normal distribution with a mean value of 0.02 and a COV of 0.3. The correlation coefficient between ΔQ_{t_i} and S_{t_i} equals $\rho = 0.5$. Discuss on the conditional distribution of ΔQ_{t_i} on S_{t_i}.

Solution

Due to the normality of both S_{t_i} and ΔQ_{t_i}, we introduce a normal variable S_0 and a constant k such that

$$\Delta Q_{t_i} = k S_{t_i} + S_0 \tag{5.54}$$

The correlation coefficient between ΔQ_{t_i} and S_{t_i} is obtained as $k\sqrt{\dfrac{\mathbb{V}(S_{t_i})}{\mathbb{V}(\Delta Q_{t_i})}}$ according to Eq. (5.54), which is equal to $\rho = 0.5$. Thus, k is obtained as 0.025. Furthermore, the mean value and standard deviation of S_0 are obtained as 0.0125 and 0.0052, respectively. Conditional on a realization of S_{t_i}, s, Eq. (5.54) becomes $\Delta Q_{t_i} = ks + S_0$, indicating that $\Delta Q_{t_i}|s$ also follows a normal distribution with a mean value of $0.025s + 0.0125$ and a standard deviation of 0.0052. Clearly, the standard deviation of ΔQ_{t_i} is reduced conditional on a realization of S_{t_i}, s. Figure 5.12 shows the conditional PDF of ΔQ_{t_i} on $s = 0.2$ and 0.4 respectively. For the purpose of comparison, the unconditional PDF of ΔQ_{t_i} is also presented in Fig. 5.12. It can be seen that due to the positive correlation coefficient between ΔQ_{t_i} and S_{t_i}, if s is smaller than its mean value (0.3 herein), then the conditional PDF of ΔQ_{t_i} shifts leftwards; similarly, if s is greater than $\mathbb{E}(S_{t_i})$, the conational PDF moves rightwards.

More generally, other types of copula function can also be used to construct the joint distribution of correlated random variables. Here we discuss the correlation between shock deterioration (ΔQ_{t_i}) and load effect (S_{t_i}) only, as an extension of the case in Example 5.12. A schematic diagram for their joint probability distribution, denoted by $h_{S_{t_i}, \Delta Q_{t_i}}(x, y)$, is as in Fig. 4.3, if interpreting S, R in Fig. 4.3 as ΔQ_{t_i}, S_{t_i} respectively. Due to their mutual correlation, with a given value of S_{t_i}, s, the conditional PDF of ΔQ_{t_i}, $f_{\Delta Q_{t_i}|s}$, differs from the marginal distribution $f_{\Delta Q_{t_i}}$, and vice versa. With the marginal distributions of S_{t_i} and ΔQ_{t_i}, denoted by $F_{S_{t_i}}$ and $F_{\Delta Q_{t_i}}$ respectively, according to the Sklar's Theorem (c.f. Eq. (2.326)), there always exists a copula function C that satisfies $H_{S_{t_i}, \Delta Q_{t_i}}(x, y) = C(F_{S_{t_i}}(x), F_{\Delta Q_{t_i}}(y))$, where $H_{S_{t_i}, \Delta Q_{t_i}}(x, y)$ is the joint CDF of S_{t_i} and ΔQ_{t_i}. Thus, the conditional CDF of ΔQ_{t_i} on $S_{t_i} = s$ is given by (c.f. Example 3.13)

$$F_{\Delta Q_{t_i}|s}(\tau) = \lim_{\Delta s \to 0} \frac{\mathbb{P}(\Delta Q_{t_i} \leq \tau \cap s - \Delta s < S_{t_i} \leq s)}{\mathbb{P}(s - \Delta s < S_{t_i} \leq s)} = \frac{1}{f_{S_{t_i}}(s)} \cdot \frac{\partial H_{S_{t_i}, \Delta Q_{t_i}}(s, \tau)}{\partial x}$$

(5.55)

where $f_{S_{t_i}}$ is the PDF of S_{t_i}.

5.3 Stochastic Load Process

As mentioned earlier, the load process can be categorized into two types: discrete and continuous, as shown in Fig. 5.10. In this section, some frequently-used models for both types of load process will be discussed.

Recall that in Fig. 5.1b, if the structural resistance is time-variant (constant), then the time-dependent reliability can be converted into a classic reliability problem, where the maximum load effect over the considered reference period should be determined first.

For both types of load process (discrete and continuous), we are interested in the probabilistic behaviour of $S_{max} = \max_{0 \leq t \leq T} S(t)$, where T is the duration of the reference period of interest (c.f. Eq. (5.17)). The distribution of S_{max} will be discussed in the following.

5.3.1 Discrete Load Process

5.3.1.1 Bernoulli, Poisson, Renewal and Markov Load Processes

In the presence of a discrete load process, if the sequence of load effect is denoted by $S_1, S_2, \ldots S_k$, then $S_{max} = \max_{i=1}^{k}\{S_i\}$. Let $F_{S_{max}}(x)$ be the CDF of S_{max}, and it follows,

$$F_{S_{max}}(x) = \mathbb{P}(S_{max} \leq x) = \mathbb{P}(S_1 \leq x \cap S_2 \leq x \cap \cdots \cap S_k \leq x)$$

(5.56)

In this Section, four models of discrete process are discussed, namely Bernoulli, Poisson, Renewal and Markov load processes.

Fig. 5.13 Illustration of a
Bernoulli pulse process

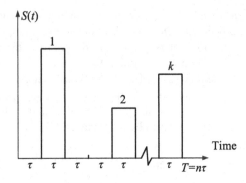

(a) Bernoulli Process

For a reference period of $[0, T]$, we divide it into n identical sections, with which the
duration of each interval is $\tau = \frac{T}{n}$. We assume that in each interval, at most one load
event may occur, and the probability of load occurrence is p, as shown in Fig. 5.13.
The number of load events, N, is a binomial random variable (c.f. Sect. 2.2.1.2),
whose distribution is as follows,

$$\mathbb{P}(N = k) = C_n^k p^k (1 - p)^{n-k}, \quad k = 0, 1, 2, \ldots n \tag{5.57}$$

Assuming that each S_i is statistically independent and identically distributed, the
conditional CDF of $S_{\max}$ on $N = k$ is

$$\mathbb{P}(S_{\max} \leq x | N = k) = [F_S(x)]^k \tag{5.58}$$

where $F_S(x)$ is the CDF of S_i. Furthermore, using the law of total probability, it
follows,

$$\mathbb{P}(S_{\max} \leq x) = \sum_{k=0}^{n} \mathbb{P}(S_{\max} \leq x | N = k) \cdot \mathbb{P}(N = k) = \sum_{k=0}^{n} [F_S(x)]^k \cdot C_n^k p^k (1 - p)^{n-k} \tag{5.59}$$

Recall that for two real numbers a and b, $(a + b)^n = \sum_{k=0}^{n} C_n^k a^k b^{n-k}$ holds. Thus,
Eq. (5.59) becomes

$$\mathbb{P}(S_{\max} \leq x) = F_{S_{\max}}(x) = [1 - p(1 - F_S(x))]^n \tag{5.60}$$

Specifically, if $p = 1$ in Eq. (5.60) (i.e., a load event occurs in each interval), then

$$F_{S_{\max}}(x) = [F_S(x)]^n \tag{5.61}$$

On the other hand, if $p \to 0$ in Eq. (5.60), then

$$F_{S_{\max}}(x) \approx 1 - np(1 - F_S(x)) \tag{5.62}$$

with which $\mathbb{P}(S_{\max} > x) \approx np(1 - F_S(x))$.

(b) Poisson Process

The Poisson process has been previously discussed in Sect. 3.4.1. If the considered reference period is $[0, T]$, we divide it into n identical sections, where n is a sufficiently large integer. Correspondingly, the probability of load occurrence within each interval is $\lambda \frac{T}{n}$, where λ is the occurrence rate. The PMF of the number of loads within $[0, T]$, N, is given by

$$\mathbb{P}(N = k) = \frac{(\lambda T)^k \exp(-\lambda T)}{k!}, \quad k = 0, 1, 2, \ldots \tag{5.63}$$

With $S_{\max} = \max\{S_1, S_2, \ldots\}$ as before, we assume that each S_i is statistically independent and identically distributed as before. With this, using the law of total probability, similar to Eq. (5.59),

$$\mathbb{P}(S_{\max} \le x) = \sum_{k=0}^{\infty} \frac{(\lambda T)^k \exp(-\lambda T)}{k!} \cdot [F_S(x)]^k \tag{5.64}$$

Recall that for a real number x, one has $\exp(x) = \sum_{i=1}^{\infty} \frac{x^i}{i!}$. Thus, Eq. (5.64) becomes

$$F_{S_{\max}}(x) = \mathbb{P}(S_{\max} \le x) = \sum_{k=0}^{\infty} \frac{(\lambda T \cdot F_S(x))^k \exp(-\lambda T)}{k!}$$
$$= \exp(-\lambda T) \cdot \exp(\lambda T \cdot F_S(x)) = \exp[-\lambda T(1 - F_S(x))] \tag{5.65}$$

Remark

It was mentioned in Sect. 2.2.3 that the Poisson distribution is a special case of the binomial distribution. This fact can be reflected through the comparison between Eqs. (5.60) and (5.65). In fact, if letting $np = \lambda T$ and $n \to \infty$, Eq. (5.60) becomes

$$F_{S_{\max}}(x) = \exp\{n \ln[1 - p(1 - F_S(x))]\} \approx \exp\{-np(1 - F_S(x))\} = \exp[-\lambda T(1 - F_S(x))] \tag{5.66}$$

which implies the consistency between Eqs. (5.60) and (5.65).

Example 5.13

Consider a building subjected to wind loads, whose wind-resisting resistance is $r = 50$ m/s. The annual probability of extreme wind is $p = 0.1$. Given the occurrence of a wind event, the wind speed has a normal distribution with a mean value of 30 m/s and a COV of 0.25. Compute the probability of failure in a service period of 20 years. Use (a) a Bernoulli process and (b) a Poisson process to model the extreme wind occurrence.

Solution

(a) When the wind occurrence is modelled by a Bernoulli process, according to Eq. (5.60),

$$\mathbb{P}_f(0, T) = 1 - \mathbb{P}(S_{\max} \leq r) = 1 - \left[1 - p(1 - F_S(r))\right]^T \quad (5.67)$$

with which

$$\mathbb{P}_f(0, 20) = 1 - \left[1 - 0.1 \times \left(1 - \Phi\left(\frac{50 - 30}{30 \times 0.25}\right)\right)\right]^{20} = 0.007633 \quad (5.68)$$

(b) If the Poisson process is used to describe the wind occurrence, according to Eq. (5.65),

$$\mathbb{P}_f(0, T) = 1 - \mathbb{P}(S_{\max} \leq r) = 1 - \exp[-\lambda T(1 - F_S(r))] \quad (5.69)$$

with which

$$\mathbb{P}_f(0, 20) = 1 - \exp\left[-0.1 \times 20 \times \left(1 - \Phi\left(\frac{50 - 30}{30 \times 0.25}\right)\right)\right] = 0.007631 \quad (5.70)$$

The comparison between $\mathbb{P}_f(0, 20)$ associated with the two wind occurrence models demonstrates that both models can well represent the occurrence sequence of stationary rare events.

Example 5.14

Reconsider Example 5.15. If the annual probability of extreme wind increases from $p = 0.1$ at present to 0.2 by 50 years, compute the probability of failure in a service period of 20 years. Use a Poisson process to model the extreme wind occurrence.

Solution

The time-variant occurrence rate is $\lambda(t) = 0.1 + \frac{0.1t}{50}$. Similar to Eq. (5.63), for a reference period of $[0, T]$, the PMF of the number of wind events is as follows,

$$\mathbb{P}(N = k) = \frac{\left(\int_0^T \lambda(t)dt\right)^k \exp\left(-\int_0^T \lambda(t)dt\right)}{k!}, \quad k = 0, 1, 2, \ldots \quad (5.71)$$

with which

$$\mathbb{P}(S_{\max} \leq x) = \sum_{k=0}^n \frac{\left(F_S(x) \cdot \int_0^T \lambda(t)dt\right)^k \exp\left(-\int_0^T \lambda(t)dt\right)}{k!} = \exp\left[-\int_0^T \lambda(t)dt \cdot (1 - F_S(x))\right] \quad (5.72)$$

Thus, the probability of failure is estimated by

$$\mathbb{P}_f(0, T) = 1 - \mathbb{P}(S_{\max} \leq r) = 1 - \exp\left[-\int_0^T \lambda(t)dt \cdot (1 - F_S(r))\right] \quad (5.73)$$

Now, with $T = 20$ and $\lambda(t) = 0.1 + \frac{0.1t}{50}$, it follows,

$$\mathbb{P}_f(0, 20) = 1 - \exp\left[-\int_0^{20} \left(0.1 + \frac{0.1t}{50}\right) dt \cdot \left(1 - \Phi\left(\frac{50 - 30}{30 \times 0.25}\right)\right)\right] = 0.00915$$

$$(5.74)$$

Compared with the result from Example 5.15, it can be seen that the increase of wind occurrence rate leads to a greater failure probability.

(c) Renewal Process

We have discussed the renewal process in Sect. 3.4.2. Let $\{T_1, T_2, \ldots\}$ be a sequence of independent and identically distributed non-negative variables, and define $X_0 = 0$, $X_j = \sum_{i=1}^{j} T_i$, $j = 1, 2, \ldots$, then X_j is the time of the jth renewal, and the number of events within a reference period of $[0, T]$ is determined by $\max\{j : X_j \leq T\}$ for $j = 0, 1, 2, \ldots$. With this, similar to Eq. (5.64), the probability of $S_{\max} \leq x$ is computed by

$$\mathbb{P}(S_{\max} \leq x) = F_{S_{\max}}(x) = \sum_{k=0}^{\infty} \mathbb{P}\left(\max\{j : X_j \leq T\} = k\right) \cdot [F_S(x)]^k \quad (5.75)$$

The expression of $\mathbb{P}\left(\max\{j : X_j \leq T\} = k\right)$ in Eq. (5.75) would depend on the PDF of each T_i. An example is given in the following.

Example 5.15

Reconsider Example 5.15. If the occurrence of extreme winds is modelled by a renewal process, and the interval between two successive events (in years) follows a Gamma distribution with a shape parameter of a and a scale parameter of b, (1) Derive the probability of failure within a reference period of T years. (2) Compute the failure probability with $T = 20$, $a = 4$ and $b = 2.5$.

Solution

(1) Let, as before, $\{T_1, T_2, \ldots\}$ denote the sequence of time intervals between two successive wind events, and $X_0 = 0$, $X_j = \sum_{i=1}^{j} T_i$, $j = 1, 2, \ldots$. Then the PDFs of T_i and X_i ($i = 1, 2, \ldots$) are respectively (c.f. Eq. (2.212)),

$$f_{T_i}(t) = \frac{(t/b)^{a-1}}{b\Gamma(a)} \exp(-t/b); \quad t \geq 0$$

$$f_{X_i}(t) = \frac{(t/b)^{ai-1}}{b\Gamma(ai)} \exp(-t/b); \quad t \geq 0 \quad (5.76)$$

Now we consider the expression of $\mathbb{P}\left(\max\{j : X_j \leq T\} = k\right)$. First, for the case of $k = 0$, one has

$$\mathbb{P}\left(\max\{j : X_j \leq T\} = k\right) = \mathbb{P}(X_1 > T) = 1 - \mathbb{P}(T_1 \leq T) = 1 - F_{T_1}(T) \quad (5.77)$$

where $F_{T_1}(x)$ is the CDF of T_1. Next, for $k = 1, 2, \ldots,$

$$\mathbb{P}\left(\max\{j : X_j \leq T\} = k\right) = \mathbb{P}(X_k \leq T \cap X_{k+1} > T) = \mathbb{P}(X_k \leq T \cap X_k + T_{k+1} > T)$$

$$= \mathbb{P}(X_k \leq T \cap T_{k+1} > T - X_k)$$

$$= \int_0^T \left[1 - F_{T_{k+1}}(T - t)\right] \cdot f_{X_k}(t)dt$$

$$(5.78)$$

where $F_{T_{k+1}}(t)$ is the CDF of T_{k+1}. Thus, Eq. (5.75) becomes

$$\mathbb{P}(S_{\max} \leq x) = 1 - F_{T_1}(T) + \sum_{k=1}^{\infty} \left\{\int_0^T \left[1 - F_{T_{k+1}}(T - t)\right] \cdot f_{X_k}(t)dt\right\} \cdot [F_S(x)]^k$$

$$(5.79)$$

with which

$$\mathbb{P}_f(0, T) = 1 - \mathbb{P}(S_{\max} \leq r) = F_{T_1}(T) - \sum_{k=1}^{\infty} \left\{\int_0^T \left[1 - F_{T_{k+1}}(T - t)\right] \cdot f_{X_k}(t)dt\right\} \cdot [F_S(r)]^k$$

$$(5.80)$$

(2) Substituting $T = 20$, $a = 4$ and $b = 2.5$ into Eq. (5.80) gives

$$\mathbb{P}_f(0, 20) = F_{T_1}(20) - \sum_{k=1}^{\infty} \left\{\int_0^{20} \left[1 - F_{T_{k+1}}(20 - t)\right] \cdot f_{X_k}(t)dt\right\} \cdot \left[\Phi\left(\frac{50 - 30}{30 \times 0.25}\right)\right]^k = 0.00621$$

$$(5.81)$$

(d) Markov Process

The Markov process was introduced in Sect. 3.4.4. An important feature of a Markov process that each state is only dependent on the previous state on the temporal scale. Now we consider using a Markov process to model the load sequence. For a reference period of $[0, T]$, we divide it into n identical sections, and let S_k denote the maximum load effect within the kth interval, $k = 1, 2, \ldots n$. Furthermore, we assume that each S_i is identically distributed, with a CDF of $F_S(x)$.

For each S_k, it simply follows $\mathbb{P}(S_k \leq x) = F_S(x)$. Now, taking into account the dependency of S_{k+1} on S_k, we introduced a function $\gamma(x)$ such that

$$\mathbb{P}(S_{k+1} \leq x | S_k \leq x) = \gamma(x) \cdot \mathbb{P}(S_k \leq x) = \gamma(x) \cdot F_S(x) \qquad (5.82)$$

Clearly, if $\gamma(x) \equiv 1$ in Eq. (5.82), then S_{k+1} is independent of S_k; if $\gamma(x) \geq 1$, then S_{k+1} is positively dependent on S_k; similarly, if $\gamma(x) \leq 1$, then S_{k+1} is negatively dependent on S_k. Based on Eq. (5.82), the CDF of $S_{\max}$ is determined by

$$F_{S_{\max}}(x) = \mathbb{P}(S_{\max} \leq x) = \mathbb{P}(S_1 \leq x \cap S_2 \leq x \cap \cdots \cap S_n \leq x)$$

$$= \mathbb{P}(S_1 \leq x) \cdot \prod_{k=1}^{n-1} \mathbb{P}(S_{k+1} \leq x | S_k \leq x) = F_S(x) \cdot \left[\gamma(x) \cdot F_S(x)\right]^{n-1}$$

$$= \gamma^{n-1}(x) \cdot F_S^n(x)$$

$$(5.83)$$

Comparing Eqs. (5.61) and (5.83), the CDF of $S_{\max}$ would be affected by the item $\gamma^{n-1}(x)$. This will be demonstrated in Example 5.16. Furthermore, it can be seen that the temporal correlation in the load process has been incorporated in Eq. (5.83). More details on the temporal correlation of the load occurrence process will be discussed in Sect. 5.3.1.2.

Example 5.16

The annual maximum wind speed follows a normal distribution with a mean value of 30 m/s and a COV of 0.25. Modelling the wind speed sequence on the yearly scale as a Markov process, the linear correlation coefficient between the maximum wind speeds of two adjacent years is ρ. Find the CDF of the maximum wind speed within a reference period of 20 years. Use $\rho = 0.4$, $\rho = 0$ and $\rho = -0.4$ respectively.

Solution

For the annual maximum wind speeds S_k and S_{k+1} in the kth and $(k+1)$th years, we can introduce a normal variable S_0 such that (c.f. Eq. (5.54))

$$S_{k+1} = \rho S_k + S_0, \quad k = 1, 2, \ldots \tag{5.84}$$

The mean value and standard deviation of S_0 are $30(1 - \rho)$ and $7.5\sqrt{1 - \rho^2}$ respectively according to Eq. (5.84).

Next, with the Bayes' theorem (c.f. Sect. 2.1.2), Eq. (5.82) becomes

$$\mathbb{P}(S_{k+1} \leq x | S_k \leq x) = \mathbb{P}(\rho S_k + S_0 \leq x | S_k \leq x) = \frac{\mathbb{P}(S_0 \leq x - \rho S_k \cap S_k \leq x)}{\mathbb{P}(S_k \leq x)}$$

$$= \frac{\int_{-\infty}^{x} F_{S_0}(x - \rho \tau) f_S(\tau) d\tau}{F_S(x)} = \gamma(x) F_S(x) \tag{5.85}$$

where $F_S(x)$ and $F_{S_0}(x)$ are the CDFs of S_k and S_0 respectively. With this, $\gamma(x)$ is obtained as follows,

$$\gamma(x) = \frac{\int_{-\infty}^{x} F_{S_0}(x - \rho \tau) f_S(\tau) d\tau}{F_S^2(x)} \tag{5.86}$$

For illustration purpose, Fig. 5.14a shows the function $\gamma(x)$ for different values of ρ. It can be seen that the cases of $\rho = -0.4, 0, 0.4$ correspond to $\gamma(x) \leq 1, = 1$ and ≥ 1 respectively. Furthermore, the CDF of the maximum wind speed over 20 years is obtained, according to Eq. (5.83), as follows,

$$F_{S_{\max}}(x) = \mathbb{P}(S_{\max} \leq x) = \gamma^{19}(x) \cdot \Phi^{20}\left(\frac{x - 20}{20 \times 0.25}\right) \tag{5.87}$$

Figure 5.14b presents the CDFs of $S_{\max}$ associated with the three values of ρ. A positive ρ shifts the CDF leftwards, while a negative value of ρ moves the CDF rightwards.

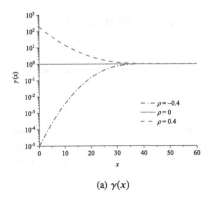

(a) $\gamma(x)$ (b) $F_{S_{max}}(x)$

Fig. 5.14 $\gamma(x)$ and $F_{S_{max}}(x)$ in Example 5.16

5.3.1.2 Temporal Correlation in Load Occurrence

Practically, temporal correlation often exists in the load process due to common causes. Such a correlation may arise from the similarity between the load intensities and/or the dependency in the load occurrence process [10, 26, 61]. One example is the tropical cyclone wind load: the common underlying climatological causes may lead to the inter-correlation between two successive cyclone events [10, 26, 61]. Moreover, the multi-year and multi-decadal oscillations in the cyclone events also suggest the existence of such a temporal correlation in the cyclone load process [47]. In this Section, the modelling of temporal correlation in the load occurrence will be discussed; the correlation in the load intensities will be addressed in Sect. 5.3.1.3.

The Poisson process has been widely used to account for the uncertainty associated with the occurrence time of a discrete load process, as has been discussed before (c.f. Sect. 5.3.1.1). For a service period of $[0, T]$, given the occurrence of totally n events, the load occurrence times are denoted by $t_1, t_2, \ldots t_n$, respectively. With this, the sequence of time interval between two subsequent events, $\Delta = \{\Delta_1, \Delta_2, \ldots \Delta_n\}$, is obtained by $\Delta_i = t_i - t_{i-1}$ for $i = 1, 2, \ldots n$ ($t_0 = 0$). From a view of a renewal process, given the value of t_{i-1}, Δ_i follows an exponential distribution with a CDF F_{Δ_i} being

$$F_{\Delta_i}(t) = 1 - \exp\left[-\int_0^t \lambda(t_{i-1} + \tau)d\tau\right], \quad t \geq 0 \tag{5.88}$$

for $i = 1, 2, \ldots$, where $\lambda(\tau)$ is the occurrence rate at time τ. For a stationary load process, Eq. (5.88) simply becomes $F_{\Delta_i}(t) = 1 - \exp(-\lambda t)$ if $\lambda(\tau) = \lambda$ for $\forall \tau$ (c.f. Eq. (2.162)).

Now we consider the temporal correlation in the load occurrence. Two methods will be discussed in the following.

Method 1

First, we discuss the modelling approach developed by [61]. The correlation in the load occurrence can be reflected by the correlation between the annual numbers of the events, which follow a Poisson distribution. Mathematically, let N_i and N_j denote the numbers of events within the ith and the jth year, respectively, with a correlation coefficient of ρ_{ij} between them.

To begin with, we assume that N_i and N_j are identically distributed with a mean value of λ. Recall Eq. (2.155) implied that the sum of a sequence of statistically independent and identically distributed binomial random variables (totally n variables) with a mean value of p follows a Poisson distribution with a mean value of np if n is sufficiently large. Motivated by this fact, we introduce two sequences of binomial random variables $\{X_{i1}, X_{i2}, \ldots X_{in}\}$ and $\{X_{j1}, X_{j2}, \ldots X_{jn}\}$ which have an identical mean value of λ/n and satisfy

$$N_i = \sum_{k=1}^{n} X_{ik}, \quad N_j = \sum_{k=1}^{n} X_{jk} \tag{5.89}$$

Furthermore, in order to reflect the correlation between N_i and N_j, we assume that for each $k = 1, 2, \ldots n$, X_{ik} is identically dependent on X_{jk} but is independent of the other elements in $\{X_{j1}, X_{j2}, \ldots X_{jn}\}$. By noting that $\mathbb{P}(X_{ik} = 1) = \mathbb{P}(X_{jk} = 1) = p = \frac{\lambda}{n}$, a dependence factor γ is introduced such that (c.f. Eq. (5.82))

$$\mathbb{P}(X_{jk} = 1 | X_{ik} = 1) = \gamma p = \gamma \cdot \frac{\lambda}{n} \tag{5.90}$$

With this, according to the law of total probability,

$$\mathbb{P}(X_{jk} = 1) = \mathbb{P}(X_{jk} = 1 | X_{ik} = 1) \cdot \mathbb{P}(X_{ik} = 1) + \mathbb{P}(X_{jk} = 1 | X_{ik} = 0) \cdot \mathbb{P}(X_{ik} = 0) \tag{5.91}$$

which yields

$$\mathbb{P}(X_{jk} = 1 | X_{ik} = 0) = \frac{\mathbb{P}(X_{jk} = 1) - \mathbb{P}(X_{jk} = 1 | X_{ik} = 1) \cdot \mathbb{P}(X_{ik} = 1)}{\mathbb{P}(X_{ik} = 0)} = \frac{p}{1-p}(1 - \gamma p) \tag{5.92}$$

Thus, one may simulate a sample pair of (N_i, N_j) by first generate a sequence of $\{X_{i1}, X_{i2}, \ldots X_{in}\}$ and then the sequence $\{X_{j1}, X_{j2}, \ldots X_{jn}\}$ according to the conditional probability of X_{jk} on X_{ik}, where the dependence factor γ is to be discussed in the following.

Recall that the correlation coefficient ρ_{ij} between N_i and N_j can by definition be estimated as follows,

$$\rho_{ij} = \frac{\mathbb{E}(N_i N_j) - \mathbb{E}(N_i)\mathbb{E}(N_j)}{\sqrt{\mathbb{V}(N_i)\mathbb{V}(N_j)}} \tag{5.93}$$

where

$$\mathbb{E}(N_i N_j) = \mathbb{E}\left(\sum_{k=1}^{n} X_{ik} \sum_{k=1}^{n} X_{jk}\right) = \mathbb{E}\left[\sum_{k=1}^{n} X_{ik} X_{jk} + \sum_{k=1}^{n}\left(X_{ik} \sum_{l \neq k} X_{jl}\right)\right]$$

$$= \mathbb{E}\left[\sum_{k=1}^{n} X_{ik} X_{jk}\right] + \sum_{k=1}^{n}\left[\mathbb{E}(X_{ik}) \sum_{l \neq k} \mathbb{E}(X_{jl})\right]$$

(5.94)

According to Eq. (5.90), $\mathbb{E}\left[\sum_{k=1}^{n} X_{ik} X_{jk}\right] = \sum_{k=1}^{n} \mathbb{E}\left[X_{ik} X_{jk}\right] = n\gamma p^2 = \gamma\lambda p$.
With this,

$$\mathbb{E}(N_i N_j) = n\gamma p^2 + n(p \cdot (n-1)p) = \gamma\lambda p + \lambda^2 - \gamma p \qquad (5.95)$$

With Eqs. (5.93) and (5.95), one has

$$\rho_{ij} = \frac{\gamma\lambda p - \gamma p}{\lambda(1-p)} = \frac{p}{1-p}(\gamma - 1) \qquad (5.96)$$

which further gives

$$\gamma = 1 + \rho_{ij}\frac{1-p}{p} \approx 1 + \frac{\rho_{ij}}{p} \qquad (5.97)$$

Equation (5.97) implies that the dependence factor can be uniquely determined by the values of ρ_{ij} and $p = \frac{\lambda}{n}$.

Next, we consider the generalized case where the mean values of N_i and N_j are λ_i and λ_j respectively. The linear correlation coefficient between N_i and N_j is denoted by ρ_{ij} as before. We use a similar equation as in Eq. (5.89), with each X_{ik} and X_{jk} following a binomial distribution with a mean value of λ_i/n and λ_j/n respectively. Referring to Eq. (5.90), a dependence factor γ' is introduced which satisfies

$$\mathbb{P}(X_{jk} = 1 | X_{ik} = 1) = \gamma' \cdot \frac{\lambda_j}{n} \qquad (5.98)$$

With this, using the law of total probability again (c.f. Eq. (5.91)), it follows,

$$\mathbb{P}(X_{jk} = 1 | X_{ik} = 0) = \frac{\mathbb{P}(X_{jk} = 1) - \mathbb{P}(X_{jk} = 1 | X_{ik} = 1) \cdot \mathbb{P}(X_{ik} = 1)}{\mathbb{P}(X_{ik} = 0)} = \frac{\frac{\lambda_j}{n} - \gamma'\frac{\lambda_j}{n}\frac{\lambda_i}{n}}{1 - \frac{\lambda_i}{n}}$$

(5.99)

where γ' can be determined provided ρ_{ij}. Since

$$\mathbb{E}(N_i N_j) = \mathbb{E}\left[\sum_{k=1}^{n} X_{ik} X_{jk}\right] + \sum_{k=1}^{n}\left[\mathbb{E}(X_{ik}) \sum_{l \neq k} \mathbb{E}(X_{jl})\right]$$

$$= n\gamma'\frac{\lambda_i}{n}\frac{\lambda_j}{n} + n\frac{\lambda_i}{n} \cdot (n-1)\frac{\lambda_j}{n} = \lambda_i\lambda_j\left(\frac{\gamma'-1}{n} + 1\right)$$

(5.100)

With Eqs. (5.93) and (5.100), one has

$$\rho_{ij} = \frac{\lambda_i \lambda_j \left(\frac{\gamma'-1}{n} + 1 \right) - \lambda_i \lambda_j}{\sqrt{n \cdot \frac{\lambda_i}{n} \left(1 - \frac{\lambda_i}{n} \right)} \sqrt{n \cdot \frac{\lambda_j}{n} \left(1 - \frac{\lambda_j}{n} \right)}} = \frac{\lambda_i \lambda_j (\gamma' - 1)}{n \sqrt{\lambda_i \lambda_j}} \tag{5.101}$$

which further gives

$$\gamma' = 1 + \rho_{ij} \frac{n}{\sqrt{\lambda_i \lambda_j}} \tag{5.102}$$

Clearly, Eq. (5.97) is a specific case of Eq. (5.102).

Remark

(1) Substituting Eq. (5.102) into Eq. (5.98) gives

$$\mathbb{P}(X_{jk} = 1 | X_{ik} = 1) = \gamma' \cdot \frac{\lambda_j}{n} = \left(1 + \rho_{ij} \frac{n}{\sqrt{\lambda_i \lambda_j}} \right) \cdot \frac{\lambda_j}{n} < 1 \tag{5.103}$$

which further yields

$$\rho_{ij} < \sqrt{\frac{\lambda_i}{\lambda_j}} \tag{5.104}$$

This indicates that the method in Eq. (5.89) cannot describe a perfectly correlated pair of N_i and N_j if $\lambda_i \leq \lambda_j$. This observation is consistent with that from Example 3.19.

(2) For two independent Poisson variables N_1 and N_2 with mean values of λ_1 and λ_2 respectively, $N_1 + N_2$ also follows a Poisson distribution with a mean value of $\lambda_1 + \lambda_2$. This fact was previously discussed in Example 2.23. However, if N_1 and N_2 are mutually correlated with a correlation coefficient of ρ_{12}, then $N_1 + N_2$ does not necessarily follow a Poisson distribution. This can be easily verified by noting that the mean value and variance of $N_1 + N_2$ are not identical if $\rho_{12} \neq 0$. As such, it can be seen that Method 1 has ignored the temporal correlation of the load events within each year.

Example 5.17

Reconsider Example 2.31. Suppose that the damage loss conditional on the occurrence of one cyclone event has a mean value of μ_D and a standard deviation of σ_D. The cyclone load process is modelled by Method 1 (c.f. Eq. (5.89)). The mean value of the annual load numbers is λ, and the correlation coefficient between the annual cyclone numbers of two adjacent years is ρ_1. Let D_c denote the linearly accumulative damage for a reference period of $[0, T]$. Derive the mean value and variance of D_c.

Solution

The cumulative damage D_c is estimated by

$$D_c = \sum_{i=1}^{T} \sum_{j=1}^{N_i} D_{ij} \tag{5.105}$$

where N_i is the number of cyclone events in the ith year, and D_{ij} is the damage due to load j in the ith year. Note that $\mathbb{E}(D_{ij}) = \mu_D$ and $\mathbb{V}(D_{ij}) = \sigma_D^2$.

Based on Eq. (5.105), the mean and variance of D_c can be obtained through conditioning on the random number of cyclone events, $N_1, N_2, \ldots N_T$ (c.f. Sect. 2.3.2). First, the mean value of D_c is

$$
\mathbb{E}(D_c) = \mathbb{E}\left[\sum_{i=1}^{T} \sum_{j=1}^{N_i} \mathbb{E}(D_{ij}) \middle| (N_1, N_2, \ldots N_T) \right] = \mathbb{E}\left[\sum_{i=1}^{T} N_i \mathbb{E}(D) \right]
$$

$$
= \sum_{i=1}^{T} \lambda \mathbb{E}(D) = \lambda T \mu_D
$$

(5.106)

Equation (5.106) implies that $\mathbb{E}(D_c)$ does not depend on the temporal correlation of the load process. Also, the result in Eq. (5.106) is consistent with that of Example 2.47.

Next, we consider the variance of D_c. According to Eq. (5.105),

$$
\mathbb{V}(D_c) = \mathbb{E}\left[\mathbb{E}\left(\sum_{i=1}^{T} \sum_{j=1}^{N_i} D_{ij} \right)^2 \middle| (N_1, N_2, \ldots N_T) \right] - (\lambda T \mu_D)^2
$$

$$
= \mathbb{E}\left[\left(\sum_{i=1}^{T} \sum_{j=1}^{N_i} \mathbb{E}(D_{ij}) \right)^2 + \mathbb{V}\left(\sum_{i=1}^{T} \sum_{j=1}^{N_i} D_{ij} \right) \middle| (N_1, N_2, \ldots N_T) \right] - (\lambda T \mu_D)^2
$$

$$
= \mathbb{E}\left[\left(\mu_D \sum_{i=1}^{T} N_i \right)^2 + \sigma_D^2 \sum_{i=1}^{T} N_i \middle| (N_1, N_2, \ldots N_T) \right] - (\lambda T \mu_D)^2
$$

$$
= \mu_D^2 \mathbb{E}\left(\sum_{i=1}^{T} N_i \right)^2 + \sigma_D^2 \lambda T - (\lambda T \mu_D)^2 = \mu_D^2 \mathbb{V}\left(\sum_{i=1}^{T} N_i \right) + \sigma_D^2 \lambda T
$$

$$
= (\mu_D^2 + \sigma_D^2) \lambda T + 2\mu_D^2 \sum_{i=1}^{T} \sum_{j=i+1}^{T} \mathbb{C}(N_i, N_j)
$$

(5.107)

where $\mathbb{C}(N_i, N_j)$ is the covariance of N_i and N_j. To compute $\mathbb{C}(N_i, N_j)$, we let, as before,

$$
N_i = \sum_{k=1}^{n} X_{ik}, \quad N_j = \sum_{k=1}^{n} X_{jk}
$$

(5.89)

According to Eq. (2.261), it follows,

$$
\mathbb{C}(N_i, N_j) = \sum_{k=1}^{n} \sum_{l=1}^{n} \mathbb{C}(X_{ik}, X_{jl}) = \sum_{k=1}^{n} \mathbb{C}(X_{ik}, X_{jk}) = \sum_{k=1}^{n} \left(\mathbb{E}(X_{ik} X_{jk}) - p^2 \right)
$$

(5.108)

where $p = \frac{\lambda}{n}$. Note that with the law of total probability, for $j > i$,

$$\mathbb{P}(X_{jk} = 1|X_{ik} = 1)$$

$$= \mathbb{P}(X_{jk} = 1|X_{(j-1)k} = 1) \cdot \mathbb{P}(X_{(j-1)k} = 1|X_{ik} = 1) + \mathbb{P}(X_{jk} = 1|X_{(j-1)k} = 0)$$
$$\cdot \mathbb{P}(X_{(j-1)k} = 0|X_{ik} = 1)$$

$$= \gamma_1 p \cdot \mathbb{P}(X_{(j-1)k} = 1|X_{ik} = 1) + \frac{p}{1-p}(1 - \gamma_1 p)[1 - \mathbb{P}(X_{(j-1)k} = 1|X_{ik} = 1)]$$

$$= \mathbb{P}(X_{(j-1)k} = 1|X_{ik} = 1)\left[\gamma_1 p - \frac{p}{1-p}(1 - \gamma_1 p)\right] + \frac{p}{1-p}(1 - \gamma_1 p)$$

$$(5.109)$$

where $\gamma_1 = 1 + \rho_1 \frac{1-p}{p}$ according to Eq. (5.97). Thus,

$$\mathbb{P}(X_{jk} = 1|X_{ik} = 1) = \mathbb{P}(X_{(j-1)k} = 1|X_{ik} = 1)\rho_1 + p(1 - \rho_1) \approx \mathbb{P}(X_{(j-1)k} = 1|X_{ik} = 1)\rho_1$$
$$(5.110)$$

with which

$$\mathbb{P}(X_{jk} = 1|X_{ik} = 1) = \rho_1^{j-i} \tag{5.111}$$

Subsequently, Eq. (5.108) is rewritten as

$$\mathbb{C}(N_i, N_j) = \sum_{k=1}^{n} \mathbb{P}(X_{jk} = 1|X_{ik} = 1) \cdot \mathbb{P}(X_{ik} = 1) - np^2 = np\rho_1^{j-i} = \lambda\rho_1^{j-i}$$
$$(5.112)$$

with which Eq. (5.107) becomes

$$\mathbb{V}(D_c) = (\mu_D^2 + \sigma_D^2)\lambda T + 2\mu_D^2 \lambda \sum_{i=1}^{T} \sum_{j=i+1}^{T} \rho_1^{j-i}$$

$$= (\mu_D^2 + \sigma_D^2)\lambda T + \frac{2\mu_D^2 \lambda \rho_1}{1 - \rho_1}\left(T - 1 - \frac{\rho_1^T - \rho_1}{\rho_1 - 1}\right) \tag{5.113}$$

The impact of temporal correlation in the load occurrence on the variance of D_c is reflected in Eq. (5.113). Clearly, $\mathbb{V}(D_c)$ increases if the correlation between the annual numbers of load becomes larger. Specially, if $\rho_1 = 0$ in Eq. (5.113), then

$$\mathbb{V}(D_c) = (\mu_D^2 + \sigma_D^2)\lambda T \tag{5.114}$$

which corresponds to the case of load occurrences being independent (c.f. Example 2.47).

Method 2

Now we discuss another approach for the modelling of the temporal correlation in the load process [55]. Recall Eq. (5.88), which depicts the time interval of two successive events from an independent load process. Here, we model the time interval sequence Δ as a correlated Markov process. That is, each Δ_{i+1} is an exponential random variable and is correlated with Δ_i for $i = 1, 2, \ldots$. Mathematically, if the linear correlation coefficient between Δ_i and Δ_{i+1} is ϱ_i, then that between Δ_i and Δ_j is

$$\rho_{ij} = \begin{cases} \prod_{k=j}^{i-1} \varrho_k, & i > j \\ 1, & i = j \\ \rho_{ji}, & i < j \end{cases} \qquad (5.115)$$

Specifically, if $\varrho_i \equiv \varrho$ for $\forall i$, ρ_{ij} in Eq. (5.115) simply becomes ϱ^{i-j} when $i > j$. Let N denote the number of load events within time interval $[0, T]$, which is determined according to

$$N = \max\left\{k : \sum_{j=0}^{k} \Delta_j \le T\right\} \qquad (5.116)$$

where $\Delta_0 = 0$. With this, one can use the following procedure to generate a sequence of Δ, $\{\delta_1, \delta_2, \ldots \delta_n\}$, for the case of a stationary and correlated process with the help of the Nataf transformation method (c.f. Sect. 3.3).

(1) Transforming Δ into a correlated sequence of standard normal distribution $\mathbf{Y}$, determine the correlation matrix of $\mathbf{Y}$, $\boldsymbol{\rho}' = [\rho'_{ij}]$, and solve the lower triangular matrix $\mathbf{L} = [l_{ij}]$ which satisfies $\mathbf{L}\mathbf{L}^\mathsf{T} = \boldsymbol{\rho}'$ (c.f. Eqs. (3.40) and (3.41)).
(2) Generate n independent standard normal realizations $z_1, z_2, \ldots z_n$.
(3) Let $\delta_i = F_{\Delta_i}^{-1}\left[\Phi\left(\sum_{j=1}^{i} l_{ij}z_j\right)\right]$ for $i = 1, 2, \ldots n$, where F_{Δ_i} is the CDF of Δ_i.

For the case of a non-stationary process, however, the aforementioned procedure is not applicable since the correlation matrix $\boldsymbol{\rho}'$ is dependent on the realization of Δ and thus cannot be determined prior to the sampling of Δ. This is explained by the fact that the COV of Δ_i depends on t_{i-1} and thus varies with i. In such a case, a recursion-based method is as follows to sample a realization of $\mathbf{\Delta}$, $\{\delta_1, \delta_2, \ldots \delta_n\}$.

(1) Sample δ_1. This is achieved by first generating a standard normal variable z_1, and let $\delta_1 = F_{\Delta_1}^{-1}(\Phi(z_1))$.
(2) Let $t_1 = t_0 + \delta_1 = \delta_1$, and calculate the COV of δ_2 with Eq. (5.88).
(3) Find ρ'_{21} and the matrix $\mathbf{L} = [l_{ij}]$ satisfying $\mathbf{L}\mathbf{L}^\mathsf{T} = \boldsymbol{\rho}'$.
(4) Generate a standard normal variable z_2, and let $\delta_2 = F_{\Delta_2}^{-1}[\Phi(l_{21}z_1 + l_{22}z_2)]$.
(5) One can further sample $\delta_3, \delta_4, \ldots \delta_n$ based on the recursion process as in Steps (2)–(4).

The modelling of the correlated load process as in Eq. (5.116) is guaranteed by the convergence of $\lim_{T\to\infty} \frac{N}{T}$. To illustrate this statement, we consider the case of a stationary process, where the time intervals are identically exponentially distributed with a mean value of $\frac{1}{\lambda}$ and the correlation coefficient for two subsequent intervals is ϱ. First, it can be shown that $\lim_{T\to\infty} N = \infty$ with probability 1. This is explained by noting that the probability of $\lim_{T\to\infty} N$ being finite with a value less than x is

$$\lim_{T \to \infty} \mathbb{P}(N < x) = 1 - \lim_{T \to \infty} \mathbb{P}(N \geq x) = \lim_{T \to \infty} \mathbb{P}(X_N \leq T) = 0 \qquad (5.117)$$

where $X_N = \sum_{i=1}^{N} \Delta_i$. Next, it is easy to see that

$$\frac{X_N}{N} \leq \frac{T}{N} < \frac{X_{N+1}}{N} \qquad (5.118)$$

Consider the lower bound of Eq. (5.118),

$$\lim_{N \to \infty} \mathbb{E}\left(\frac{X_N}{N}\right) = \lim_{N \to \infty} \frac{\sum_{i=1}^{N} \mathbb{E}(\Delta_i)}{N} = \frac{1}{\lambda} \qquad (5.119)$$

and

$$\lim_{N \to \infty} \mathbb{V}\left(\frac{X_N}{N}\right) = \lim_{N \to \infty} \frac{1}{N^2} \mathbb{V}\left(\sum_{i=1}^{N} \Delta_i\right) = \lim_{N \to \infty} \frac{1}{N^2} \left[\sum_{i=1}^{N} \mathbb{V}(\Delta_i) + 2 \sum_{1 \leq i < j \leq n} \mathbb{C}(\Delta_i, \Delta_j)\right]$$

$$= \lim_{N \to \infty} \frac{1/\lambda^2}{N^2} \left[N + 2 \sum_{1 \leq i < j \leq n} \varrho^{j-i}\right] = 0$$

$$(5.120)$$

Similarly, for the upper bound of Eq. (5.118),

$$\lim_{N \to \infty} \mathbb{E}\left(\frac{X_{N+1}}{N}\right) = 1/\lambda, \quad \lim_{N \to \infty} \mathbb{V}\left(\frac{X_{N+1}}{N}\right) = 0 \qquad (5.121)$$

As such, both $\frac{X_N}{N}$ and $\frac{X_{N+1}}{N}$ converge to $1/\lambda$ as $T \to \infty$. Thus, with Eq. (5.118), $\lim_{T \to \infty} \frac{T}{N} = 1/\lambda$, or equivalently, $\lim_{T \to \infty} \frac{N}{T} = \lambda$.

Finally, it is noted that the modelling technique in Eq. (5.116) can be further extended to the case where each Δ_i follows an arbitrary (positive) distribution type but not necessarily an exponential distribution. One could imagine this extension by recalling the relationship between a Poisson process and a renewal process.

Example 5.18

Consider a structure subjected to repeated loads. Within a reference period of 50 years, the total number of load events, N, has a mean value of 100 and a COV of 0.3. Let n_{cr} denote the maximum number of events that the structure can withstand (i.e., the structure fails if $N > n_{cr}$).

(1) Modelling the load process using Method 1, find the relationship between the structural failure probability within 50 years and n_{cr}.

(2) Modelling the load process using Method 2, find the relationship between the structural failure probability within 50 years and n_{cr}. Assume a stationary and correlated load process.

Solution

(1) Let N_i denote the number of load events in the ith year, $i = 1, 2, \ldots 50$, with which $N = \sum_{i=1}^{50} N_i$. We model each N_i as a Poisson random variable with a mean value of 2.

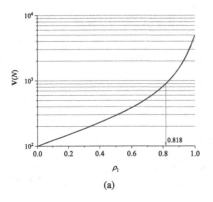
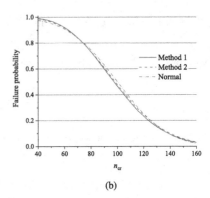

Fig. 5.15 Dependence of failure probability on n_{cr}

Thus, the variance of N_i is also 2. Let ρ_1 denote the correlation coefficient between N_i and N_{i+1} (i.e., the numbers of load within two adjacent years). To determine the value of ρ_1, we consider the variance of N. According to Eq. (5.112), $\mathbb{C}(N_i, N_j) = \lambda \rho_1^{j-i}$, where $\lambda = 2$/year. Thus, according to Eq. (2.264),

$$\mathbb{V}(N) = \mathbb{V}\left(\sum_{i=1}^{50} N_i\right) = \sum_{i=1}^{50} \mathbb{V}(N_i) + 2\sum_{i=1}^{50}\sum_{j>i}^{50} \mathbb{C}(N_i, N_j)$$

$$= 2 \times 50 + 2 \times 2 \times \sum_{i=1}^{50}\sum_{j>i} \rho_1^{j-i} = 100 + \frac{4\rho_1}{1-\rho_1}\left(50 - 1 - \frac{\rho_1^{50} - \rho_1}{\rho_1 - 1}\right)$$

$$\tag{5.122}$$

Since $\mathbb{V}(N)=(100 \times 0.3)^2 = 900$, ρ_1 is obtained numerically as 0.818 according to Eq. (5.122), as shown in Fig. 5.15a.

Let $N_i = \sum_{k=1}^{n} X_{ik}$ for $i = 1, 2, \ldots 50$ as in Eq. (5.89). We use a simulation-based approach to simulate a realization of N. For each simulation run, the procedure is as follows,

(a) Simulate samples for $X_{11}, X_{12}, \ldots X_{1n}$, denoted by $x_{11}, x_{12}, \ldots x_{1n}$ respectively. Note that X_{1k} is a binomial variable with a mean value of $p = \frac{\lambda}{n}$.

(b) For $i = 2, 3, \ldots 50$, generate samples for $X_{i1}, X_{i2}, \ldots X_{in}$, denoted by $x_{i1}, x_{i2}, \ldots x_{in}$ respectively, given the realizations $x_{(i-1)1}, x_{(i-1)2}, \ldots x_{(i-1)n}$. According to Eqs. (5.90) and (5.92), if $x_{(i-1)k} = 1$, then the binomial variable X_{ik} has a mean value of $\gamma_1 p$, where $\gamma_1 = 1 + \rho_1 \frac{1-p}{p}$ with Eq. (5.97). On the other hand, if $x_{(i-1)k} = 0$, then X_{ik} has a mean value of $\frac{p}{1-p}(1 - \gamma_1 p)$ or equivalently, $p(1 - \rho_1)$.

(c) A sample for the number of load events, N, is obtained as $\sum_{i=1}^{50}\sum_{j=1}^{n} x_{ij}$.

Figure 5.15(b) shows the dependence of failure probability on the value of n_{cr}, using $n = 60$ and 10^5 simulation runs for each n_{cr}. For comparison purpose, the failure probability assuming a normal distribution for N is also presented in Fig. 5.15b. The difference between the two lines is due to the selection of different load process models.

(2) Using Method 2 to describe the load process, we first need to determine the values of λ and ϱ, where $\frac{1}{\lambda}$ is the mean value of each time interval between two adjacent load events, and ϱ is the correlation coefficient between two adjacent intervals. By trial and error, it is found that $\lambda = 1.92$ and $\varrho = 0.82$ (note that λ is not necessarily 2). Subsequently, the dependence of failure probability on n_{cr} is computed via a simulation-based approach, as shown in Fig. 5.15b. It can be seen that the failure probabilities associated with the two load models (Methods 1 and 2) are close to each other.

5.3.1.3 Temporal Correlation in Load Intensity

The temporal correlation in loads may arise from not only the load occurrence times but also the intensities associated with two load events. Conceptually, the correlation between load intensities is expected to decrease with the time separation of the two events. That being the case, some frequently-used models, such as the Gaussian, exponential and triangle models, can be used to describe such a correlation decay.

For a discrete load process, a correlation model for load intensity should capture the main characteristics of correlation decay as a function of time separation, as well as the intermittence in loads considering the fact that the occurrence of significant loads is often temporally non-continuous [55]. Illustratively, the traditional Gaussian, exponential and triangle models for correlation decay should take the following forms,

$$\rho_{ij} = \begin{cases} \exp\left(-\dfrac{|t_i - t_j|}{L_c}\right) \cdot \dfrac{\iota(t_i)\iota(t_j)}{t_i t_j}, & \text{Exponential} \\[2ex] \exp\left(-\dfrac{(t_i - t_j)^2}{L_c^2}\right) \cdot \dfrac{\iota(t_i)\iota(t_j)}{t_i t_j}, & \text{Gaussian} \\[2ex] \left(1 - \dfrac{|t_i - t_j|}{L_c}\right) \cdot \mathbb{I}(|t_i - t_j| \le L_c) \cdot \dfrac{\iota(t_i)\iota(t_j)}{t_i t_j}, & \text{Triangle} \end{cases} \quad (5.123)$$

where ρ_{ij} is the load intensity correlation between the ith and the jth events occurring at times t_i and t_j respectively, L_c is the correlation length, and $\iota(t) = t$ if there occurs a load event at time t and 0 otherwise. The intermittence of loads is reflected in Eq. (5.123). The Nataf transformation method can be used to simulate a sample sequence for the load intensity, conditional on the occurrence times of each load event.

Example 5.19

Reconsider Example 5.15. Let $\mathbb{P}_f(0, 20)$ denote the structural failure probability in a service period of 20 years, and L_c the correlation length of the wind speeds. Use a Poisson process to model the extreme wind occurrence and an exponential correlation decay model (c.f. Eq. (5.123)). Find the relationship between $\mathbb{P}_f(0, 20)$ and L_c.

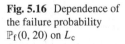

Fig. 5.16 Dependence of
the failure probability
$\mathbb{P}_f(0, 20)$ on L_c

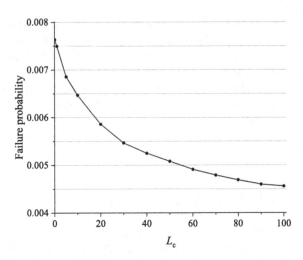

Solution

We use a simulation-based approach to compute the dependence of $\mathbb{P}_f(0, 20)$ on the correlation length L_c. For each simulation run, the procedure is as follows,

    (a) Simulate the number of loads, n, and sample a sequence of the load occurrence times, denoted by $t_1, t_2, \ldots t_n$ respectively.

    (b) Construct the correlation matrix of the wind speeds, $\boldsymbol{\rho}$ according to Eq. (5.123), and find the lower triangle matrix $\mathbf{L} = [l_{ij}]$ satisfying $\mathbf{LL}^\mathsf{T} = \boldsymbol{\rho}$ via Cholesky decomposition of $\boldsymbol{\rho}$ (note that the wind speed is a normal variable in this example).

    (c) The sampled wind speed vector is $\mathbf{v} = \mu + \sigma \mathbf{Lz}$, where $\mathbf{v}$ is a vector consisting of n standard normal variables, μ and σ are the mean value and standard deviation of the wind speed respectively (30 and 7.5 m/s in Example 5.15).

    (d) If all the n elements in $\mathbf{v}$ does not exceed the resistance $r = 50$ m/s, then the structure is deemed as survival; otherwise, failure occurs.

Figure 5.16 presents the failure probability $\mathbb{P}_f(0, 20)$ for different values of L_c using 10^7 samples for each case. It can be seen that a greater value of L_c (i.e., a slower decay rate of the intensity correlation) leads to a smaller failure probability. When L_c is small enough, the load process is approximately independent and the failure probability is 0.007636, consistent with that from Example 5.15.

5.3.2   Continuous Load Process

In this Section, we discuss the continuous load process. The statistical characteristics of a general continuous process is first addressed in Sect. 5.3.2.1, followed by the discussion of a continuous load process in Sect. 5.3.2.2 for use in structural reliability assessment.

5.3.2.1 Characteristics of a Continuous Stochastic Process

We have introduced in Sect. 3.4 the definition of a continuous stochastic process, that is, a collection of random variables on a continuous time scale time scale. For a continuous process $\{X(t), t \in A\}$, where A is the index set of the process, if for any $t_1, t_2, \ldots t_n \in A$ and $\tau > 0$, $t_1 + \tau, t_2 + \tau, \ldots t_n + \tau \in A$, and the joint CDF of $X(t_1), X(t_2), \ldots X(t_n)$ and that of $X(t_1 + \tau), X(t_2 + \tau), \ldots X(t_n + \tau)$ are the same, then $X(t)$ is a *strictly stationary process*. In a wider sense, if $\mathbb{E}(X(t)) =$ constant and $\mathbb{C}(X(t), X(t + \tau))$ depends on the time lag τ only, then $X(t)$ is a *weakly stationary process* or a *covariance stationary process* [1]. It is noticed, however, that the condition $\mathbb{C}(X(t), X(t + \tau))$ depending on τ only cannot guarantee $\mathbb{E}(X(t))$ is a constant.

Example 5.20

Consider a continuous load process $X(t) = A \sin(\omega t) + B \cos(\omega t)$, where A and B are two independent standard normal variables, and ω is a constant. Show that $X(t)$ is a covariance stationary process.

Solution

It is easy to verify that

$$\mathbb{E}(X(t)) = \mathbb{E}(A) \cdot \sin(\omega t) + \mathbb{E}(B) \cdot \cos(\omega t) = 0 = \text{constant} \tag{5.124}$$

and that

$$
\begin{aligned}
\mathbb{C}(X(t_1), X(t_2)) &= \mathbb{C}(A \sin(\omega t_1) + B \cos(\omega t_1), A \sin(\omega t_2) + B \cos(\omega t_2)) \\
&= \mathbb{C}(A \sin(\omega t_1), A \sin(\omega t_2)) + \mathbb{C}(B \cos(\omega t_1), B \cos(\omega t_2)) \\
&= \sin(\omega t_1) \sin(\omega t_2) \mathbb{V}(A) + \cos(\omega t_1) \cos(\omega t_2) \mathbb{V}(B) = \cos(\omega |t_1 - t_2|)
\end{aligned}
\tag{5.125}
$$

Thus, $X(t)$ is a covariance stationary process.

Now we analyze a covariance (weakly) stationary stochastic process $X(t)$ on the frequency domain, which would be based on the Fourier transform method (c.f. Sect. 4.2.4). Treating $X(t)$ as the sum of harmonic stochastic processes with different frequencies and amplitudes, the purpose of frequency analysis or spectral analysis is to identify the contribution of each oscillation in relation to a certain frequency.

Let $X(t)$ be a zero-mean periodic function with a period of T. Similar to Eq. (4.124), the Fourier expansion of $X(t)$ takes the form of,

$$X(t) = \frac{A_0}{2} + \sum_{j=1}^{\infty} \left[A_j \cos\left(\frac{2j\pi}{T}\right) + B_j \sin\left(\frac{2j\pi}{T}\right) \right] \tag{5.126}$$

where A_j and B_j are the Fourier coefficients,

$$A_j = \frac{2}{T} \int_{-T/2}^{T/2} X(t) \cos \frac{2j\pi}{T} t \, dt, \quad j = 0, 1, 2, \ldots$$

$$B_j = \frac{2}{T} \int_{-T/2}^{T/2} X(t) \sin \frac{2j\pi}{T} t \, dt, \quad j = 1, 2, \ldots$$

(5.127)

Let $i = \sqrt{-1}$ denote the imaginary unit as before, and $C_j = \frac{1}{2}(A_j - iB_j)$, $C_{-j} = \overline{C_j} = \frac{1}{2}(A_j + iB_j)$, where the overline indicates the conjugate of the complex. With this, Eq. (5.126) is rewritten as follows,

$$X(t) = \sum_{j=-\infty}^{\infty} C_j \exp\left(i\frac{2j\pi}{T} t\right)$$

(5.128)

That is, C_j is the amplitude of the oscillation with a frequency of $\frac{2j\pi}{T}$. According to Eq. (5.127), the term C_j in Eq. (5.128) is obtained as,

$$C_j = \frac{1}{T} \int_{-T/2}^{T/2} X(t) \exp\left(-i\frac{2j\pi}{T} t\right) dt, \quad j = 0, \pm 1, \pm 2, \ldots$$

(5.129)

We let $\omega_j = \frac{2j\pi}{T} = j\omega_1$. With Eq. (5.128), the following equation holds for a sufficiently large T,

$$X(t) = \frac{T}{2\pi} \sum_{j=-\infty}^{\infty} \left[C_j \exp\left(i\omega_j t\right) \cdot \frac{2\pi}{T} \right] \xrightarrow{\text{large } T} \frac{T}{2\pi} \sum_{j=-\infty}^{\infty} \left[C_j \exp\left(i\omega_j t\right) \cdot d\omega_j \right]$$

$$\xrightarrow{\text{let } C_j = C(\omega_j)} \frac{T}{2\pi} \int_{-\infty}^{\infty} C(\omega) \exp(i\omega t) d\omega$$

(5.130)

where $C(\omega) = \frac{1}{T} \int_{-T/2}^{T/2} X(t) \exp(-i\omega t) \, dt$ according to Eq. (5.129). Equation (5.130) implies that $X(t)$ and $T \cdot C(\omega)$ is a Fourier transform pair when $T \to \infty$.

Consider the average power (or "energy") of the process $X(t)$ within one period, which is defined by

$$\text{Average power} = \frac{1}{T} \int_{-T/2}^{T/2} X^2(t) dt$$

(5.131)

Substituting Eqs. (5.129) and (5.130) into Eq. (5.131) gives

$$\text{Average power} = \frac{1}{T} \int_{-T/2}^{T/2} X^2(t) dt = \frac{1}{T} \int_{-T/2}^{T/2} X(t) \cdot X(t) dt$$

$$= \frac{1}{2\pi} \int_{-T/2}^{T/2} \int_{-\infty}^{\infty} X(t) \exp(i\omega t) \cdot C(\omega) d\omega dt$$

$$= \frac{T}{2\pi} \int_{-\infty}^{\infty} \overline{C(\omega)} \cdot C(\omega) d\omega = \frac{T}{2\pi} \int_{-\infty}^{\infty} |C(\omega)|^2 d\omega$$

(5.132)

Taking the expectation of both sides of Eq. (5.132) yields

$$\mathbb{E}(\text{Average power}) = \int_{-\infty}^{\infty} \frac{T}{2\pi} \mathbb{E}\left(|C(\omega)|^2\right) d\omega \qquad (5.133)$$

Now we define the *power spectral density function* (PSDF) of $X(t)$, $\mathbb{S}(\omega)$, as follows,

$$\mathbb{S}(\omega) = \frac{T}{2\pi} \mathbb{E}\left(|C(\omega)|^2\right) \qquad (5.134)$$

With this, it is easy to see, from Eq. (5.133), that $\mathbb{E}(\text{Average power}) = \int_{-\infty}^{\infty} \mathbb{S}(\omega) d\omega$. That is, $\mathbb{S}(\omega)$ represents a portion of the average power associated with an oscillation frequency of ω.

Let $\mathbb{R}(\tau)$ denote the *autocorrelation function* of $X(t)$, i.e., $\mathbb{R}(\tau) = \mathbb{C}(X(t), X(t + \tau))$. One can show that, interestingly, $\mathbb{R}(\tau)$ and $\mathbb{S}(\omega)$ is a Fourier transform pair. The proof is as follows. Note that $\mathbb{R}(\tau) = \mathbb{E}(X(t)X(t + \tau)) = \mathbb{E}(\overline{X(t)}X(t + \tau))$. Thus,

$$
\begin{aligned}
\mathbb{S}(\omega) &= \frac{T}{2\pi} \mathbb{E}\left(C(\omega) \cdot \overline{C(\omega)}\right) = \frac{T}{2\pi} \mathbb{E}\left(\frac{1}{T}\int_{-T/2}^{T/2} X(t_1) \exp(-i\omega t_1)\, dt_1 \cdot \frac{1}{T}\int_{-T/2}^{T/2} X(t_2)\exp(i\omega t_2)\, dt_2\right) \\
&= \frac{1}{2\pi T}\int_{-T/2}^{T/2}\int_{-T/2}^{T/2} \mathbb{E}[X(t_1)X(t_2)]\exp(i\omega(t_2 - t_1))\, dt_1 dt_2 \\
&\xrightarrow{\text{let } t_2 - t_1 = \tau} \frac{1}{2\pi T}\int_{-T/2}^{T/2}\left[\int_{-T/2}^{T/2} \mathbb{E}[X(t_1)X(t_1 + \tau)]\exp(i\omega\tau)\, d\tau\right] dt_1 \\
&\xrightarrow{T \to \infty} \frac{1}{2\pi}\int_{-\infty}^{\infty} \mathbb{R}(\tau)\exp(i\omega\tau)\, d\tau
\end{aligned}
$$

$$(5.135)$$

which completes the proof.

It is noticed from Eq. (5.135) that both $\mathbb{R}(\tau)$ and $\mathbb{S}(\omega)$ are even functions. Furthermore, with Eq. (5.135),

$$\mathbb{R}(\tau) = \int_{-\infty}^{\infty} \mathbb{S}(\omega)\exp(-i\omega\tau)\, d\omega \qquad (5.136)$$

Assigning $\tau = 0$ in Eq. (5.136) yields

$$\mathbb{V}[X(t)] = \int_{-\infty}^{\infty} \mathbb{S}(\omega) d\omega \qquad (5.137)$$

implying that the integral of $\mathbb{S}(\omega)$ over $(-\infty, \infty)$ equals the variance of $X(t)$.

Example 5.21

For a weakly stationary process $X(t)$ with an autocorrelation function of (1) $\mathbb{R}(\tau) = \exp(-k|\tau|)$, (2) $\mathbb{R}(\tau) = \exp(-k\tau^2)$, (3) $\mathbb{R}(\tau) = \left|1 - \frac{\tau}{k}\right| \cdot \mathbb{I}(|\tau| < k)$, where $k > 0$ is a constant and $\tau \in (-\infty, \infty)$, find the PSDF of $X(t)$ for the three cases.

Solution

(1) According to Eq. (5.135),

$$
\begin{aligned}
\mathbb{S}(\omega) &= \frac{1}{2\pi} \int_{-\infty}^{\infty} \mathbb{R}(\tau) \exp\left(i\omega\tau\right) d\tau = \frac{1}{2\pi} \int_{-\infty}^{\infty} \exp(-k|\tau|) \exp\left(i\omega\tau\right) d\tau \\
&= \frac{1}{2\pi} \int_{-\infty}^{\infty} \exp(-k|\tau|) \cos(\omega\tau) d\tau = \frac{1}{\pi} \int_{0}^{\infty} \exp(-k\tau) \cos(\omega\tau) d\tau \quad (5.138) \\
&= \frac{k}{\pi(k^2 + \omega^2)}
\end{aligned}
$$

(2) For the Gaussian autocorrelation function, with Eq. (5.135),

$$
\begin{aligned}
\mathbb{S}(\omega) &= \frac{1}{2\pi} \int_{-\infty}^{\infty} \exp(-k\tau^2) \exp\left(i\omega\tau\right) d\tau = \frac{1}{2\pi} \int_{-\infty}^{\infty} \exp(-k\tau^2) \cos(\omega\tau) d\tau \\
&= \frac{1}{\pi} \int_{0}^{\infty} \exp(-k\tau^2) \cos(\omega\tau) d\tau = \frac{1}{2\sqrt{k\pi}} \exp\left(-\frac{\omega^2}{4k}\right)
\end{aligned}
$$

$$(5.139)$$

(3) For the triangle autocorrelation function, applying Eq. (5.135) gives,

$$
\begin{aligned}
\mathbb{S}(\omega) &= \frac{1}{2\pi} \int_{-\infty}^{\infty} \mathbb{R}(\tau) \exp\left(i\omega\tau\right) d\tau = \frac{1}{2\pi} \int_{-k}^{k} \left|1 - \frac{\tau}{k}\right| \exp\left(i\omega\tau\right) d\tau \\
&= \frac{1}{\pi} \int_{0}^{k} \left(1 - \frac{\tau}{k}\right) \cos(\omega\tau) d\tau = \frac{1 - \cos(k\omega)}{\pi k \omega^2}
\end{aligned}
$$

$$(5.140)$$

Example 5.22

Let $X(t)$ be a stationary stochastic process with a PSDF of $\mathbb{S}(\omega)$. Show that the PSDF of the jth derivative of $X(t)$, $X^{(j)}(t)$, is $\omega^{2j}\mathbb{S}(\omega)$.

Solution

From a view of the method of induction, we only need to consider the case of $j = 1$, that is, to show that the PSDF of $\dot{X}(t)$ is $\omega^2 \mathbb{S}(\omega)$.

Let $\mathbb{S}_1(\omega)$ denote the PSDF of $\dot{X}(t)$. With Eq. (5.130),

$$
\dot{X}(t) = \frac{T}{2\pi} \int_{-\infty}^{\infty} [(i\omega)C(\omega)] \exp(i\omega t) d\omega \qquad (5.141)
$$

Thus, according to Eq. (5.134),

$$
\mathbb{S}_1(\omega) = \frac{T}{2\pi} \mathbb{E}\left(|(i\omega)C(\omega)|^2\right) = \omega^2 \frac{T}{2\pi} \mathbb{E}\left(|C(\omega)|^2\right) = \omega^2 \mathbb{S}(\omega) \qquad (5.142)
$$

which completes the proof.

5.3.2.2 Continuous Stochastic Process for Loads

Consider a continuous stochastic load process $S(t)$. For a reference period of $[0, T]$, we are interested in the probabilistic behaviour of the maxima of the load process, denoted by $S_{max} = \max\{S(t)\}, t \in [0, T]$.

Note that

$$
\begin{aligned}
F_{S_{max}}(x) &= \mathbb{P}(S_{max} \le x) = \mathbb{P}(S(t) \le x, \forall t \in [0, T]) \\
&= \mathbb{P}(S(0) \le x \cap S(t) \le x \forall t \in (0, T]) \\
&= \mathbb{P}(S(t) \le x \forall t \in (0, T] | S(0) \le x) \cdot \mathbb{P}(S(0) \le x)
\end{aligned} \tag{5.143}
$$

Typically, $\mathbb{P}(S(0) \le x)$ is fairly close to 1. Based on this, we have

$$
F_{S_{max}}(x) = \mathbb{P}(S(t) \le x \forall t \in (0, T] | S(0) \le x) \tag{5.144}
$$

To estimate $\mathbb{P}(S(t) \le x \forall t \in (0, T] | S(0) \le x)$, we use an up-crossing analysis. Let $v^+(x, t)$ denote the up-crossing rate of the stochastic process $S(t)$ with respect to the threshold x within $(t, t + \mathrm{d}t]$, where $\mathrm{d}t \to 0$. Conditional on $S(0) \le x$, the probability of $S(t) \le x$ for $\forall t \in (0, T]$ equals the probability that no up-crossing occurs within $[0, T]$. When x is large enough, the up-crossing rate is small, based on which a Poisson process can be assumed for the occurrence of up-crossings. According to Eqs. (2.171) and (5.144), it follows,

$$
F_{S_{max}}(x) = \mathbb{P}(S(t) \le x \forall t \in (0, T] | S(0) \le x) = \exp\left(-\int_0^T v^+(x, t)\mathrm{d}t\right) \tag{5.145}
$$

Specifically, for a stationary process $S(t)$, $v^+(x, t)$ is independent of time, written as $v^+(x)$, and Eq. (5.145) becomes

$$
F_{S_{max}}(x) = \exp(-v^+(x)T) \tag{5.146}
$$

We further consider a stationary Gaussian process $S(t)$ (i.e., $S(t)$ valued at any time is a normal variable) with a mean value of μ_S and a standard deviation of σ_S, as shown in Fig. 5.17. The term $v^+(x)$ in Eq. (5.146) is evaluated by

$$
\begin{aligned}
v_x^+ \mathrm{d}t &= \mathbb{P}(S(t) < x \cap S(t + \mathrm{d}t) > x) = \mathbb{P}(S(t) < x \cap S(t) + \dot{S}(t)\mathrm{d}t) > x) \\
&= \int_0^\infty \int_{x - \dot{s}\mathrm{d}t}^x f_{S, \dot{S}}(s, \dot{s})\mathrm{d}s\mathrm{d}\dot{s}
\end{aligned} \tag{5.147}
$$

where $f_{S, \dot{S}}(s, \dot{s})$ is the joint PDF of $S(t)$ and $\dot{S}(t)$. Since $S(t)$ is a stationary Gaussian process, it can be shown that $S(t)$ and $\dot{S}(t)$ are mutually independent. With this,

Fig. 5.17 Crossing rate of a stochastic process $S(t)$ with respect to threshold x

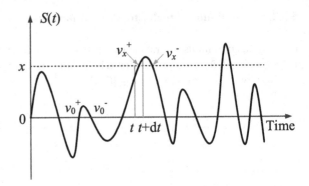

$$f_{S,\dot{S}}(s,\dot{s}) = \frac{1}{2\pi\sigma_S\sigma_{\dot{S}}} \exp\left[-\frac{1}{2}\left(\frac{s^2}{\sigma_S^2} + \frac{\dot{s}^2}{\sigma_{\dot{S}}^2}\right)\right] \tag{5.148}$$

where $\sigma_{\dot{S}}$ is the standard deviation of $\dot{S}(t)$.

Substituting Eq. (5.148) into Eq. (5.147) yields

$$v_x^+ dt \approx \int_0^\infty \dot{s}\,dt f_{S,\dot{S}}(x,\dot{s})\,ds\,d\dot{s} = \frac{dt}{2\pi}\frac{\sigma_{\dot{S}}}{\sigma_S}\exp\left(-\frac{x^2}{2\sigma_S^2}\right) \tag{5.149}$$

with which

$$v_x^+ = \frac{1}{2\pi}\frac{\sigma_{\dot{S}}}{\sigma_S}\exp\left(-\frac{x^2}{2\sigma_S^2}\right) \tag{5.150}$$

Thus, Eq. (5.146) becomes

$$F_{S_{\max}}(x) = \exp\left[-\frac{T}{2\pi}\frac{\sigma_{\dot{S}}}{\sigma_S}\exp\left(-\frac{x^2}{2\sigma_S^2}\right)\right] \tag{5.151}$$

As x is large enough,

$$F_{S_{\max}}(x) \approx 1 - \frac{T}{2\pi}\frac{\sigma_{\dot{S}}}{\sigma_S}\exp\left(-\frac{x^2}{2\sigma_S^2}\right) \tag{5.152}$$

Equation (5.152) implies that the maxima of a stationary Gaussian process at the upper tail follows a Rayleigh distribution (c.f. Sect. 2.2.6).

Example 5.23

For a stationary Gaussian process $S(t)$ with a mean value of μ_S and a PSDF of $\mathbb{S}(\omega)$, one can simulate a realization of $S(t)$ by letting

$$S(t) = \mu_S + \lim_{n\to\infty}\sigma_S\cdot\sqrt{\frac{2}{n}}\sum_{j=1}^n \cos\left(\Omega_j t + \Psi_j\right) \tag{5.153}$$

where n is a sufficiently large integer, σ_S is the standard deviation of $S(t)$ satisfying $\sigma_S^2 = \int_{-\infty}^{\infty} \mathbb{S}(\omega)d\omega$, Ω_j is a random variable whose PDF is $\frac{\mathbb{S}(\omega)}{\sigma_S^2}$, and Ψ_j is a uniform variable within $[0, 2\pi]$.

Now we consider a zero-mean stationary Gaussian load process $S(t)$. The PSDF of $S(t)$ is as in Eq. (5.139) with $k = 1$. Use a simulation-based approach to discuss the accuracy of Eq. (5.152). Use $T = 10$ and $T = 50$ respectively.

Solution

In Eq. (5.139), assigning $k = 1$ gives

$$\mathbb{S}(\omega) = \frac{1}{2\sqrt{\pi}} \exp\left(-\frac{\omega^2}{4}\right) \qquad (5.154)$$

Thus, the standard deviation of $S(t)$, σ_S, is computed as

$$\sigma_S = \sqrt{\int_{-\infty}^{\infty} \mathbb{S}(\omega)d\omega} = \sqrt{\int_{-\infty}^{\infty} \frac{1}{2\sqrt{\pi}} \exp\left(-\frac{\omega^2}{4}\right) d\omega} = 1 \qquad (5.155)$$

The standard deviation of $\dot{S}(t)$, $\sigma_{\dot{S}}$, is calculated according to Eq. (5.142) as follows,

$$\sigma_{\dot{S}} = \sqrt{\int_{-\infty}^{\infty} \omega^2 \mathbb{S}(\omega)d\omega} = \sqrt{\int_{-\infty}^{\infty} \frac{1}{2\sqrt{\pi}} \omega^2 \exp\left(-\frac{\omega^2}{4}\right) d\omega} = \sqrt{2} \qquad (5.156)$$

With this, the CDF of $S_{\max}$ can be explicitly obtained according to Eq. (5.152), as shown in Fig. 5.18. For comparison purpose, a simulation-based approach is used to approximate the CDF of $S_{\max}$. Using Eq. (5.153), note that Ω_j has a PDF of $\frac{1}{2\sqrt{\pi}} \exp\left(-\frac{\omega^2}{4}\right)$, implying that Ω_j is a normal variable having a mean value of 0 and a standard deviation of $\sqrt{2}$. With $n = 500$ and 10^4 simulation runs for each x, the simulated CDF of $S_{\max}$ is also presented in

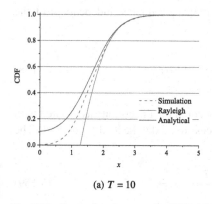

(a) $T = 10$

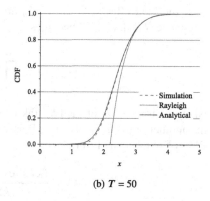

(b) $T = 50$

Fig. 5.18 Comparison between the CDFs of $S_{\max}$

Fig. 5.18. Note that the legend "Analytical" refers to Eq. (5.151), and "Rayleigh" refers to Eq. (5.152) in Fig. 5.18. It can be seen that the CDF given by Eq. (5.152) is accurate at the upper tail for both cases of $T = 10$ and $T = 50$.

5.4 Time-Dependent Reliability Considering a Discrete Load Process

Recall the definition of time-dependent reliability in Eq. (5.14),

$$\mathbb{L}(0, T) = \mathbb{P}\{R(t) > S(t), \forall t \in [0, T]\} \tag{5.14}$$

The reliability assessment in the presence of a discrete load process will be discussed in this section. The reliability with a continuous load process will be later addressed in Sect. 5.5.

5.4.1 Formulation of Time-Dependent Reliability for Some Simple Cases

We start from the simplest case where the structural resistance is time-invariant, denoted by R. Given n load effects $S_1, S_2, \ldots S_n$, the time-dependent reliability conditional on $R = r$ is

$$\mathbb{L}(0, T) = \mathbb{P}\left(S_1 \leq r \cap S_2 \leq r \cap \cdots \cap S_n \leq r\right) = \mathbb{P}\left(\cap_{i=1}^{n} S_i \leq r\right) = \mathbb{P}\left(\max_{i=1}^{n} S_i \leq r\right) \tag{5.157}$$

If the load effects, $S_1, S_2, \ldots S_n$ are a stationary sequence independent of the resistance R, Eq. (5.157) becomes

$$\mathbb{L}(0, T) = \prod_{i=1}^{n} F_{S_i}(r) = [F_S(r)]^n \tag{5.158}$$

where F_S is the CDF of each S_i. Taking into account the uncertainty associated with the number of loads, N, we use a Poisson process to model the load occurrence. The PMF of N takes the form of

$$\mathbb{P}(N = k) = \frac{(\lambda T)^k \exp(-\lambda T)}{k!}, \quad k = 0, 1, 2, \ldots \tag{5.159}$$

where λ is the occurrence rate of the load events. With Eqs. (5.158) and (5.159), using the law of total probability gives

$$\mathbb{L}(0, T) = \sum_{k=0}^{\infty} [F_S(r)]^k \cdot \frac{(\lambda T)^k \exp(-\lambda T)}{k!} = \exp(-\lambda T) \sum_{k=0}^{\infty} \frac{[\lambda T \cdot F_S(r)]^k}{k!}$$

(5.160)

$$= \exp[-\lambda T (1 - F_S(r))]$$

Equation (5.160) is consistent with Eq. (5.65), by noting that $\mathbb{L}(0, T) = \mathbb{P}(S_{\max} \le r) = F_{S_{\max}}(r)$. Furthermore, considering the uncertainty associated with the resistance R and using the law of total probability again, it follows,

$$\mathbb{L}(0, T) = \int_0^{\infty} \exp[-\lambda T (1 - F_S(r))] f_R(r) dr$$

(5.161)

where $f_R(r)$ is the PDF of R.

Next, we further consider the time-variation of the structural resistance, denoted by $R(t)$. As in Eq. (5.31), we let $R(t) = R_0 \cdot G(t)$, where R_0 is the initial resistance and $G(t)$ is the deterioration function. Conditional on $R_0 = r$ and $G(t) = g(t)$, similar to Eq. (5.157), one has

$$\mathbb{L}(0, T) = \mathbb{P}(r \cdot g(t_1) \ge S_1 \cap r \cdot g(t_2) \ge S_2 \cap \cdots \cap r \cdot g(t_n) \ge S_n) = \prod_{i=1}^{n} F_S[r \cdot g(t_i)]$$

(5.162)

where t_i is the time at which S_i occurs, $i = 1, 2, \ldots n$. Taking into consideration the randomness associated with the load occurrence times, it follows,

$$\mathbb{L}(0, T) = \underbrace{\int_0^T \cdots \int_0^T}_{n\text{-fold}} \prod_{i=1}^{n} F_S[r \cdot g(t_i)] \cdot f_{\mathbf{T}}(\mathbf{t}) d\mathbf{t}$$

(5.163)

where $f_{\mathbf{T}}(\mathbf{t})$ is the joint PDF of $\{t_1, t_2, \ldots t_n\}$. Modelling the load occurrence as a stationary Poisson process, $f_{\mathbf{T}}(\mathbf{t}) = \left(\frac{1}{T}\right)^n$. Thus,

$$\mathbb{L}(0, T) = \underbrace{\int_0^T \cdots \int_0^T}_{n\text{-fold}} \prod_{i=1}^{n} F_S[r \cdot g(t_i)] \cdot \left(\frac{1}{T}\right)^n d\mathbf{t} = \left[\frac{1}{T} \int_0^T F_S[r \cdot g(t)] dt\right]^n$$

(5.164)

The number of load events has a PMF of that in Eq. (5.159). With this, applying the law of total probability, one has

$$\mathbb{L}(0, T) = \sum_{k=0}^{\infty} \left[\frac{1}{T} \int_0^T F_S[r \cdot g(t)] dt \right]^k \cdot \frac{(\lambda T)^k \exp(-\lambda T)}{k!}$$

$$= \exp(-\lambda T) \sum_{k=0}^{\infty} \frac{\left[\lambda \cdot \int_0^T F_S[r \cdot g(t)] dt \right]^k}{k!} = \exp \left[-\lambda \left(T - \int_0^T F_S[r \cdot g(t)] dt \right) \right]$$

(5.165)

Similar to Eq. (5.161), if the PDF of R_0 is $f_{R_0}(r)$, then [32]

$$\mathbb{L}(0, T) = \int_0^{\infty} \exp \left[-\lambda \left(T - \int_0^T F_S[r \cdot g(t)] dt \right) \right] f_{R_0}(r) dr \qquad (5.166)$$

Note that in Eq. (5.166), if $g(t) \equiv 1$ (i.e., the resistance deterioration is not considered), then the reliability simply becomes that in Eq. (5.161).

Example 5.24

Recall the derivation of time-dependent reliability in Eq. (5.166). If the load occurrence is modelled by a non-stationary Poisson process with an occurrence rate of $\lambda(t)$ at time t, how will Eq. (5.166) be generalized?

Solution

Using a non-stationary Poisson process for the load occurrence, the PMF of the number of loads within $[0, T]$, N, becomes

$$\mathbb{P}(N = k) = \frac{\left(\int_0^T \lambda(\tau) d\tau \right)^k \exp \left(- \int_0^T \lambda(\tau) d\tau \right)}{k!}, \quad k = 0, 1, 2, \ldots \qquad (5.167)$$

Let T_1 denote the occurrence time of one load event. The CDF of T_1 is determined by

$$F_{T_1}(t) = \mathbb{P}[N(0, t) = 1 | N(0, T) = 1] = \frac{\mathbb{P}[N(0, t) = 1 \cap N(t, T) = 0]}{\mathbb{P}[N(0, T) = 1]}$$

$$= \frac{\left(\int_0^t \lambda(\tau) d\tau \right) \cdot \exp \left(- \int_0^t \lambda(\tau) d\tau \right) \cdot \exp \left(- \int_t^T \lambda(\tau) d\tau \right)}{\left(\int_0^T \lambda(\tau) d\tau \right) \cdot \exp \left(- \int_0^T \lambda(\tau) d\tau \right)}$$

$$= \frac{\int_0^t \lambda(\tau) d\tau}{\int_0^T \lambda(\tau) d\tau}$$

(5.168)

In Eq. (5.168), it is easy to verify that $F_{T_1}(0) = 0$ and $F_{T_1}(T) = 1$. Furthermore, based on Eq. (5.168), the PDF of T_1 is obtained as follows,

$$f_{T_1}(t) = \frac{dF_{T_1}(t)}{dt} = \frac{\lambda(t)}{\int_0^T \lambda(\tau) d\tau} \qquad (5.169)$$

Thus, Eq. (5.164) is revised as follows,

$$L(0, T) = \left[\int_0^T \frac{\lambda(t)}{\int_0^T \lambda(\tau)d\tau} \cdot F_S[r \cdot g(t)]dt \right]^n = \left[\frac{\int_0^T \lambda(t) \cdot F_S[r \cdot g(t)]dt}{\int_0^T \lambda(\tau)d\tau} \right]^n$$

$$(5.170)$$

Subsequently, with Eqs. (5.167) and (5.170),

$$L(0, T) = \sum_{k=0}^{\infty} \left[\frac{\int_0^T \lambda(t) \cdot F_S[r \cdot g(t)]dt}{\int_0^T \lambda(\tau)d\tau} \right]^k \cdot \frac{\left(\int_0^T \lambda(\tau)d\tau \right)^k \exp\left(-\int_0^T \lambda(\tau)d\tau \right)}{k!}$$

$$= \exp\left(-\int_0^T \lambda(\tau)d\tau \right) \sum_{k=0}^{\infty} \frac{\left(\frac{\int_0^T \lambda(t) \cdot F_S[r \cdot g(t)]dt}{\int_0^T \lambda(\tau)d\tau} \cdot \int_0^T \lambda(\tau)d\tau \right)^k}{k!}$$

$$= \exp\left(-\int_0^T \lambda(\tau) \left[1 - F_S[r \cdot g(t)] \right] d\tau \right)$$

$$(5.171)$$

Furthermore, incorporating the uncertainty associated with R_0 gives

$$L(0, T) = \int_0^{\infty} \exp\left(-\int_0^T \lambda(\tau) \left[1 - F_S[r \cdot g(t)] \right] d\tau \right) f_{R_0}(r)dr \qquad (5.172)$$

Clearly, if $\lambda(\tau) \equiv \lambda$, then Eq. (5.172) simply becomes Eq. (5.166).

Example 5.25

Recall the derivation of time-dependent reliability in Eq. (5.166). If the load effect is time-variant with a CDF of $F_S(\cdot, t)$ at time t [52], how will Eq. (5.166) be generalized?

Solution

We subdivide the time interval $[0, T]$ into n identical sections, where n is large enough so that the duration of each interval, $\Delta = \frac{T}{n}$ is negligible. With this, it is reasonable to assume that the CDF of the load effect within the ith interval is constant, denoted by $F_S\left(\cdot, \frac{Ti}{n} \right)$. Conditional on $R_0 = r$, The probability that the structure survives within the ith interval is

$$L_i = \mathbb{P}(\text{no load occurs}) + \mathbb{P}(\text{load occurs}) \times \mathbb{P}\left[r \cdot g\left(\frac{Ti}{n} \right) > S_i \right]$$

$$= \left(1 - \lambda \frac{T}{n} \right) + \lambda \frac{T}{n} F_S\left(r \cdot g\left(\frac{Ti}{n} \right), \frac{Ti}{n} \right)$$

$$(5.173)$$

Taking the logarithmic form of both sides gives

$$\ln L_i = \ln\left\{ 1 - \lambda \frac{T}{n} \left[1 - F_S\left(r \cdot g\left(\frac{Ti}{n} \right), \frac{Ti}{n} \right) \right] \right\} \approx \lambda \frac{T}{n} \left[1 - F_S\left(r \cdot g\left(\frac{Ti}{n} \right), \frac{Ti}{n} \right) \right]$$

$$(5.174)$$

Note that the time-dependent reliability over $[0, T]$ equals the probability that the structure survives within each of the n intervals. Thus,

$$L(0, T) = \exp\left(\ln\prod_{i=1}^{n} L_i\right) = \exp\left\{\ln\prod_{i=1}^{n}\lambda\frac{T}{n}\left[1 - F_S\left(r \cdot g\left(\frac{Ti}{n}\right), \frac{Ti}{n}\right)\right]\right\}$$

$$= \exp\left\{\sum_{i=1}^{n}\lambda\frac{T}{n}\left[1 - F_S\left(r \cdot g\left(\frac{Ti}{n}\right), \frac{Ti}{n}\right)\right]\right\} = \exp\left\{\lambda\int_0^T [1 - F_S(r \cdot g(t), t)]\,dt\right\}$$

$$\tag{5.175}$$

Furthermore, taking into account the uncertainty associated with the initial resistance, Eq. (5.175) becomes

$$L(0, T) = \int_0^\infty \exp\left\{\lambda\int_0^T [1 - F_S(r \cdot g(t), t)]\,dt\right\} f_{R_0}(r)dr \tag{5.176}$$

One can easily verify that Eq. (5.176) is a generalized form of Eq. (5.166) incorporating the time-variation of the load effects on the temporal scale.

5.4.2   Hazard Function and Time-Dependent Reliability

The time-dependent reliability in Eqs. (5.161), (5.166), (5.172) and (5.176) can be also formulated with the help of the hazard function, as will be discussed in this section.

The *hazard function*, $h(t)$, is defined as the probability of structural failure at time t provided that the structure has survived up to time t. Mathematically, for a short time duration $dt \to 0$, one has,

$$h(t)dt = \mathbb{P}(t < \text{failure time} \le t + dt | \text{failure time} > t) \tag{5.177}$$

Let T_f be the time of structural failure as before, and F_{T_f} the CDF of T_f. Recall that Eq. (5.16) gives

$$F_{T_f}(\tau) = \mathbb{P}(T_f \le \tau) = \mathbb{P}(\text{failure within } [0, \tau]) = \mathbb{P}_f(0, T) = 1 - L(0, \tau) \tag{5.16}$$

Thus,

$$h(t)dt = \mathbb{P}(t < T_f \le t + dt | T_f > t) = \frac{\mathbb{P}(t < T_f \le t + dt)}{\mathbb{P}(T_f > t)} = -\frac{L(0, t + dt) - L(0, t)}{L(0, t)} \tag{5.178}$$

which further yields

$$h(t) = -\frac{L(0, t + dt) - L(0, t)}{L(0, t)dt} = -d\ln L(0, t) \tag{5.179}$$

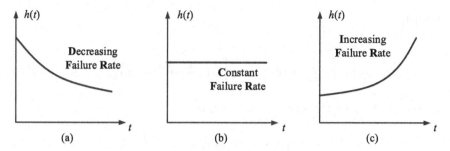

Fig. 5.19 Three types of hazard functions: DFR, CFR and IFR

Integrating both sides from 0 to t gives

$$\mathbb{L}(0, t) = \exp\left[-\int_0^t h(\tau)d\tau\right] \tag{5.180}$$

Note that the CDF of the failure time, $F_{T_f}(t)$, is closely related to the time-dependent reliability $\mathbb{L}(0, t)$ according to Eq. (5.16). With this, if the PDF of the failure time is $f_{T_f}(t)$, with Eq. (5.179), one has

$$h(t) = -d \ln[1 - F_{T_f}(t)] = \frac{f_{T_f}(t)}{1 - F_{T_f}(t)} \tag{5.181}$$

Remark

(1) If the deterioration process in Example 5.10 is modelled as independent (i.e. Case 2), then the hazard function $h(t)$ is obtained as $h(t) = h'(t) = \frac{1}{dt}\mathbb{P}(t < T_f \le t + dt)$, which subsequently overestimates the structural failure probability. The overestimation is also reflected through the comparison between Cases 1 and 2 in Example 5.10.

(2) We can categorize the hazard function $h(t)$ into three types based on its time-variant characteristics, as shown in Fig. 5.19, namely decreasing failure rate (DFR), constant failure rate (CFR) and increasing failure rare (IFR). The DFR usually occurs at the very early stage of the structural service life, mainly due to the reduction of quantity uncertainties resulted from proven survival (see, e.g., Example 2.1.2, and also Sect. 5.4.4). The CFR is often associated with the middle stage of the service life, and the IFR is often observed at the later stage where the structural resistance deterioration is significant.

Example 5.26

If the hazard function $h(t)$ in Eq. (5.180) takes a form of $h(t) = at^b + c$ with a, b and c being three non-negative constants, (1) derive the time-dependent reliability $\mathbb{L}(0, t)$; (2) Discuss the distribution type of the failure time.

Solution

(1) According to Eq. (5.180), it follows,

$$\mathbb{L}(0, t) = \exp\left[-\int_0^t h(\tau)\mathrm{d}\tau\right] = \exp\left[-\left(\frac{a}{b+1}t^{b+1} + ct\right)\right], \quad t \geq 0 \qquad (5.182)$$

(2) With Eq. (5.16), the CDF of the failure time T_f is

$$F_{T_f}(t) = 1 - \mathbb{L}(t) = 1 - \exp\left[-\left(\frac{a}{b+1}t^{b+1} + ct\right)\right] \qquad (5.183)$$

In Eq. (5.183), if $b = c = 0$, then the failure time T_f follows an exponential distribution. Correspondingly, the hazard function $h(t) = a$ is time-variant, leading to a uniform hazard on the time scale, consistent with the fact that the exponential distribution can be used to describe the time lag of a memoryless process (c.f. Sect. 2.2.3). If $b = 1$ and $c = 0$, then Eq. (5.183) becomes the CDF of a Rayleigh distribution (c.f. Sect. 2.2.6); if $b \neq 0$ and $c = 0$, then the failure time follows a Weibull distribution (c.f. Sect. 2.2.8.3).

Example 5.27

The Weibull distribution and the lognormal distribution have been used in the literature the describe the probabilistic behaviour of the fatigue service life (e.g., [21]). Comment on the suitability of the two distribution types by analysing the hazard function.

Solution

For the case of a Weibull distribution, according to Eq. (5.4.2), the hazard function takes a form of $h(t) = at^b$, where $a > 0, b = 1$. This corresponds to an IFR (c.f. Fig. 5.19c).

When using a lognormal distribution for the failure time, the CDF is $F_{T_f}(x) = \Phi\left(\frac{\ln x - \lambda}{\varepsilon}\right)$, and the PDF is $f_{T_f}(x) = \frac{1}{\sqrt{2\pi}x\varepsilon}\exp\left[-\frac{1}{2}\left(\frac{\ln x - \lambda}{\varepsilon}\right)^2\right]$ (c.f. Eqs. (2.201) and (2.202)). According to Eq. (5.181),

$$h(t) = \frac{f_{T_f}(t)}{1 - F_{T_f}(t)} = \frac{\exp\left[-\frac{1}{2}\left(\frac{\ln x - \lambda}{\varepsilon}\right)^2\right]}{\sqrt{2\pi}x\varepsilon\left[1 - \Phi\left(\frac{\ln x - \lambda}{\varepsilon}\right)\right]} \qquad (5.184)$$

Illustratively, Fig. 5.20 shows the hazard function with $\lambda = 1$ and $\epsilon = 0.2$ (corresponding to a mean value of 2.77 and a standard deviation of 0.56). An IFR is observed within the first stage of $[0, 6.84]$, but subsequently, the failure rate decreases with time. This is inconsistent with the general expectation for a degrading structure.

Now we reconsider the time-dependent reliability over a time interval of $[0, T]$, as discussed in Sect. 5.4.1. In the presence of a discrete load process as shown in Fig. 5.21, we use a Poisson process to model the occurrence of these load events as before.

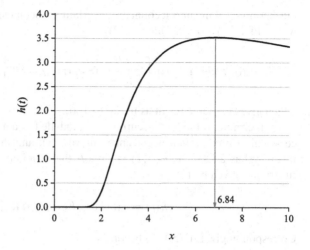

Fig. 5.20 The hazard function associated with a lognormally distributed failure time

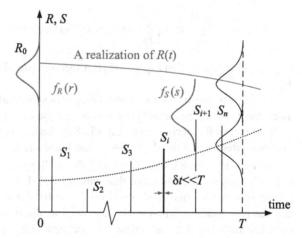

Fig. 5.21 Illustration of time-dependent reliability in the presence of a discrete load process

If the sequence of load effect is described as a stationary Poisson process with a constant occurrence rate of λ, and each load effect is statistically independent and identically distributed with a CDF of $F_S(s)$, the hazard function $h(t)$ is given in Eq. (5.185) conditional on the realizations of the initial resistance $R_0 = r$ and the deterioration function $G(t) = g(t)$.

$$h(t) = \frac{1}{dt} \mathbb{P}(t < T_f \leq t + dt | T_f > t) = \frac{1}{dt} \mathbb{P}(t < T_f \leq t + dt) = \lambda\{1 - F_S[r_0 \cdot g(t)]\}$$
(5.185)

Thus, according to Eq. (5.180),

$$\mathbb{L}(0, T) = \exp\left[\lambda \int_0^T F_S[r \cdot g(t)]dt - \lambda T\right]$$
(5.186)

Furthermore, taking into account the uncertainty associated with the initial resistance, whose PDF is $f_{R_0}(r)$, it follows [32]

$$\mathbb{L}(0, T) = \int_0^\infty \exp\left[\lambda \int_0^T F_S[r \cdot g(t)]\mathrm{d}t - \lambda T\right] \cdot f_{R_0}(r)\mathrm{d}r \qquad (5.187)$$

which gives the same result as in Eq. (5.166).

Furthermore, if the load occurrence is modelled as a non-stationary Poisson process with a time-variant occurrence rate of $\lambda(t)$, and the time-variant load effect intensity has a CDF of $F_S(\cdot, t)$ at time t, then similar to Eq. (5.185), the hazard function is given by [25]

$$h(t) = \lambda(t)\{1 - F_S[r_0 \cdot g(t), t]\} \qquad (5.188)$$

Correspondingly, Eq. (5.187) becomes

$$\mathbb{L}(0, T) = \int_0^\infty \exp\left\{\int_0^T \lambda(t)\left[F_S(r \cdot g(t), t) - 1\right]\mathrm{d}t\right\} \cdot f_{R_0}(r)\mathrm{d}r \qquad (5.189)$$

One can easily verify that Eq. (5.187) is a specific case of Eq. (5.189) with $F_S(\cdot, t) \equiv F_S(\cdot)$.

We now have introduced several formulas for structural time-dependent reliability assessment; the comparison between them can be found in Table 5.1.

Note that in the derivation of Eq. (5.189), the key is to find the hazard function $h(t)$ first. Recall that the definition of the hazard function is $h(t) = \frac{1}{\mathrm{d}t}\mathbb{P}(t < T_\mathrm{f} \le t + \mathrm{d}t|T_\mathrm{f} > t)$, which equals $h'(t) = \frac{1}{\mathrm{d}t}\mathbb{P}(t < T_\mathrm{f} \le t + \mathrm{d}t)$ if the structural failure within $[t, \mathrm{d}t]$ is independent of that within $[0, t]$. For instance, in Eq. (5.185), the relationship holds only conditional on the realizations of the initial resistance R_0 and the deterioration function $G(t)$, as well as the independence of the load process. Illustratively, if the uncertainty associated with R_0 and/or $G(t)$ is considered, then the events of both $T_\mathrm{f} \in [0, t]$ and $T_\mathrm{f} \in [t, \mathrm{d}t]$ are dependent on the probability information of R_0 and/or $G(t)$ and thus are correlated mutually. In such a case, one cannot simply let $h(t) = h'(t)$. In fact, in the definition of $h(t)$, the known information that the structure survives up to time t provides a "proof" of the structure's subsequent performance within $[t, \mathrm{d}t]$, and thus will have an impact on the probability of $T_\mathrm{f} \in [t, \mathrm{d}t]$ from a view of Baysian updating (c.f. Sect. 2.1.2). We use Example 5.28 in the following to demonstrate this point.

Example 5.28

Consider the reliability and hazard function of a structure with a resistance following a lognormal distribution with a mean value of $3S_\mathrm{n}$ and a COV of 0.2, where S_n is the nominal load effect. The resistance deterioration is not considered herein. The load process is modelled by a stationary Poisson process, with an occurrence rate of 0.5/year. Conditional on the occurrence of a load event, the load effect follows an Extreme Type I distribution with a mean value of S_n and a COV of 0.4. Discuss the relationship between $h(t)$ and $h'(t)$ for reference periods up to 50 years.

Table 5.1 Comparison between several formulas for structural time-dependent reliability

Equation	Formula	Comment
Equation (5.161)	$\mathbb{L}(0, T) = \int_0^\infty \exp[-\lambda T(1 - F_S(r))] f_R(r) dr$	No resistance deterioration, stationary load, time-invariant load effect
Equation (5.166) or (5.187)	$\mathbb{L}(0, T) =$ $\int_0^\infty \exp\left[-\lambda \left(T - \int_0^T F_S[r \cdot g(t)] dt\right)\right] f_{R_0}(r) dr$	Resistance deterioration, stationary occurrence, time-invariant load effect
Equation (5.172)	$\mathbb{L}(0, T) =$ $\int_0^\infty \exp\left(- \int_0^T \lambda(\tau) [1 - F_S[r \cdot g(t)]] d\tau\right) f_{R_0}(r) dr$	Resistance deterioration, non-stationary occurrence, time-invariant load effect
Equation (5.176)	$\mathbb{L}(0, T) =$ $\int_0^\infty \exp\left\{\lambda \int_0^T [1 - F_S(r \cdot g(t), t)] dt\right\} f_{R_0}(r) dr$	Resistance deterioration, stationary occurrence, time-variant load effect
Equation (5.189)	$\mathbb{L}(0, T) =$ $\int_0^\infty \exp\left\{\int_0^T \lambda(t) [F_S(r \cdot g(t), t) - 1] dt\right\} \cdot f_{R_0}(r) dr$	Resistance deterioration, non-stationary occurrence, time-variant load effect

Solution

Using Eq. (5.189), the time-dependent reliability for reference periods up to 50 years are calculated, based on which the hazard function $h(t)$ is obtained according to Eq. (5.179) and is presented in Fig. 5.22. For comparison purpose, the hazard function without the effect of Bayesisn updating, $h'(t)$, is also plotted in Fig. 5.22. It can be seen that the values of both $h(t)$ and $h'(t)$ are equal at the initial time as there is no updating effect when $t = 0$; however, the difference between $h(t)$ and $h'(t)$ becomes larger as the time t increases due to the fact that the structural survival up to time t is positively evident of the structure's subsequent performance. Since $h(t) \leq h'(t)$, according to Eq. (5.180), the structural reliability may be underestimated in some cases if using $h'(t)$ instead of $h(t)$ (i.e., ignoring the Bayesian updating effect of the survival information up to time t), as $h'(t)$ overestimates the risk of structural failure. For example, with Eq. (5.189), one has

Fig. 5.22 The comparison between $h(t)$ and $h'(t)$ reflects the effect of Bayesian updating

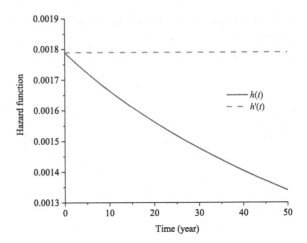

$$\mathbb{L}(0, T) = \int_0^\infty \exp\left\{\int_0^T \lambda(t)\left[F_S(r \cdot g(t), t) - 1\right]dt\right\} \cdot f_{R_0}(r)dr$$

$$> \exp\left\{\int_0^T \lambda(t)\int_0^\infty \left[F_S(r \cdot g(t), t) - 1\right] \cdot f_{R_0}(r)drdt\right\}$$

(5.190)

The use of the hazard function to derive the time-dependent reliability (c.f. Eq. (5.189)) can be further extended to the case of a series or parallel system, as will be discussed in Examples 5.29 and 5.4.2.

Example 5.29

Consider the reliability of a k-component series system within a reference period of $[0, T]$. Assume that for component i ($i = 1, 2, \ldots k$), the resistance deterioration with time is described by $R_i(t) = R_{i,0} \cdot g_i(t)$, where $R_i(t)$ is the resistance of component i at time t, $R_{i,0}$ is the initial resistance of component i, and $g_i(t)$ is the deterioration function of component i. The joint PDF of the $R_{i,0}$'s is $f_{\mathbf{R}}(\mathbf{r})$. Use a non-stationary Poisson process to model the load sequence with a time-variant occurrence rate of $\lambda(t)$. Derive the time-dependent reliability for the series system [63].

Solution

Suppose that a sequence of n discrete loads with intensities of $S_1, S_2, \ldots S_n$ occur at times $t_1, t_2, \ldots t_n$, and S_i results in a structural response of $c_j \cdot S_i$ for the jth component. Let F_i denote the event that the ith component fails within the considered time interval, with which the system reliability, $\mathbb{L}_{ss}(T)$, is evaluated by

$$\mathbb{L}_{ss}(T) = 1 - \mathbb{P}\left(F_1 \bigcup F_2 \bigcup \cdots \bigcup F_k\right)$$

(5.191)

Conditional on $R_{i,0} = r_{i,0}$ for $i = 1, 2, \ldots k$, Eq. (5.191) becomes

$$\mathbb{L}_{ss}(T) = \mathbb{P}\left\{\bigcap_{j=1}^{n}\bigcap_{i=1}^{k} r_{i,0} \times g_i(t_j) > c_i S_j\right\} \tag{5.192}$$

Modelling the load occurrence as a non-stationary Poisson process with a time-variant occurrence rate of $\lambda(t)$, the hazard function of the system is

$$h_{ss}(t) = \lambda(t)\left\{1 - F_S\left(\min_{i=1}^{k}\frac{r_{i,0}g_i(t)}{c_i}, t\right)\right\} \tag{5.193}$$

where $F_S(\cdot, t)$ is the CDF of load effect at time t. With this, according to Eq. (5.180), the system reliability in Eq. (5.192) becomes

$$\mathbb{L}_{ss}(T) = \exp\left[-\int_0^T \lambda(t)\left\{1 - F_S\left(\min_{i=1}^{k}\frac{r_{i,0}g_i(t)}{c_i}, t\right)\right\}dt\right] \tag{5.194}$$

Finally, taking into account the uncertainties associated with the initial resistances of each component, it follows,

$$\mathbb{L}_{ss}(T) = \int\ldots\int \exp\left[-\int_0^T \lambda(t)\left\{1 - F_S\left(\min_{i=1}^{k}\frac{r_{i,0}g_i(t)}{c_i}, t\right)\right\}dt\right] \cdot f_{\mathbf{R}}(\mathbf{r})d\mathbf{r} \tag{5.195}$$

Example 5.30

Reconsider the k-component system in Example 5.29. We now assume that the k components consist of a parallel system. Reevaluate the system's time-dependent reliability within a reference period of $[0, T]$.

Solution

By definition, the reliability of the parallel system is given by

$$\mathbb{L}_{ps}(T) = 1 - \mathbb{P}\left(F_1 \bigcap F_2 \bigcap \cdots \bigcap F_k\right) \tag{5.196}$$

The hazard function $h_{ps}(t)$ is

$$h_{ps}(t) = \lambda(t)\left\{1 - F_S\left(\max_{i=1}^{k}\frac{r_{i,0}g_i(t)}{c_i}, t\right)\right\} \tag{5.197}$$

Thus, Eq. (5.196) becomes

$$\mathbb{L}_{ps}(T) = \exp\left[-\int_0^T \lambda(t)\left\{1 - F_S\left(\max_{i=1}^{k}\frac{r_{i,0}g_i(t)}{c_i}, t\right)\right\}dt\right] \tag{5.198}$$

which further is rewritten as follows in the presence of the uncertainties associated with the component initial resistances.

$$
\mathbb{L}_{ps}(T) = \int \cdots \int \exp\left[-\int_0^T \lambda(t) \left\{1 - F_S\left(\max_{i=1}^k \frac{r_{i,0} g_i(t)}{c_i}, t\right)\right\} dt\right] \cdot f_{\mathbf{R}}(\mathbf{r}) d\mathbf{r}
$$

$$(5.199)$$

5.4.3 Simplified Approach for Time-Dependent Reliability Assessment

Note that the time-dependent reliability in Eq. (5.189) (and also those in Eqs. (5.195) and (5.199)) involves a two-fold integral; if the uncertainty associated with the deterioration process is also considered, then at least a three-fold integral would be involved. In an attempt to improve the calculation efficiency of Eq. (5.189), some simplified/approximate approaches can be used, as will be discussed in this section.

5.4.3.1 Method 1

Suppose that the deterioration function takes a linear form of $g(t) = 1 - kt$, and that the load effect S follows an Extreme Type I distribution with a location parameter of u and a scale parameter of α. If the mean value of the load intensity increases linearly with time according to $\mathbb{E}(S(t)) = \mathbb{E}(S(0)) + \kappa_m t$, where κ_m is a parameter reflecting the changing rate, then the time-dependent reliability is estimated as follows [58] for a stationary load process with an occurrence rate of λ.

$$
\mathbb{L}(0, T) = \int_0^\infty \exp\left(\lambda \cdot \xi_1\right) \cdot f_{R_0}(r) dr \tag{5.200}
$$

where

$$
\xi_1 = -\exp\left(-\frac{r - \mathbb{E}(S(0))}{\alpha}\right) \frac{\alpha}{kr + \kappa_m} \left[\exp\left(\frac{T(kr + \kappa_m)}{\alpha}\right) - 1\right] \tag{5.201}
$$

If, alternatively, the mean occurrence rate of the loads increases linearly with time with $\lambda(t) = \lambda(0) + \kappa_\lambda t$ while the load intensity is time-variant, where κ_λ is the changing rate of the occurrence rate, the time-dependent reliability is given by [58]

$$
\mathbb{L}(0, T) = \int_0^\infty \exp\left[\lambda(0) \cdot \xi_2 + \kappa_\lambda \cdot \psi\right] \cdot f_{R_0}(r) dr \tag{5.202}
$$

where

$$\psi = -\exp\left(-\frac{r-u}{\alpha}\right)\left\{\exp\left(\frac{krT}{\alpha}\right)\left[\frac{\alpha T}{kr} - \left(\frac{\alpha}{kr}\right)^2\right] + \left(\frac{\alpha}{kr}\right)^2\right\} \qquad (5.203)$$

and

$$\xi_2 = -\exp\left(-\frac{r-u}{\alpha}\right)\frac{\alpha}{kr}\left[\exp\left(\frac{krT}{\alpha}\right) - 1\right] \qquad (5.204)$$

Example 5.31

Consider the time-dependent reliability problem of an aging structure, whose deterioration is associated with imprecise information due to the fact that the deterioration may be a multifarious process involving multiple deterioration mechanisms. The example is adopted from [59, 64]. The structure was initially designed at the limit state of $0.9R_n = 1.2D_n + 1.6L_n$, in which R_n is the nominal resistance, D_n and L_n are the nominal dead load and live load, respectively. It is assumed that $D_n = L_n$. The dead load D is assumed to be deterministic and equals to D_n. The live load is modelled as a Poisson process, and the magnitude of the live load follows an Extreme Type I distribution with a standard deviation of $0.12L_n$ and a time-variant mean of $(0.4 + 0.005t)L_n$ in year t. The occurrence rate of the live load is 1.0/year. The initial resistance of the structure, denoted by R_0, is assumed to be deterministic and equals to $1.05R_n$. In year t, the resistance deteriorates to $R(t)$, given by $R(t) = R_0 \cdot (1 - \overline{G}(t))$, in which $\overline{G}(t)$ is the complement of the degradation function and is linearly with time. At the end of 40 years, the COV of $\overline{G}(40)$ is 0.4, and the mean value of $\overline{G}(40)$ is denoted by $\mu_{\overline{G}(40)}$. However, the distribution type of $\overline{G}(40)$ is unknown.

Compute the lower and upper bounds of the time-dependent probability of failure for reference periods up to 40 years (c.f. Sect. 4.5.3.1). Use the values of $\mu_{\overline{G}(40)} = 0.2$ and $\mu_{\overline{G}(40)} = 0.4$ respectively.

Solution

Taking into account the uncertainty associated with the function $\overline{G}(t)$, suppose that in year T, the PDF of $\overline{G}(T)$ is $f_{\overline{G}}(g)$. With this, the time-dependent reliability, $\mathbb{L}(0, T)$, is given by

$$\mathbb{L}(0, T) = \int_0^1 \exp\left[-\int_0^T \lambda(1 - F_S[r(t|g) - D, t])dt\right] \cdot f_{\overline{G}}(g)dg \qquad (5.205)$$

where $r(t|g)$ is the resistance at time t given that $\overline{G}(T)$ equals g, λ is the occurrence rate of the load, and F_S is the CDF of each live load effect. It is noted that $\overline{G}(T)$ should not be less than 0 for structures without maintenance or repair measures because the resistance process in non-increasing, nor be greater than 1 since the resistance of a structure never becomes a negative value, accounting for the integration limits of 0 and 1 in Eq. (5.205). For the case where the mean of load effect increases linearly with time (i.e., $\mu_S(t) = \mu_S(0) + \kappa_m t$), while the standard deviation of load effect, σ_L, is constant, the core of Eq. (5.205),

$$\nu(g) = \exp\left[-\int_0^T \lambda(t)(1 - F_S[r(t|g) - D, t])dt\right] \qquad (5.206)$$

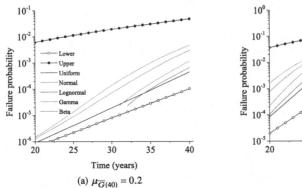
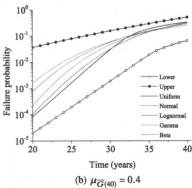

(a) $\mu_{\overline{G}(40)} = 0.2$ (b) $\mu_{\overline{G}(40)} = 0.4$

Fig. 5.23 The lower and upper bounds of the time-dependent failure probability

can be simplified as follows according to Eq. (5.200),

$$v(g) = \exp(-\lambda \cdot \xi_1),$$ (5.207)

in which

$$\xi_1 = \exp\left(\frac{m_0 + D - r_0}{a}\right) \frac{aT}{r_0 g + \kappa_m T} \left[\exp\left(\frac{r_0 g + \kappa_m T}{a}\right) - 1\right],$$ (5.208)

where $a = \frac{\sqrt{6}\sigma_L}{\pi}$, and $m_0 = \mu_S(0) - 0.5772a$.

The lower and upper bounds of the time-dependent probability of failure for reference periods up to 40 years are computed using the linear programming-based method in Sect. 4.5.3.1, and are plotted in Fig. 5.23. As a comparison, Fig. 5.23 also shows the probabilities of failure with additional assumptions of the distribution type of $\overline{G}(40)$, i.e., several commonly-used distributions including normal, lognormal, Gamma, Beta and uniform distributions. It can be seen from Fig. 5.23 that for both cases of $\mu_{\overline{G}(40)}$, the lower and upper bounds establish an envelope to enclose the failure probabilities associated with additional assumptions for the distribution type of $\overline{G}(40)$.

5.4.3.2 Method 2

We divide the reference period $[0, T]$ into m identical sections on a yearly scale (m is the integer that is closest to T). For example, for a reference period of 50 years, we let $m = 50$ and the service period $[0, 50]$ is divided into 50 sections. Assume that the resistance is approximately constant within each year. Consider the CDF of the maximum load effect within the jth year, $F_{max}(\cdot, j)$ for $j = 1, 2, \ldots m$. We further divide the jth year into n identical intervals (n is sufficiently large), with which the probability that one load event occurs within the ith interval (for the jth year) is $\frac{1}{n}\lambda \left(j - 1 + \frac{i}{n} \right)$, and the CDF of the load effect is $F_S \left(s, j - 1 + \frac{i}{n} \right)$. Thus

$$F_{\max}(s, j) = \lim_{n\to\infty} \prod_{i=1}^{n} \left\{ 1 - \lambda \left(j - 1 + \frac{i}{n} \right) \cdot \frac{1}{n} \cdot \left[1 - F_S \left(s, j - 1 + \frac{i}{n} \right) \right] \right\}$$

$$(5.209)$$

Taking the logarithmic form of both sides of Eq. (5.209) gives,

$$\ln F_{\max}(s, j) = \lim_{n\to\infty} \sum_{i=1}^{n} \left\{ -\lambda \left(j - 1 + \frac{i}{n} \right) \cdot \frac{1}{n} \cdot \left[1 - F_S \left(s, j - 1 + \frac{i}{n} \right) \right] \right\}$$

$$= -\int_{j-1}^{j} \lambda(t) \left\{ 1 - F_S(s, t) \right\} dt$$

$$(5.210)$$

Thus,

$$F_{\max}(s, j) = \exp \left[-\int_{j-1}^{j} \lambda(t) \left\{ 1 - F_S(s, t) \right\} dt \right] \qquad (5.211)$$

With this, as $T \gg 1$year, Eq. (5.189) becomes [51]

$$\mathbb{L}(0, T) = \int_{0}^{\infty} \prod_{j=1}^{m} F_{\max}[r \cdot g(j), j] \cdot f_{R_0}(r) dr \qquad (5.212)$$

which gives an expression for the time-dependent reliability with a lower-dimensional integral compared with Eq. (5.189).

5.4.4   Updating Structural Reliability with Prior Information

We have mentioned in Sect. 5.2 that the external load is one of the major contributors to the reduction of structural reliability. However, if it is known that a structure has survived from a historical load, then this information would be evident of the structural resistance from a viewpoint of statistics [8, 13, 14, 23, 42, 65]. For example, for an existing bridge with a service period of T years, the lower tail of the probability distribution of current resistance, $R(T)$, is progressively truncated due to the fact that $R(T)$ is higher than any of the historically imposed loads (c.f. Example 2.1.2). In fact, the successful historical load can be regarded as a proof load test with uncertainty for an existing service-proven structure. The use of prior information to update the structural resistance and reliability will be discussed in this Section.

5.4.4.1   Updated Resistance for Service Proven Structures

If an existing structure is subjected to a known proof load, then the distribution of
resistance is simply truncated at this known load effect (c.f. Eq. (2.37)) [13]. If the
proof load is uncertain, then the updated distribution of structural resistance is [14]

$$f_R'(r) = \frac{f_R(r)F_S(r)}{\int_0^\infty f_R(r)F_S(r)\mathrm{d}r} \tag{2.39}$$

where $f_R(r)$ and $f_R'(r)$ are the PDFs of structural resistance prior to and post loading
respectively, F_S is the CDF of the load effect.

Now we consider the structural resistance at time T with resistance deterioration
over $[0, T]$. Suppose the PDF of initial resistance, R_0, is $f_{R_0}(r)$, and the PDF of
$G(T)$ (deterioration function value at time T, c.f. Eq. (5.31)) is $f_{G(T)}(g)$. Prior to
loading, the PDF of $R(T)$ is determined by

$$
\begin{aligned}
f_{R(T)}(r) &= \frac{1}{\mathrm{d}r}\mathbb{P}[R(T) = r] = \frac{1}{\mathrm{d}r}\mathbb{P}[R_0 \cdot G(T) = r] \\
&= \frac{1}{\mathrm{d}r}\sum_\tau \mathbb{P}[R_0 \cdot G(T) = r | R_0 = \tau] \cdot \mathbb{P}(R_0 = \tau) \\
&= \frac{1}{\mathrm{d}r}\int_0^\infty \mathbb{P}\left[G(T) = \frac{r}{\tau}\right] \cdot f_{R_0}(\tau)\mathrm{d}\tau = \frac{1}{\mathrm{d}r}\int_0^\infty f_{G(T)}\left(\frac{r}{\tau}\right)\mathrm{d}r \cdot f_{R_0}(\tau)\mathrm{d}\tau \\
&= \int_0^\infty f_{R_0}(\tau) \cdot f_{G(T)}\left(\frac{r}{\tau}\right) \cdot \frac{1}{\tau}\mathrm{d}\tau
\end{aligned}
$$

$$\tag{5.213}$$

Provided the fact that the structure has survived for T years, let $S_1, S_2, \ldots S_T$
denote the T maximum load effects experienced within each year (c.f. Sect. 5.4.3.2).
If the deterioration function in Eq. (5.32) is adopted, the updated distribution of
structural resistance at current time T is computed as follows [23]

$$
\begin{aligned}
f_{R(T)}'(r) &= \frac{1}{\mathrm{d}r}\mathbb{P}\{R(T) = r | \text{survival within } [0, T]\} = \frac{1}{\mathrm{d}r}\mathbb{P}\{R_0 \cdot G(T) = r | \text{survival within } [0, T]\} \\
&= \frac{1}{\mathrm{d}r} \cdot \frac{\mathbb{P}\{R_0 \cdot G(T) = r \cap \text{survival within } [0, T]\}}{\mathbb{P}\{\text{survival within } [0, T]\}} \\
&= \frac{1}{\mathrm{d}r} \cdot \frac{\sum_\tau \mathbb{P}[R_0 \cdot G(T) = r \cap \text{survival within } [0, T] | R_0 = \tau] \cdot \mathbb{P}(R_0 = \tau)}{\sum_\tau \sum_g \mathbb{P}[\text{survival within } [0, T] | R_0 = \tau, G(T) = g] \cdot \mathbb{P}(R_0 = \tau)\mathbb{P}(G(T) = g)} \\
&= \frac{1}{\mathrm{d}r} \cdot \frac{\int_0^\infty f_{R_0}(\tau)f_{G(T)}\left(\frac{r}{\tau}\right)\frac{1}{\tau}\mathrm{d}r \cdot \mathbb{P}\left\{\bigcap_{i=1}^T\left[\tau \cdot \left(1 - (1-g) \cdot \left(\frac{i}{T}\right)^\alpha\right) \geq S_i\right]\right\}\mathrm{d}\tau}{\int_0^1 \int_0^\infty f_{R_0}(\tau)f_{G(T)}(g) \cdot \mathbb{P}\left\{\bigcap_{i=1}^T\left[\tau \cdot \left(1 - (1-g) \cdot \left(\frac{i}{T}\right)^\alpha\right) \geq S_i\right]\right\}\mathrm{d}\tau\mathrm{d}g} \tag{5.214} \\
&= \frac{\int_0^\infty f_{R_0}(\tau)f_{G(T)}\left(\frac{r}{\tau}\right) \cdot \frac{1}{\tau} \cdot \prod_{i=1}^T F_{S,i}\left[\tau \cdot \left(1 - (1-\frac{r}{\tau}) \cdot \left(\frac{i}{T}\right)^\alpha\right)\right]\mathrm{d}\tau}{\int_0^1 \int_0^\infty f_{R_0}(\tau)f_{G(T)}(g) \cdot \prod_{i=1}^T F_{S,i}\left[\tau \cdot \left(1 - (1-g) \cdot \left(\frac{i}{T}\right)^\alpha\right)\right]\mathrm{d}\tau\mathrm{d}g}
\end{aligned}
$$

where $F_{S,i}$ is the CDF of maximum load effect experienced within the ith year, $i = 1, 2, \ldots T$. It can be seen that Eq. (5.214) works for both stationary ($F_{S,i}$ is time-invariant) and non-stationary ($F_{S,i}$ is time-variant) load processes.

5.4.4.2 Updated Reliability of Service-Proven Structures for Subsequent Years

For a structure that has survived for T years, we consider the reliability for a subsequent T' years, denoted by $\mathbb{L}(T, T + T')$. Using the Bayes' theorem, it follows,

$$\mathbb{L}(T'|T) = \frac{\mathbb{L}(0, T + T')}{\mathbb{L}(0, T)} \tag{5.215}$$

We let $S_{T+1}, S_{T+2}, \ldots S_{T+T'}$ denote the maximum load effect experienced within the $(T + 1)$th, $(T + 2)$th,...$(T + T')$th years respectively. Similar to the denominator of Eq. (5.214), the numerator and denominator of Eq. (5.215) can be evaluated by using the law of total probability as follows,

$$\mathbb{L}(0, T) = \int_0^1 \int_0^\infty f_{R_0}(\tau) f_{G(T)}(g) \cdot \prod_{i=1}^{T} F_{S,i}\left[\tau \cdot \left(1 - (1 - g) \cdot \left(\frac{i}{T}\right)^\alpha\right)\right] d\tau dg$$

$$\mathbb{L}(0, T + T') = \int_0^1 \int_0^\infty f_{R_0}(\tau) f_{G(T)}(g) \cdot \prod_{i=1}^{T+T'} F_{S,i}\left[\tau \cdot \left(1 - (1 - g) \cdot \left(\frac{i}{T}\right)^\alpha\right)\right] d\tau dg \tag{5.216}$$

where $F_{S,i}$ is the CDF of S_i. Thus, Eq. (5.215) becomes [23]

$$\mathbb{L}(T'|T) = \frac{\int_0^1 \int_0^\infty f_{R_0}(\tau) f_{G(T)}(g) \cdot \prod_{i=1}^{T+T'} F_{S,i}\left[\tau \cdot \left(1 - (1 - g) \cdot \left(\frac{i}{T}\right)^\alpha\right)\right] d\tau dg}{\int_0^1 \int_0^\infty f_{R_0}(\tau) f_{G(T)}(g) \cdot \prod_{i=1}^{T} F_{S,i}\left[\tau \cdot \left(1 - (1 - g) \cdot \left(\frac{i}{T}\right)^\alpha\right)\right] d\tau dg} \tag{5.217}$$

Based on Eq. (5.217), the reliability index $\beta(T'|T)$ is computed as $\Phi^{-1}[\mathbb{L}(T'|T)]$.

For comparison purpose, if the historical service information is not considered, the reliability for subsequent T' years is simply

$$\mathbb{L}(T, T + T') = \int_0^1 \int_0^\infty \prod_{i=T+1}^{T+T'} F_{S,i}\left[\tau \cdot \left(1 - (1 - g) \cdot \left(\frac{i}{T}\right)^\alpha\right)\right] \cdot f_{R_0}(\tau) f_{G(T)}(g) d\tau dg \tag{5.218}$$

Corresponding to Eq. (5.218), the reliability index is $\beta(T, T + T') = \Phi^{-1}[\mathbb{L}(T', T + T')]$.

Example 5.32

Consider the resistance (moment-bearing capacity) of a bridge girder [23]. The initial resistance has a lognormal distribution with a mean value of 4022 kN·m and a standard

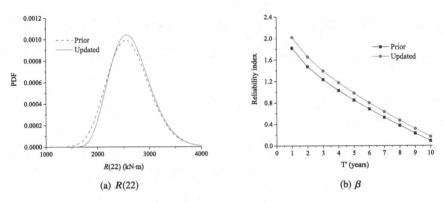

Fig. 5.24 Updating the resistance and reliability with historical service information

deviation of 479 kN·m (c.f. Fig. 1.3). The bridge has successfully served for 22 years. The deterioration function valued at the current time, $G(22)$, follows a Beta distribution with a mean value of 0.65 and a COV of 0.1. The dead load is deterministically 1169 kN·m, and the nominal live load L_n equals 1215 kN·m. The annual maximum load effect follows an Extreme Type I distribution with a mean value of $0.46L_n$ and a standard deviation of $0.18L_n$. Use a linear deterioration model for $G(t)$. (1) Compute the updated resistance at the current time with the successful service information. (2) Calculate the reliability indices for subsequent service periods up to 10 years with and without the historical information.

Solution

The updated estimate of current resistance $R(22)$ of the service-proven bridge girder is obtained from Eq. (5.214) with service history divided equally into 22 time intervals ($T = 22$). The result is shown in Fig. 5.24a, where the effect of service load history on the estimate of current resistance can be seen. The mean value of current resistance, $R(22)$, increases from 2615 kN·m to 2653 kN·m (by 1.5%). The COV of $R(22)$ decreases from 0.156 to 0.146.

The updated reliability index of the service-proven bridge girder for subsequent periods up to 10 years is obtained from Eq. (5.217) with $T = 22$ and $T' = 1, 2, \ldots 10$. The result is presented in Fig. 5.24b. For comparison purpose, the prior time-dependent reliability index is calculated with Eq. (5.218) and is also plotted in Fig. 5.24(b). The reliability indices for subsequent service years increase significantly due to the updating effect of the prior service loads.

Example 5.33

Reconsider Example 5.32. Now we use a Gamma process to model the resistance deterioration process as in Eq. (5.35). Assume that $\overline{G}(22) = 1 - G(22)$ has a mean value of 0.35 and a COV of 0.25. Also, the Poisson process is used to describe the load occurrence with an occurrence rate of $\lambda = 1$/year. Reevaluate the updated distribution of $R(22)$ with prior service information [65].

Solution

Similar to Eq. (5.214), the updated PDF of $R(T)$ is estimated by

$$f'_{R(T)}(r) = \frac{\frac{1}{dr} \cdot \mathbb{P}\{R_0 \cdot G(T) = r \cap \text{survival within } [0, T]\}}{\mathbb{P}\{\text{survival within } [0, T]\}} \tag{5.219}$$

Modelling the resistance degradation as a Gamma process, let N denote the number of loads, which is a Poisson random variable having a mean value of λT. Furthermore, suppose that the N loads occur at times $t_1, t_2, \ldots t_N$ with load effects of $S_1, S_2, \ldots S_N$ respectively. With this, the denominator of Eq. (5.219) is estimated according to Eq. (5.35) as follows,

$$\mathbb{P}\{\text{survival within } [0, T]\} = \mathbb{P}\left\{ \bigcap_{i=1}^{N} \left[R_0 \cdot \left(1 - \sum_{k=1}^{i} D_k \right) \geq S_i \right] \right\} \tag{5.220}$$

where D_k is the resistance deterioration (normalized by the initial resistance) within the interval of $[t_{k-1}, t_k]$. The numerator of Eq. (5.219) is

$$\frac{1}{dr} \cdot \mathbb{P}\{R_0 \cdot G(T) = r \cap \text{survival within } [0, T]\}$$

$$= \frac{1}{dr} \int_0^\infty f_{R_0}(\tau) \cdot \mathbb{P}\left\{ \bigcap_{i=1}^{N} \left[\tau \cdot \left(1 - \sum_{k=1}^{i} D_k \right) \geq S_i \right] \bigcap \left(\frac{r}{\tau} < 1 - \sum_{k=1}^{N} D_k \leq \frac{r + dr}{\tau} \right) \right\} d\tau \tag{5.221}$$

A purely analytical solution would have existed if the estimates of the probabilities in Eqs. (5.220) and (5.221) can be solved numerically. However, multiple-fold integral is involved in Eqs. (5.220) and (5.221) so that it is difficult to find the numerical solution directly. Therefore, we use a simulation-based approach herein to estimate the posterior PDF of the current resistance $R(T)$. For each simulation run, the procedure is as follows,

(a) Simulate a realization of the initial resistance R_0, denoted by r_0, and the number of load events, n.

(b) Simulate the time sequence for the n loads, $t_1, t_2, \ldots t_n$.

(c) Generate the samples for $D_1, D_2, \ldots D_n$, denoted by $d_1, d_2, \ldots d_n$ respectively.

(d) Simulate the load effect corresponding to each t_i, $s_1, s_2, \ldots s_n$.

(e) If $r_0 \cdot \left(1 - \sum_{k=1}^{i} D_k \right) \geq s_i$ holds for $\forall i = 1, 2, \ldots n$, then the structure is deemed as "survival". In this case, record r_0.

The simulation-based PDF of $R(22)$ is presented in Fig. 5.25. For comparison purpose, the result associated with a fully-correlated deterioration process (c.f. Example 5.32) is also plotted in Fig. 5.25. It can be seen that the use of a fully-correlated deterioration process may overestimate the current resistance. The mean value of $R(22)$ prior to updating is 4022kN·m; the updated $R(22)$ has a mean value of 4061 kN·m with a Gamma process-based deterioration, and 4050 kN·m based on a fully-correlated deterioration.

Fig. 5.25 Updated PDFs of $R(22)$ with different deterioration models

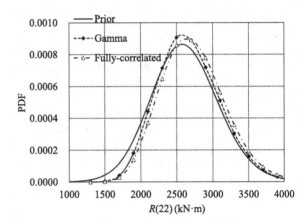

$R(22)$ (kN·m)

Example 5.34

Consider a structure that has a time-variant resistance of $R(t) = R_0 \cdot g(t)$, where R_0 is the initial resistance and $g(t)$ is the deterioration function. The PDF of R_0 is $f_{R_0}(r)$. The sequence of load effect is a stationary Poisson process with a constant occurrence rate of λ, and each load effect is statistically independent and identically distributed with a CDF of $F_S(s)$. Given that the structure has survived for T years, derive the structural reliability for a subsequent T' years [60].

Solution

Let $\mathbb{L}(T'|T)$ denote the time-dependent reliability for a subsequent period of t' years given the successful service history of T years. According to Eq. (5.215), $\mathbb{L}(T'|T) = \frac{\mathbb{L}(0,T+T')}{\mathbb{L}(0,T)}$. Furthermore, with Eq. (5.166),

$$\mathbb{L}(0, T) = \int_0^\infty \exp\left[-\lambda\left(T - \int_0^T F_S[r \cdot g(t)]dt\right)\right] f_{R_0}(r)dr$$

$$\mathbb{L}(0, T + T') = \int_0^\infty \exp\left[-\lambda\left(T + T' - \int_0^{T+T'} F_S[r \cdot g(t)]dt\right)\right] f_{R_0}(r)dr$$

$$(5.222)$$

Thus,

$$\mathbb{L}(T'|T) = \frac{\mathbb{L}(0, T + T')}{\mathbb{L}(0, T)} = \frac{\int_0^\infty \exp\left[-\lambda\left(T + T' - \int_0^{T+T'} F_S[r \cdot g(t)]dt\right)\right] f_{R_0}(r)dr}{\int_0^\infty \exp\left[-\lambda\left(T - \int_0^T F_S[r \cdot g(t)]dt\right)\right] f_{R_0}(r)dr}$$

$$(5.223)$$

5.4.5 Synthetic Approach for Time-Dependent Reliability Assessment

This section discusses the assessment approach for structural reliability in the presence of both the temporal correlation in loads and the deterioration-load dependency [55]. The probability models introduced in Sects. 5.3.1.2 and 5.3.1.3 are used herein. The resistance deterioration model as illustrated in Fig. 5.11 is adopted.

5.4.5.1 Type-I failure

A structure is deemed to fail at time t if the load effect, $S(t)$, exceeds the resistance, $R(t)$. This type failure is referred to as Type-I failure in this Section, which is actually the classical definition of structural failure (c.f. Eq. (5.1)). With this, similar to Eq. (5.26), the instantaneous failure probability at time t, $\mathbb{P}_{f,ins}(t)$, is given by

$$\mathbb{P}_{f,ins}(t) = \mathbb{P}(R(t) - S(t) \le 0) = \int_0^\infty F_R(x, t) f_S(x, t) dx \qquad (5.224)$$

where $F_R(x, t)$ is the instantaneous CDF of R at time t, and $f_S(x, t)$ is the instantaneous PDF of S at time t. Equation (5.224) holds under the assumption of statistically independent R and S.

Now we consider the structural failure probability within time interval $[0, T]$. If there are n load events with intensities of $S_1, S_2, \ldots S_n$ occurring at times $t_1, t_2, \ldots t_n$, respectively, the failure probability $p_{f,1}(T)$ can be expressed as

$$\mathbb{P}_{f,1}(T) = 1 - \mathbb{P}\left(R_1 > S_1 \bigcap R_2 > S_2 \bigcap \cdots \bigcap R_n > S_n\right) \qquad (5.225)$$

where R_i is the resistance corresponding to time t_i for $i = 1, 2, \ldots n$,

$$R_i = R_0 - \left(\sum_{k=1}^i G_k + \sum_{k=1}^i D_k\right) \qquad (5.226)$$

in which G_k is the gradual deterioration during time interval $[t_{k-1}, t_k)$, and D_k is the shock deterioration at time t_k. Conditional on $\{t_1, t_2, \ldots t_n\}$ and $\{S_1, S_2, \ldots S_n\}$, Eq. (5.225) becomes [55]

$$\mathbb{P}_{f,1}(T) = 1 - \int_0^{\gamma_1} \cdots \int_0^{\gamma_n} dF_{D_n|S_n}(x_n) dF_{D_{n-1}|S_{n-1}}(x_{n-1}) \ldots dF_{D_1|S_1}(x_1) \qquad (5.227)$$

where $\gamma_i = R_0 - \sum_{k=1}^i G_k - \sum_{k=1}^{i-1} x_k - S_i$, and $F_{D_i|S_i}$ is the conditional CDF of D_i on S_i for $i = 1, 2, \ldots n$.

5.4.5.2 Type-II failure

The second type of failure occurs if the cumulative deterioration reaches or exceeds the permissible level, g_a, over the considered reference period, i.e.,

$$\mathbb{P}_{f,2}(T) = 1 - \mathbb{P}\left[\left(\sum_{i=1}^{n+1} G_i + \sum_{i=1}^{n} D_i\right) \geq g_a\right] = \int \cdots \int_{\Omega} \mathrm{d}F_{D_n|S_n}(x_n)\mathrm{d}F_{D_{n-1}|S_{n-1}}(x_{n-1})\ldots \mathrm{d}F_{D_1|S_1}(x_1)$$

(5.228)

where $\mathbb{P}_{f,2}(T)$ is the probability of Type-II failure within a reference period of T years, $\Omega = \{x_1,\ldots x_n | \sum_{i=1}^{n} x_i \geq g_a - \sum G_i\}$, and G_{n+1} is the gradual deterioration within time interval $[t_n, T]$. Since the second type failure mode is related to the cumulative deterioration condition only, which is essentially a monotonic process, one may simply focus on the deterioration state at the end of the service period of interest.

Example 5.35

Consider a structure subjected to cumulative damage caused by repeated shocks [53]. Based on the concept of linear combination, the damage at any time equals the sum of all the damage increments that occurred before this time, that is, $D_c = \sum_{j=1}^{N} Q(t_j)$, where D_c is the cumulative damage index (referred to as *cumulative damage* in this example), $\{t_j\}$ is the time sequence at which the loads occur, N is the number of shock events described by a Poisson random variable, $Q(t)$ is the damage increment at time t given the occurrence of a shock event. Suppose that $Q(t_j)$ is statistically independent of each other, and that there is no compounding effect from the loads on the structure.

Assume that both the Poisson occurrence rate and mean value of each damage vary linearly with time, i.e.,

$$\lambda(t) = \lambda(0)(1 + \kappa_\lambda t); \quad \mu(t) = \mu(0)(1 + \kappa_\mu t)$$

(5.229)

where κ_λ and κ_μ are two constants indicating the changing rate of $\lambda(t)$ and $\mu(t)$, respectively. Let $\lambda(0) = 0.2$, $\mu(0) = 0.01$, $\kappa_\lambda = \kappa_\mu = 0.01$. The COV of each shock damage is assumed to be time-invariant and equals 0.3. Show that a Gamma distribution can well describe the probabilistic behaviour of D_c.

Solution

We first estimate the mean value and variance of D_c (afterwards, the probability distribution of D_c can be uniquely determined provided that D_c follows a Gamma distribution). With the sequence of load events modelled as a Poisson process with a time-variant occurrence rate of $\lambda(t)$, we subdivide the time period $[0, T]$ into n identical sections, where n is large enough so that at most one load may occur during each time interval. The occurrence probability of a load event during the kth time interval is then $\lambda(t_k) \cdot \frac{T}{n}$ for $k = 1, 2 \ldots n$, where $t_k = \frac{T}{n} \cdot k$. Note that an equivalent expression for D_c is

$$D_c = \sum_{k=1}^{n} \tilde{Q}_k = \sum_{k=1}^{n} B_k \cdot Q(t_k)$$

(5.230)

in which $\widetilde{Q}_k$ is the damage within the kth time interval, and B_k is a Bernoulli random variable,

$$\mathbb{P}(B_k = 1) = \lambda(t_k) \cdot \frac{T}{n}; \ \mathbb{P}(B_k = 0) = 1 - \mathbb{P}(B_k = 1) \tag{5.231}$$

Thus,

$$\mathbb{E}(D_c) = \sum_{k=1}^{n} \mathbb{E}[B_k \cdot Q(t_k)] = \sum_{k=1}^{n} \lambda(t_k) \cdot \frac{T}{n} \cdot \mu(t_k) = \int_0^T \lambda(t)\mu(t)\mathrm{d}t \tag{5.232}$$

Furthermore, according to Eq. (2.307),

$$\mathbb{V}[B_k \cdot Q(t_k)] = \mathbb{E}\{\mathbb{V}[B_k \cdot Q(t_k)|B_k]\} + \mathbb{V}\{\mathbb{E}[B_k \cdot Q(t_k)|B_k]\}$$
$$= \mathbb{E}\{B_k^2\mathbb{V}[Q(t_k)]\} + \mathbb{V}\{B_k\mathbb{E}[Q(t_k)]\} = (\mu^2(t_k) + \sigma^2(t_k))\frac{\lambda(t_k)T}{n} \tag{5.233}$$

with which

$$\mathbb{V}(D_c) = \sum_{k=1}^{n} (\mu^2(t_k) + \sigma^2(t_k))\frac{\lambda(t_k)T}{n} = \int_0^T \lambda(t)\left[\mu^2(t) + \sigma^2(t)\right]\mathrm{d}t \tag{5.234}$$

where $\sigma(t)$ is the standard deviation of the damage at time t. Note that Eqs. (5.232) and (5.234) are consistent with the results from Examples 2.47 and 5.3.1.2.

Next, to verify the accuracy of Gamma distribution for the cumulative damage, the analytical and simulated CDFs of D_c are presented in Fig. 5.26. A service period of $T = 10$ years is considered in Fig. 5.26a and 100 years in Fig. 5.26b. The analytical CDF is that of a Gamma distribution with mean value and variance obtained from Eqs. (5.232) and (5.234). The simulated CDF is approximated by the histogram of sampled D_c with 100,000 Monte Carlo replications. Illustratively, each damage increment, $Q(t_k)$, is assumed to have a Gamma, Weibull or lognormal distribution in the simulation for the purpose of comparison. From Fig. 5.26 it can be seen that the Gamma distribution well captures the probabilistic behaviour of the cumulative damage D_c for both reference periods. Moreover, the distribution type of each damage increment has a negligible impact on the distribution of D_c, indicating that one may only focus on the estimate of mean value and variance of $Q(t_k)$ in practice.

Example 5.36

Reconsider Eq. (5.35). Let $f_Q(z, t)$ be the PDF of $Q(t)$. Derive the PDF and CDF of the cumulative damage D_c for a reference period of $[0, T]$ [54].

Solution

With the load occurrence process modelled as a non-stationary Poisson process with a time-variant occurrence rate of $\lambda(t)$, as before, we divide the time period $[0, T]$ into n identical sections with a sufficiently large n. We will find the distribution of the cumulative damage D_c with the help of the moment generating function, as introduced in Sect. 2.1.6.

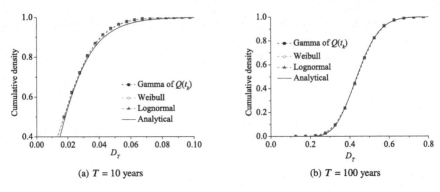

Fig. 5.26 Simulated and analytical CDFs of the cumulative damage for Example 5.35

The MGF of D_c, $\phi_D(\tau)$, is obtained according to Eq. (5.230) as follows,

$$\phi_D(\tau) = \mathbb{E}\left[\exp\left(\sum_{k=1}^{n} \tau \tilde{Q}_k\right)\right] \tag{5.235}$$

By noting the independence between $\tilde{Q}_1, \tilde{Q}_2, \dots \tilde{Q}_n$ and the Taylor expansion of an exponential function (i.e., $\exp(x) = \sum_{j=0}^{\infty} \frac{x^j}{j!}$ holds for any number x), Eq. (5.235) becomes

$$\phi_D(\tau) = \mathbb{E}\left[\prod_{k=1}^{n} \exp(\tau \tilde{Q}_k)\right] = \mathbb{E}\left\{\prod_{k=1}^{n}\left[\sum_{j=0}^{\infty} \frac{(\tau \tilde{Q}_k)^j}{j!}\right]\right\} = \prod_{k=1}^{n}\left[\sum_{j=0}^{\infty} \frac{\mathbb{E}(\tau \tilde{Q}_k)^j}{j!}\right] \tag{5.236}$$

Taking the logarithmic form for both sides of Eq. (5.236), since $\ln(1+x)$ approximates x for a small value of x, we have

$$\phi_D(\tau) = \exp\left\{\sum_{k=1}^{n} \ln\left[\sum_{j=0}^{\infty} \frac{\mathbb{E}(\tau \tilde{Q}_k)^j}{j!}\right]\right\} = \exp\left(\sum_{k=1}^{n}\sum_{j=1}^{\infty} \frac{\mathbb{E}(\tau \tilde{Q}_k)^j}{j!}\right) = \exp\left(\sum_{j=1}^{\infty} \xi_j \tau^j\right) \tag{5.237}$$

where

$$\xi_j = \frac{1}{j!}\sum_{k=1}^{n} \mathbb{E}\left[(\tilde{Q}_k)^j\right] \tag{5.238}$$

Let $\phi_{iD}(i\tau)$ denote the characteristic function of D_c. The PDF of D_c, $f_D(x)$, is obtained, by referring to Eq. (2.133), as follows,

$$f_D(x) = \frac{1}{2\pi}\int_{-\infty}^{\infty} \exp(-ix\tau)\phi_{iD}(i\tau)d\tau \tag{5.239}$$

Substituting Eq. (5.237) into Eq. (5.239), it follows,

$$f_D(x) = \frac{1}{2\pi} \int_{-\infty}^{\infty} \exp\left(\sum_{m=1}^{\infty}(-1)^m \xi_{2m}\tau^{2m}\right) \cdot \exp\left(i\sum_{m=0}^{\infty}(-1)^m \xi_{2m+1}\tau^{2m+1} - ix\tau\right) d\tau \tag{5.240}$$

By noting that for any real number x, $\exp(ix) = \cos x + i\sin x$, we let

$$\theta(\tau, x) = \sum_{m=0}^{\infty}(-1)^m \xi_{2m+1}\tau^{2m+1} - x\tau \tag{5.241}$$

with which Eq. (5.240) becomes

$$f_D(x) = \frac{1}{2\pi} \int_{-\infty}^{\infty} \exp\left(\sum_{m=1}^{\infty}(-1)^m \xi_{2m}\tau^{2m}\right)\cos\theta(\tau, x)d\tau + i\frac{1}{2\pi}\int_{-\infty}^{\infty}\exp\left(\sum_{m=1}^{\infty}(-1)^m \xi_{2m}\tau^{2m}\right)\sin\theta(\tau, x)d\tau \tag{5.242}$$

Since the term $\exp\left(\sum_{m=1}^{\infty}(-1)^m \xi_{2m}\tau^{2m}\right) \cdot \sin\theta(\tau, x)$ in Eq. (5.242) is an odd function of τ, the following equation holds,

$$\int_{-\infty}^{\infty} \exp\left(\sum_{m=1}^{\infty}(-1)^m \xi_{2m}\tau^{2m}\right) \cdot \sin\theta(\tau, x)d\tau = 0 \tag{5.243}$$

Thus, Eq. (5.242) becomes

$$f_D(x) = \frac{1}{2\pi} \int_{-\infty}^{\infty} \exp\left(\sum_{m=1}^{\infty}(-1)^m \xi_{2m}\tau^{2m}\right) \cdot \cos\theta(\tau, x)d\tau \tag{5.244}$$

Recall Eq. (5.238), with which one has

$$\xi_j = \lim_{n\to\infty}\frac{1}{j!}\sum_{k=1}^{n}\mathbb{E}\left(B_k Q^j(t_k)\right) = \lim_{n\to\infty}\frac{1}{j!}\sum_{k=1}^{n}\mathbb{E}\left[(Q(t_k))^j\right] \cdot \lambda(t_k) \cdot \frac{T}{n} \tag{5.245}$$

By definition,

$$\mathbb{E}\left[(Q(t_k))^j\right] = \int_{0}^{\infty} z^j \cdot f_Q(z, t_k)dz \tag{5.246}$$

Substituting Eq. (5.246) into Eq. (5.245), we have

$$\xi_j = \frac{1}{j!}\int_{0}^{T}\int_{0}^{\infty} z^j \lambda(t) f_Q(z, t)dzdt \tag{5.247}$$

With this,

$$\xi_{2m} = \frac{1}{(2m)!}\int_{0}^{T}\int_{0}^{\infty} z^{2m}\lambda(t) f_Q(z, t)dzdt \tag{5.248}$$

and further,

$$\sum_{m=1}^{\infty}(-1)^m \xi_{2m}\tau^{2m} = \int_{0}^{T}\int_{0}^{\infty}\sum_{m=1}^{\infty}\frac{(-1)^m}{(2m)!}z^{2m} \cdot \lambda(t) f_Q(z, t)dzdt \tag{5.249}$$

We let $\delta(\tau) = \sum_{m=1}^{\infty}(-1)^m \xi_{2m} \tau^{2m}$ for simplicity. By noting that $\cos x = 1 + \sum_{m=1}^{\infty} \frac{(-1)^m}{(2m)!} x^2$ for any real number x, Eq. (5.249) becomes

$$\delta(\tau) = \int_0^T \int_0^\infty \cos(z\tau) \cdot \lambda(t) f_Q(z, t) dz dt - \int_0^T \lambda(t) dt \qquad (5.250)$$

Similarly, since $\sin x = \sum_{m=0}^{\infty} \frac{(-1)^m}{(2m+1)!} x^{2m+1}$ holds for any real number x, the item $\theta(\tau, x)$ in Eq. (5.241) becomes

$$\theta(\tau, x) = \int_0^T \int_0^\infty \sin(z\tau)\lambda(t) f_Q(z, t) dz dt - x\tau \qquad (5.251)$$

Substituting Eqs. (5.250) and (5.251) into Eq. (5.244), it follows,

$$f_D(x) = \frac{1}{\pi} \int_0^\infty \exp(\delta(\tau)) \cdot \cos\theta(\tau, x) d\tau = \frac{1}{\pi} \int_0^\infty \exp(\delta(\tau)) \cdot \cos[\theta_0(\tau) - x\tau] d\tau \qquad (5.252)$$

where $\theta_0(\tau) = \theta(\tau, x) + x\tau$. Equation (5.252) presents an explicit form of the PDF of $D(T)$. Further, the CDF of D_c, $F_D(x)$, is obtained as

$$F_D(x) = \int_0^x f_D(y) dy = \frac{1}{\pi} \int_0^\infty \frac{\exp(\delta(\tau))}{\tau} \cdot [\sin\theta_0(\tau) - \sin(\theta_0(\tau) - x\tau)] d\tau \qquad (5.253)$$

5.4.5.3 Reliability Analysis Considering both Failure Mechanisms

Consider the reliability of a structure within service period $[0, T]$ subjected to both failure modes as in Sects. 5.4.5.1 and 5.4.5.2. Due to the complexity of calculating the multi-fold integral in Eqs. (5.227) and (5.228), a simulation-based method can be utilized to compute the probability of failure. For each simulation run, the procedure is as follows.

(1) Sample the number of load events, n, and the time sequence $\{t_1, t_2, \ldots t_n\}$ according to the simulation technique presented in Sect. 5.3.1.2.
(2) Simulate the load intensities $\{S_1, S_2, \ldots S_n\}$ corresponding to the time sequence.
(3) Generate a sample sequence for shock deterioration $\{D_1, D_2, \ldots D_n\}$, and calculate the gradual deterioration sequence $\{G_1, G_2, \ldots G_n, G_{n+1}\}$.
(4) The structure is deemed to fail if either of the following two criteria is satisfied: (a) any load intensity S_i exceeds the corresponding resistance for $i = 1, 2 \ldots n$, and (b) the cumulative deterioration, $\sum_{i=1}^{n+1} G_i + \sum_{i=1}^{n} D_i$, exceeds the permissible level g_a.

Performing the simulation procedure for m times, if the structure fails for r times, then the structural failure probability is approximated by r/m.

Note that in Step (4) of the above procedure, if criterion (a) is considered only, one can obtain the failure probability associated with the first failure type; the probability of second-type failure is estimated if one takes into account criterion (b) only.

Example 5.37

Consider the reliability of a structure subjected to a sequence of load events described by Method 2 in Sect. 5.3.1.2 [55]. The design criterion is given by $0.9R_n = 1.2D_n + 1.6L_n$, in which R_n, L_n and D_n are the nominal resistance, live load and dead load, respectively. Assume that $\mathcal{D}_n = \mathcal{L}_n$. The initial resistance and dead load are assumed to be deterministic. The initial resistance is taken as $r_0 = 1.05\mathcal{R}_n$, and the dead load equals $1.0\mathcal{D}_n$. The live load effect has a mean of $0.5L_n$ and a COV of 0.3, following an Extreme value Type I distribution. The live load occurrence rate, λ, is assumed to be 0.5/year. The correlation between two subsequent time intervals is set to be 0.5. The correlation length for load intensity, L_c, is assumed to be 2.8854, with which the correlation declines to 0.5 with a time separation of 2 years (c.f. the exponential model in Eq. (5.123)).

The resistance degrades due to the combined effects of both the gradual and shock deteriorations. The former is assumed to be deterministic and cause a 15% reduction of initial resistance linearly over a reference period of 50 years. The latter is modelled to follow a lognormal distribution with a mean value of $0.01r_0$ and a COV of 0.5, and have a linear correlation of 0.8 with the load intensity. Use a Gaussian copula for the dependency between load effect and shock deterioration. Let F_1, F_2 and F_{12} respectively denote the type-I, type-II and combined-failures (i.e., $F_{12} = F_1 \bigcup F_2$).

(1) Compute the probabilities of F_1, F_2 and F_{12} for reference periods up to 50 years. Use $g_a = 0.45r_0$ and $0.5r_0$ respectively.

(2) For the two cases in (1), compute the correlation coefficient between $\mathcal{F}_1$ and $\mathcal{F}_2$, where

$$\mathcal{F}_i = \begin{cases} 1, & F_i \text{ occurs} \\ 0, & F_i \text{ does not occur} \end{cases} \text{ for } i = 1, 2.$$

Solution

(1) The structural failure probabilities for reference periods up to 50 years are obtained using a simulation-based approach, as detailed above. Totally 10^6 simulation runs are used to approximate the failure probability. The results are presented in Fig. 5.27. The permissible level of cumulative damage, g_a, equals $0.5r_0$ in Fig. 5.27a and $0.45r_0$ in Fig. 5.27b. The failure probability increases as the service period becomes longer, which is characteristic of structural aging and accumulation of risk. The probability of F_{12} is greater than that associated with a single failure mode (F_1 or F_2) as expected. For the case of $g_a = 0.5r_0$, the probabilities of type-I and type-II failures are comparable. For reference periods up to 46 years, the probability of type-II failure is greater than that associated with F_1. However, at the latter stage, the probability of type-II failure becomes greater. For the case of $g_a = 0.45r_0$, the type-II failure governs for all service periods.

(2) For both cases of Figs. 5.27(a) and (b), $\mathbb{P}(F_1 \bigcap F_2) > \mathbb{P}(F_1)\mathbb{P}(F_2)$, indicating that F_1 and F_2 are positively correlated. In order to measure the correlation between $\mathcal{F}_1$ and $\mathcal{F}_2$, we compute their linear correlation coefficient as follows,

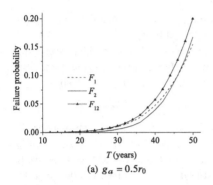

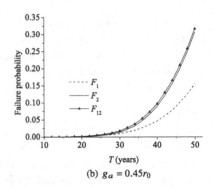

(a) $g_a = 0.5r_0$ (b) $g_a = 0.45r_0$

Fig. 5.27 Time-dependent reliability analysis for reference periods up to 50 years

Table 5.2 Correlation coefficient $\rho(\mathcal{F}_1, \mathcal{F}_2)$ for different service periods

g_a	$T = 20$ years	$T = 30$ years	$T = 40$ years	$T = 50$ years
$0.5r_0$	0.362	0.540	0.681	0.712
$0.45r_0$	0.460	0.530	0.578	0.586

$$\rho(\mathcal{F}_1, \mathcal{F}_2) = \frac{\mathbb{P}(F_1 \cap F_2) - \mathbb{P}(F_1)\mathbb{P}(F_2)}{\sqrt{\mathbb{P}(F_1)\mathbb{P}(F_2)(1 - \mathbb{P}(F_1))(1 - \mathbb{P}(F_2))}}. \tag{5.254}$$

Table 5.2 presents the correlation coefficients $\rho(\mathcal{F}_1, \mathcal{F}_2)$ associated with different reference periods. The positive correlation between $\mathcal{F}_1$ and $\mathcal{F}_2$ is clearly reflected, due to the common effects of cumulative shock deterioration. Moreover, the correlation $\rho(\mathcal{F}_1, \mathcal{F}_2)$ increases with T, which is characteristic of the cumulation of load temporal correlation and deterioration-load dependency with time.

5.5   Time-Dependent Reliability Considering a Continuous Load Process

5.5.1   Reliability Assessment with a Continuous Load Process

Recall that in Fig. 5.17, we discussed the upcrossing rate of the stochastic load process $S(t)$ with respect to the threshold x, which yields the CDF of the maxima of $S(t)$ over $[0, T]$, denoted by $F_{S_{\max}}(x)$. Now, if the structural resistance is constantly r, then the reliability is simply evaluated as $F_{S_{\max}}(r)$. If the resistance deterioration is taken into account, the time-dependent reliability will be discussed in this Section [66].

Fig. 5.28 Upcrossing rate of stochastic process $X(t)$ relative to $\Omega(t)$

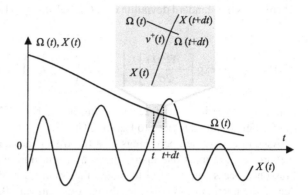

Let the stochastic process $R(t)$ denote the structural resistance, and $S(t)$ the load process as before. Based on Eq. (5.14), the time-dependent reliability over a reference period of $[0, T]$ is evaluated by

$$\mathbb{P}(Z(t) \geq 0, \forall t \in [0, T]) = \mathbb{P}(R(t) - S(t) \geq 0, \forall t \in [0, T]) = \mathbb{P}(\Omega(t) - X(t) \geq 0, \forall t \in [0, T])$$
(5.255)

where $Z(t) = R(t) - S(t)$, $\Omega(t) = R(t) - \mathbb{E}[S(t)]$ and $X(t) = S(t) - \mathbb{E}[S(t)]$.

We first consider the case where $R(t)$ is deterministic and $S(t)$ is a stationary Gaussian process with a standard deviation of σ_S. With this, the term $X(t)$ in Eq. (5.255) is also a zero-mean stationary Gaussian process with a standard deviation of $\sigma_X = \sigma_S$. Figure 5.28 presents an illustration of the upcrossing rate-based reliability problem. The upcrossing rate of $X(t)$ relative to $\Omega(t)$ at time t, $v^+(t)$, is estimated by

$$\lim_{dt \to 0} v^+(t)dt = \mathbb{P}\left\{ \Omega(t) > X(t) \bigcap \Omega(t + dt) < X(t + dt) \right\}$$

$$= \mathbb{P}\left\{ \Omega(t + dt) - \dot{X}(t)dt < X(t) < \Omega(t) \right\}$$
(5.256)

$$= \int_{\dot{\Omega}(t)}^{\infty} \left[\dot{X}(t) - \dot{\Omega}(t) \right] f_{X\dot{X}}\left[\Omega(t), \dot{X}(t) \right] d\dot{X}(t)dt$$

Rearranging Eq. (5.256) simply gives

$$v^+(t) = \int_{\dot{\Omega}(t)}^{\infty} \left(\dot{X} - \dot{\Omega} \right) f_{X\dot{X}}\left(\Omega, \dot{X} \right) d\dot{X}$$
(5.257)

Since $X(t)$ is a zero-mean stationary Gaussian process, $X(t)$ are $\dot{X}(t)$ are mutually independent. Similar to Eq. (5.148), one has

$$f_{X\dot{X}}(x, \dot{x}) = \frac{1}{2\pi \sigma_X \sigma_{\dot{X}}} \exp\left\{ -\frac{1}{2}\left(\frac{x^2}{\sigma_X^2} + \frac{\dot{x}^2}{\sigma_{\dot{X}}^2} \right) \right\}$$
(5.258)

where $\sigma_{\dot{X}}$ is the standard deviation of $\dot{X}(t)$. Substituting Eq. (5.258) into Eq. (5.257) gives

$$v^+(t) = \frac{1}{2\pi\sigma_X}\exp\left[-\frac{\Omega^2(t)}{2\sigma_X^2}\right]\cdot\left\{\sigma_{\dot{X}}\exp\left(-\frac{\dot{\Omega}^2(t)}{2\sigma_{\dot{X}}^2}\right) - \sqrt{2\pi}\dot{\Omega}(t)\left[1 - \Phi\left(\frac{\dot{\Omega}(t)}{\sigma_{\dot{X}}}\right)\right]\right\}$$
(5.259)

It is easy to verify that if $\Omega(t) = $ constant in Eq. (5.259), i.e., $\dot{\Omega}(t) = 0$, then Eq. (5.259) simply degrades to Eq. (5.150).

Assuming that the upcrossings of $X(t)$ to $\Omega(t)$ are temporally independent and are rare (e.g., at most one upcrossing may occur during a short time interval), the Poisson process can be used to model the occurrence of the upcrossings. Let N_T denote the number of upcrossings during time interval $[0, T]$, and it follows,

$$\mathbb{P}(N_T = k) = \frac{1}{k!}\left\{\int_0^T v^+(t)dt\right\}^k \exp\left\{-\int_0^T v^+(t)dt\right\}, \quad k = 0, 1, 2, \ldots$$
(5.260)

Furthermore, the structural reliability during $[0, T]$ is the probability of $N_T = 0$, with which

$$\mathbb{L}(0, T) = [1 - \mathbb{P}(0)]\exp\left\{-\int_0^T v^+(t)dt\right\}$$
(5.261)

where $\mathbb{P}(0)$ is the failure probability at initial time. Specifically, as $\mathbb{P}(0)$ is typically small enough, one has [11, 29]

$$\mathbb{L}(0, T) = \exp\left\{-\int_0^T v^+(t)dt\right\}$$
(5.262)

Equation (5.262) presents the time-dependent reliability for a reference of T years.

Next, the time-dependent reliability in the presence of a non-Gaussian load process is considered. First, recall the time-variant limit state function $Z(t)$ in Eq. (5.255), with which

$$\mathbb{L}(0, T) = \mathbb{P}(Z(t) \geq 0, \forall t \in [0, T]) = \mathbb{P}(R(t) - S(t) \geq 0, \forall t \in [0, T])$$
$$= \mathbb{P}\left\{\Phi^{-1}\left[F_{S(t)}(R(t))\right] - Q(t) \geq 0, \forall t \in [0, T]\right\}$$
(5.263)

where $Q(t) = \Phi^{-1}\left[F_{S(t)}(S(t))\right]$, and $F_{S(t)}(x)$ is the CDF of $S(t)$. With this, the term $Q(t)$ becomes a standard Gaussian process. Furthermore, an "equivalent resistance" is defined as $R^*(t) = \Phi^{-1}\left[F_{S(t)}(R(t))\right]$. In such a way, the time-dependent reliability analysis is transformed into solving a standard "first passage probability" problem. That is, Eqs. (5.259) and (5.262) apply in the presence of the "equivalent" resistance and load, $R^*(t)$ and $Q(t)$.

A key step herein is to find the correlation in $Q(t)$ provided that the correlation in $S(t)$ is known. Suppose that the correlation coefficient between $S(t_i)$ and $S(t_j)$

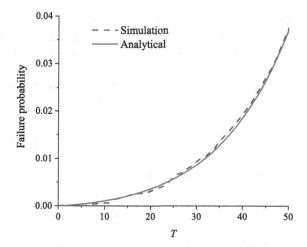

Fig. 5.29 Analytical and simulated failure probabilities in Example 5.38

is ρ_{ij}, and that between the corresponding $Q(t_i)$ and $Q(t_j)$ is ρ'_{ij}. The relationship between ρ_{ij} and ρ'_{ij} can be determined according to Eq. (3.37).

Example 5.38

Suppose that the structural resistance varies with time according to $r(t) = 8 - 0.02t$, $t \in [0, 50]$. The continuous load process $S(t)$ is a stationary Gaussian process with a mean value of 4. The PSDF of $S(t)$ is as in Eq. (5.139) with $k = 1$. Compute the time-dependent failure probability $\mathbb{P}_f(0, T)$ for T up to 50.

Solution

Similar to Example 5.23, $\sigma_S = 1$ and $\sigma_{\dot{S}} = \sqrt{2}$. We let $\Omega(t) = r(t) - \mathbb{E}[S(t)] = 4 - 0.02t$, and $X(t) = S(t) - \mathbb{E}[S(t)]$. With this, $\dot{\Omega}(t) = -0.02$, and $X(t)$ is a stationary standard Gaussian process. Thus, the time-variant upcrossing rate is estimated according to Eq. (5.259) as follows,

$$
v^+(t) = \frac{1}{2\pi} \exp\left[-\frac{(4 - 0.02t)^2}{2}\right] \cdot \left\{\sqrt{2} \exp\left(-\frac{(-0.02)^2}{4}\right) + \sqrt{2\pi} \times 0.02 \times \left[1 - \Phi\left(\frac{-0.02}{\sqrt{2}}\right)\right]\right\}
$$

$$
= 0.2291 \exp\left[-\frac{(4 - 0.02t)^2}{2}\right]
$$

$$(5.264)$$

Furthermore, with Eq. (5.262),

$$
\mathbb{L}(0, T) = \exp\left\{-\int_0^T v^+(t)\mathrm{d}t\right\} = \exp\left\{-0.2291 \int_0^T \exp\left[-\frac{(4 - 0.02t)^2}{2}\right]\mathrm{d}t\right\}
$$

$$
= \exp\left[-14.355 + 14.356 \times \mathrm{erf}(2.828 - 0.014T)\right]
$$

$$(5.265)$$

The dependence of $\mathbb{P}_f(0, T) = 1 - \mathbb{L}(0, T)$ on T is shown in Fig. 5.29. For comparison purpose, the simulated failure probability is also presented in Fig. 5.29, where the simulation procedure for the load samples can be found in Example 5.23. It can be seen that both the simulated and analytical failure probabilities agree well with each other.

Example 5.39

Reconsider Example 5.38. If the load evaluate at any time follows a lognormal distribution, recompute the time-dependent failure probability $\mathbb{P}_f(0, T)$ for T up to 50.

Solution

Since $S(t)$ has a lognormal distribution with a mean value of 4 and a standard deviation of 1, with Eq. (2.201), the CDF of $S(t)$ is

$$F_{S(t)}(x) = \Phi\left(\frac{\ln x - \lambda}{\varepsilon}\right), \quad x > 0 \tag{5.266}$$

where $\lambda = 1.356$ and $\varepsilon = 0.246$. Thus,

$$\Phi^{-1}[F_{S(t)}(x)] = \frac{\ln x - \lambda}{\varepsilon}, \quad x > 0 \tag{5.267}$$

According to Eq. (5.263),

$$\mathbb{L}(0, T) = \mathbb{P}\left\{r^* - Q(t) \geq 0, \forall t \in [0, T]\right\} \tag{5.268}$$

where $r^*(t) = \Phi^{-1}\left[F_{S(t)}(r(t))\right] = \frac{\ln r(t) - \lambda}{\varepsilon}$ and $Q(t) = \Phi^{-1}\left[F_{S(t)}(S(t))\right] = \frac{\ln S(t) - \lambda}{\varepsilon}$. In Eq. (5.268), $Q(t)$ is a stationary standard Gaussian process.

With Eq. (3.44),

$$\rho(Q(t), Q(t + \tau)) = \frac{(1 + v^2)^{\rho(S(t), S(t+\tau))} - 1}{v^2} \approx \rho(S(t), S(t + \tau)) \tag{5.269}$$

where $v = 0.25$ is the COV of $S(t)$. The error in Eq. (5.269) is less than 0.008, i.e., $|\rho[Q(t), Q(t + \tau)) - \rho(S(t), S(t + \tau))| < 0.008$. Thus, the autocorrelation function of $Q(t)$ is obtained as follows,

$$\mathbb{R}_Q(\tau) = \rho(Q(t), Q(t + \tau)) = \rho(S(t), S(t + \tau)) = \frac{\mathbb{R}_S(\tau)}{\mathbb{V}(S(t))} = \mathbb{R}_S(\tau) \tag{5.270}$$

implying that the PSDF of $Q(t)$ is also as in Eq. (5.139) with $k = 1$. Thus, similar to Example 5.23, $\sigma_{\dot{Q}} = \sqrt{2}$.

Now we can compute the time-dependent reliability with Eqs. (5.259) and (5.262), where $\Omega(t) = r^*(t) = \frac{\ln(8 - 0.02t) - \lambda}{\varepsilon}$, and $X(t) = Q(t)$. Note that $\dot{\Omega}(t) = -\frac{0.02}{\varepsilon(8 - 0.02t)}$, with which it follows,

$$v^+(t) = \frac{1}{2\pi} \exp\left[-\frac{\left(\frac{\ln(8 - 0.02t) - \lambda}{\varepsilon}\right)^2}{2}\right] \cdot \left\{\sqrt{2} \exp\left(-\frac{\left(-\frac{0.02}{\varepsilon(8 - 0.02t)}\right)^2}{4}\right)\right.$$

$$\left. + \sqrt{2\pi}\frac{0.02}{\varepsilon(8 - 0.02t)}\left[1 - \Phi\left(\frac{-\frac{0.02}{\varepsilon(8 - 0.02t)}}{\sqrt{2}}\right)\right]\right\} \tag{5.271}$$

Fig. 5.30 Time-dependent failure probabilities with lognormal and normal load processes

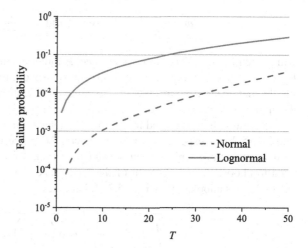

Substituting Eq. (5.271) into Eq. (5.262) gives the time-dependent reliability. Figure 5.30 shows the time-dependent failure probability $\mathbb{P}_f(0, T)$ for T up to 50. For comparsion purpose, the failure probability associated with a Gaussian load process (from Example 5.38) is also presented in Fig. 5.30. The difference between the failure probabilities with different load models is referred to as the *tail sensitivity problem*, as discussed in Sect. 4.1.

5.5.2 PSDF of the Load Process

In stochastic process based time-dependent reliability analysis in Eq. (5.262), one of the crucial ingredients is the modelling of the autocorrelation in the load process [10, 33, 55]. The autocorrelation function $\mathbb{R}(\tau)$, as discussed in Eq. (5.136), reflects the correlation structure of a stochastic process, say, $X(t)$. The PSDF is a Fourier transform of $\mathbb{R}(\tau)$, as revealed in Eqs. (5.135) and (5.136), also providing a useful tool to describe the statistical characteristics of the load process $X(t)$. In fact, the term $\sigma_{\dot{X}}$ is included in Eq. (5.259), which can be estimated by

$$\sigma_{\dot{X}} = \sqrt{\int_{-\infty}^{\infty} \omega^2 \mathbb{S}(\omega) d\omega} \tag{5.272}$$

provided that the PSDF of $X(t)$, $\mathbb{S}(\omega)$, is known (c.f. Eq. (5.142)).

However, since an improper integral is involved in Eq. (5.272), an arbitrary form of $\mathbb{S}(\omega)$ does not necessarily lead to a converged estimate of $\sigma_{\dot{X}}$. For example, if $R(\tau)$ takes the form of $R(\tau) = \sigma_X^2 \exp(-k\tau)$, where σ_X is the standard deviation of $X(t)$, then, by referring to Eq. (5.138),

$$S(\omega) = \frac{1}{\pi} \int_0^\infty R(\tau) \cos(\tau\omega) d\tau = \frac{k\sigma_X^2}{\pi(k^2 + \omega^2)} \tag{5.273}$$

with which Eq. (5.272) does not converge. Another example would be the PSDF in Eq. (5.140), which yields a non-convergent $\sigma_{\dot X}$ as well. Furthermore, even for some PSDFs that result in a converged $\sigma_{\dot X}$, the integral operation in Eq. (5.272) may halter the application of the structural reliability assessment in Eq. (5.262) (that is, a two-fold integral will be involved in Eq. (5.262) if substituting Eqs. (5.259) and (5.272) into Eq. (5.262)), especially for use in practical engineering.

In an attempt to achieve a simple and convergent form of Eq. (5.272), one would need to choose a proper PSDF of the load process. For example, if the PSDF in Eq. (5.139) is adopted, then Eq. (5.272) becomes

$$\sigma_{\dot X} = \sqrt{\int_{-\infty}^\infty \omega^2 \cdot \frac{1}{2\sqrt{k\pi}} \exp\left(-\frac{\omega^2}{4k}\right) d\omega} = \sqrt{2k} \tag{5.274}$$

which gives an explicit estimate of $\sigma_{\dot X}$ once k is given.

Another useful PSDF for $X(t)$ was suggested in [66], which takes the form of

$$S(\omega) = \frac{a}{\omega^6 + b}, \quad -\infty < \omega < +\infty \tag{5.275}$$

where a and b are two positive constants. Clearly, Eq. (5.275) satisfies the basic properties of a PSDF: it's an even function of ω (i.e., $S(-\omega) = S(\omega)$) and positive for $\forall \omega \in (-\infty, \infty)$. The application of Eq. (5.275) enables the autocorrelation function $R(\tau)$ to decrease sharply at the early stage (with relatively small τ) and converge to zero latter with a fluctuation along the horizontal axis.

With the PSDF in Eq. (5.275), according to Eq. (5.136), it follows

$$\begin{aligned} R(\tau) &= R(\tau, b) = 2a \cdot \int_0^\infty \frac{1}{\omega^6 + b} \cos(\omega\tau) d\omega \\ \sigma_X^2 &= R(0, b) = 2a \cdot \int_0^\infty \frac{1}{\omega^6 + b} d\omega = \frac{2a\pi}{3b^{5/6}} \end{aligned} \tag{5.276}$$

The integral operation involved in Eq. (5.276) can be solved in a closed form. To begin with, one has

$$R(1, b) = \frac{2a\pi}{12b^{5/6}} \exp\left(-\frac{b^{1/6}}{2}\right) \cdot \left[2\exp\left(-\frac{b^{1/6}}{2}\right) + 4\cos\left(\frac{\sqrt{3}}{2}b^{1/6} - \frac{\pi}{3}\right)\right] \tag{5.277}$$

Further, it is easy to find that

$$R(\tau, b) = \tau^5 \cdot R(1, b\tau^6) \tag{5.278}$$

As such, Eq. (5.276) provides a straightforward approach to find the parameters a and b in the density function $\mathbb{S}(\omega)$, provided that the autocorrelation function of the load process is known. In Eq. (5.278), the magnitude of $\mathbb{R}(\tau, b)$ is controlled by the term $\exp\left(-\frac{b^{1/6}\tau}{2}\right)$, which is a monotonically decreasing function of τ with a given b, while the fluctuation of $\mathbb{R}(\tau, b)$ is posed by the term $2\exp\left(-\frac{b^{1/6}\tau}{2}\right) + 4\cos\left(\frac{\sqrt{3}}{2}b^{1/6}\tau - \frac{\pi}{3}\right)$.

Furthermore, with $\mathbb{S}(\omega)$ taking the form of Eq. (5.275), according to Eq. (5.272), it follows

$$\sigma_{\dot{X}}^2 = 2a \cdot \int_0^\infty \frac{\omega^2}{\omega^6 + b}d\omega = \frac{\pi a}{3\sqrt{b}} \tag{5.279}$$

It can be seen from Eq. (5.279) that $a > 0$, $b > 0$ due to the fact that $\sigma_{\dot{X}}^2$ is a positive real number. Furthermore, with Eq. (5.279), it is easy to see that Eq. (5.259) has a simple form with only fundamental algebras involved, which is beneficial for the application of structural reliability assessment when substituting Eq. (5.259) into Eq. (5.262).

Example 5.40

Reconsider Example 5.38. The load process $S(t)$ is a stationary Gaussian process with a mean value of 4 and a standard deviation of 1. The correlation coefficient of $S(t)$ and $S(t + 1)$ equals $\exp(-1)$.

Now, if the PSDF of the load process is modelled by Eq. (5.275), recompute the time-dependent failure probability $\mathbb{P}_f(0, T)$ for T up to 50.

Solution

We let $X(t) = S(t) - \mathbb{E}[S(t)] = S(t) - 4$ as before. Since the PSDF of $X(t)$ is modelled by Eq. (5.275), we first need to determine the two parameters a and b in Eq. (5.275). With Eqs. (5.276) and (5.277), b is obtained, by referring to Fig. 5.31, as follows,

$$\rho(X(t), X(t+1)) = \frac{\mathbb{R}(1, b)}{\mathbb{V}(X(t))} = \frac{1}{4}\exp\left(-\frac{b^{1/6}}{2}\right) \cdot \left[2\exp\left(-\frac{b^{1/6}}{2}\right) + 4\cos\left(\frac{\sqrt{3}}{2}b^{1/6} - \frac{\pi}{3}\right)\right]$$
$$\Rightarrow b = 57.24$$

(5.280)

Subsequently,

$$\sigma_X^2 = \frac{2a\pi}{3b^{5/6}} = 1 \Rightarrow a = \frac{3b^{5/6}}{2\pi} = 13.92 \tag{5.281}$$

Thus, according to Eq. (5.279),

$$\sigma_{\dot{X}} = \sqrt{\frac{\pi a}{3\sqrt{b}}} = 1.388 \tag{5.282}$$

The time-dependent reliability $\mathbb{L}(0, T)$ is computed by Eq. (5.262), where $\Omega(t) = 4 - 0.02t$, and

Fig. 5.31 Relationship between $\rho(X(t), X(t+1))$ and b

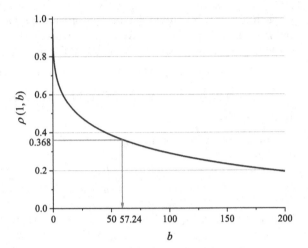

$$v^+(t) = \frac{1}{2\pi} \exp\left[-\frac{(4 - 0.02t)^2}{2}\right] \cdot \left\{1.388 \exp\left(-\frac{(-0.02)^2}{2 \times 1.388^2}\right) - \sqrt{2\pi} \times (-0.02) \times \left[1 - \Phi\left(\frac{-0.02}{1.388}\right)\right]\right\}$$

$$= 0.2249 \exp\left[-\frac{(4 - 0.02t)^2}{2}\right]$$

$$(5.283)$$

Thus,

$$P_f(0, T) = 1 - \exp\left\{-\int_0^T v^+(t)\mathrm{d}t\right\} = \exp[-14.095 + 14.096 \times \mathrm{erf}(2.828 - 0.014T)]$$

$$(5.284)$$

5.5.3 Comparison Between Reliabilities with Discrete and Continuous Load Processes

Recall that the time-dependent reliability problem has been addressed in Sects. 5.4 and 5.5.1, respectively. The former considers a discrete load process where only the significant load events that may impair the structural safety directly are incorporated, while the later is derived based on a continuous load process. The difference between the reliabilities associated with the two types of load model is discussed in this Section.

Consider a stationary Gaussian load process, as shown in Fig. 5.28. Let $X(t) = S(t) - \mathbb{E}[S(t)]$ as before, which is a zero-mean load process (c.f. Eq. (5.255)). Consider the CDF of $\max\{X(t)\}$ within a time duration of Δ, $F_{X_{\max}|\Delta}$. In Eqs. (5.259) and (5.262), assigning $\Omega(t) = x$ and $\dot{\Omega}(t) = 0$, as x is large enough, one has [33]

$$F_{X_{\max}|\Delta}(x) = \exp\left\{-\frac{\sigma_{\dot{X}}}{2\pi\sigma_X}\exp\left(-\frac{x^2}{2\sigma_X^2}\right)\Delta\right\} \approx 1 - \frac{\sigma_{\dot{X}}\Delta}{2\pi\sigma_X}\exp\left(-\frac{x^2}{2\sigma_X^2}\right)$$

$$(5.285)$$

which yields a Rayleigh distribution, consistent with Eq. (5.152).

Furthermore, with Eqs. (5.259) and (5.262), the time-dependent reliability over $[0, T]$, $\mathbb{L}_c(0, T)$, is as follows (the subscript $_c$ is to emphasize that the reliability is associated with a continuous load process),

$$\mathbb{L}_c(0, T) = \exp\left\{-\frac{\sigma_{\dot{X}}}{2\pi\sigma_X}\int_0^T \underbrace{\exp\left[-\frac{\Omega^2(t)}{2\sigma_X^2}\right]}_{\text{part 1}} \cdot \underbrace{\left\{\exp\left(-\frac{\dot{\Omega}^2(t)}{2\sigma_X^2}\right) - \frac{\sqrt{2\pi}\dot{\Omega}(t)}{\sigma_{\dot{X}}}\left[1 - \Phi\left(\frac{\dot{\Omega}(t)}{\sigma_{\dot{X}}}\right)\right]\right\}}_{\text{part 2}} dt\right\}$$

$$(5.286)$$

Now we recompute the time-dependent reliability by first discretizing the time interval $[0, T]$ into n identical sections with $n\Delta = T$. Let λ denote the occurrence rate of the load events within each interval (i.e., the probability of load occurrence equals $\lambda\Delta/n$). From a view of a discrete load process, the CDF of $\max\{X(t)\}$ within a short time interval of Δ is given by

$$F_{X_{\max}|\Delta}(x) = 1 - \lambda\Delta \cdot (1 - F_S(x)) \tag{5.287}$$

where F_S is the CDF of load effect conditional on the occurrence of one load event.

If we let the two CDFs of maximum load effect in Eqs. (5.285) and (5.287) be equal, then $F_S(x)$ in Eq. (5.287) is determined by

$$F_S(x) = 1 - \frac{\sigma_{\dot{X}}}{2\pi\lambda\sigma_X}\exp\left(-\frac{x^2}{2\sigma_X^2}\right) \tag{5.288}$$

Equation (5.288) implies that if a continuous Gaussian process is transformed to a discrete one, the CDF of the load effect conditional on the occurrence of one load event simply follows a Rayleigh distribution.

Remark

For the more generalized case of a non-Gaussian load process, $S(t)$ can be converted into a Gaussian process $Q(t)$, as discussed before. With this, for a reference period of Δ, the CDF of $\max\{S(t)\}$ is given by

$$F_{S_{\max}|\Delta}(x) = \mathbb{P}\left\{\bigcap_{0 \leq t \leq \Delta}\left(\Phi^{-1}[F_S(S(t))] < \Phi^{-1}(F_S(x))\right)\right\} \tag{5.289}$$

Let $x^* = \Phi^{-1}(F_S(x))$, with which Eq. (5.289) becomes

$$F_{S_{\max}|\Delta}(x) = \exp\left\{-\frac{\sigma_{\dot{\varrho}}\Delta}{2\pi}\exp\left(-\frac{x^{*2}}{2}\right)\right\} \approx 1 - \frac{\sigma_{\dot{\varrho}}\Delta}{2\pi}\exp\left(-\frac{x^{*2}}{2}\right)$$

$$= 1 - \frac{\sigma_{\dot{\varrho}}\Delta}{2\pi}\exp\left\{-\frac{[\Phi^{-1}(F_S(x))]^2}{2}\right\} \tag{5.290}$$

It should be noted that Eq. (5.290) is only valid when x is large enough. Eq. (5.290) implies that when the load process is non-Gaussian, the maximum load effect within a time interval does not necessarily follow a Rayleigh distribution. The distribution type in Eq. (5.290) is known as *Pseudo-Rayleigh distribution* [66].

Recall that Eq. (5.165) gives the time-dependent reliability in the presence of a discrete load process,

$$\mathbb{L}(0, T) = \exp\left[-\lambda\left(T - \int_0^T F_S[r \cdot g(t)]dt\right)\right] \tag{5.165}$$

Substituting Eq. (5.288) into Eq. (5.160) gives the expression of $\mathbb{L}_d(0, T)$ (the subscript $_d$ means that the reliability is related to a discrete load process) as follows,

$$\mathbb{L}_d(0, T) = \mathbb{P}\left[\bigcap_{0<t\leq T}(\Omega(t) - X_{\max} > 0)\right] = \exp\left[-\frac{\sigma_{\dot{X}}}{2\pi\sigma_X}\int_0^T \underbrace{\exp\left(-\frac{\Omega^2(t)}{2\sigma_X^2}\right)}_{\text{part 1}}dt\right] \tag{5.291}$$

It is noticed that the "part 1" of Eq. (5.291) is exactly the same as the "part 1" of Eq. (5.286). To further compare $\mathbb{L}_d(0, T)$ and $\mathbb{L}_c(0, T)$, we introduce a function

$$\xi(x) = \exp\left(-\frac{x^2}{2}\right) - \sqrt{2\pi}x[1 - \Phi(x)] \tag{5.292}$$

with which "part 2" of Eq. (5.286) equals $\xi\left(\frac{\dot{\Omega}(t)}{\sigma_{\dot{X}}}\right)$. Figure 5.32 shows the dependence of $\xi(x)$ on x, from which it can be seen that $\xi(x)$ is a monotonically decreasing function of x. Since $\xi(0) = 1$, $\xi(x) \geq \xi(0) = 1$ if $x \geq 0$.

Based on the function $\xi(x)$, we rewrite Eq. (5.293) as follows,

$$\mathbb{L}_c(0, T) = \exp\left\{-\frac{\sigma_{\dot{X}}}{2\pi\sigma_X}\int_0^T \underbrace{\exp\left[-\frac{\Omega^2(t)}{2\sigma_X^2}\right]}_{\text{part 1}}\cdot\underbrace{\xi\left(\frac{\dot{\Omega}(t)}{\sigma_{\dot{X}}}\right)}_{\text{part 2}}dt\right\} \tag{5.293}$$

Fig. 5.32 Plot of the function $\xi(x)$ in Eq. (5.292)

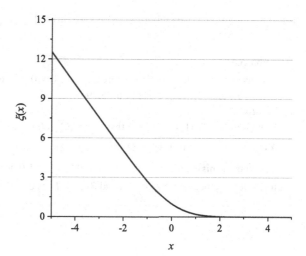

Using the mean value theorem for integrals (e.g., [5]), there exists a real number $x_0 \in \left[\min_{t=0}^{T} \frac{\dot{\Omega}(t)}{\sigma_{\dot{X}}}, \max_{t=0}^{T} \frac{\dot{\Omega}(t)}{\sigma_{\dot{X}}} \right]$ such that

$$\mathbb{L}_c(0, T) = \exp \left\{ -\xi(x_0) \cdot \frac{\sigma_{\dot{X}}}{2\pi \sigma_X} \underbrace{\int_0^T \exp \left[-\frac{\Omega^2(t)}{2\sigma_X^2} \right] dt}_{\text{part 1}} \right\} = [\mathbb{L}_d(0, T)]^{\xi(x_0)}$$

(5.294)

Note that $\frac{\dot{\Omega}(t)}{\sigma_{\dot{X}}}$ is typically non-positively if taking into account the structural resistance deterioration, with which $x_0 \leq 0$ and $\xi(x_0) \geq 1$. Thus,

$$\mathbb{L}_c(0, T) = [\mathbb{L}_d(0, T)]^{\xi(x_0)} \leq \mathbb{L}_d(0, T) \tag{5.295}$$

Equation (5.295) implies that the choice of a discrete load model overestimates the structural safety or equivalently, underestimates the failure probability, if the realistic load process is continuous. In fact, based on Eq. (5.295), since the failure probability $\mathbb{P}_{f,d}(0, T) = 1 - \mathbb{L}_d(0, T)$ is typically small enough for well-designed structures, one has

$$\mathbb{P}_{f,c}(0, T) = 1 - \mathbb{L}_c(0, T) = 1 - [1 - \mathbb{P}_{f,d}(0, T)]^{\xi(x_0)} \approx \xi(x_0)\mathbb{P}_{f,d}(0, T) \tag{5.296}$$

which implies that, theoretically, the failure probability is underestimated by a factor of $\frac{1}{\xi(x_0)}$ if the continuous load process is modelled as a discrete one. It is noticed, however, that the difference between $\mathbb{P}_{f,c}(0, T)$ and $\mathbb{P}_{f,d}(0, T)$ would be fairly negligible for many practical cases where $\xi(x_0)$ is close to 1.0. This point will be demonstrated in Example 5.41.

Example 5.41

Reconsider Example 5.38. For a reference period of $[0, 50]$, discuss the difference between $\mathbb{P}_{f,c}(0, T)$ and $\mathbb{P}_{f,d}(0, T)$ as in Eq. (5.296).

Solution

From Example 5.41, it is known that $\sigma_{\dot{S}} = \sqrt{2}$, $\Omega(t) = r(t) - \mathbb{E}[S(t)] = 4 - 0.02t$, and $X(t) = S(t) - \mathbb{E}[S(t)]$. Thus, $\dot{\Omega}(t) = -0.02$ and $\sigma_{\dot{X}} = \sigma_{\dot{S}} = \sqrt{2}$. Since $\frac{\dot{\Omega}(t)}{\sigma_{\dot{X}}} = -\frac{0.02}{\sqrt{2}}$ is a constant in this example, $x_0 = -\frac{0.02}{\sqrt{2}}$, and $\xi(x_0) = 1.018$ Thus, according to Eq. (5.296), the difference between $\mathbb{P}_{f,c}(0, T)$ and $\mathbb{P}_{f,d}(0, T)$ is only 1.8%.

Problems

5.1 In Problem 4.20, if each bar has an identical circular cross-section, and its diameter degrades with time according to $D(t) = D(0) \cdot \frac{40}{40+t}$, where t is in years and $D(0)$ is the initial diameter,
(1) What is the resistance deterioration model of each bar?
(2) What is the system failure probability at the end of 10 years?

5.2 For a specific type of load within a reference period of T years, its occurrence is modeled by a Poisson process with a time-variant occurrence rate $\lambda(t) = \lambda_0 + \kappa_\lambda t$, and the load intensity, conditional on occurrence, is a Weibull variable with a constant shape parameter α and a time-variant scale parameter $u(t) = u_0 + \kappa_u t$. Derive the CDF of the maximum load within time interval $[0, T]$.

5.3 In Problem 5.2, if the load is applied to a structure whose resistance is lognormally distributed with mean μ_R and standard deviation σ_R, what is the structural failure probability for a reference period of T years? Use the following values to evaluate your result: $T = 10$, $\lambda_0 = 1$, $u_0 = 0.5$, $\kappa_\lambda = \kappa_u = 0.1$, $\alpha = 1.5$, $\mu_R = 4$, $\sigma_R = 1$.

5.4 Suppose that the initial resistance of a structure is deterministically 3. The resistance degrades linearly with time, which is modeled by a Gamma process. The deterioration function at the end of 50 years has a mean of 0.7 and a standard deviation of 0.3. The load occurrence is a Poisson process with a constant occurrence rate of 0.2/year. The load effect, conditional on occurrence, is an Extreme Type I variable with a mean of 0.5 and a COV of 0.6.
(1) Use a simulation-based approach to find the time-dependent failure probabilities for reference periods up to 50 years.
(2) Show that simply considering the maximum of the load effect and the minimum of resistance over a reference period of interest may introduce a significant error in the estimate of structural failure probability.

5.5 In Problem 4.22, if the resistances of each bar independently degrades by 10% linearly over 10 year, what is the time-dependent failure probability of the truss for a reference period of 10 years?

5.6 In Problem 5.5, we further know that the resistance deterioration of each bar is a linear process and is fully correlated on the temporal scale, and that the deterioration function at the end of 10 years is a Beta variable with mean 0.9 and COV 0.2.
(1) If the deterioration process of each bar is mutually independent, what is the time-dependent failure probability of the truss for a reference period of 10 years?
(2) If the deterioration processes of all the bars are correlated, and the linear correlation coefficient of the deterioration functions evaluated at the end of 10 years is 0.7 for any two bars, what is the failure probability of the truss for a reference period of 10 years?

5.7 Recall Example 5.11. If in Eq. (5.43), $\hat{d}_i(t_i) = k \cdot t_i^m \cdot |t_i - t_{i-1}|$, and $\epsilon(t_i)$ is a Gamma variable with mean 1 and standard deviation $\frac{p \cdot t_i^q}{\sqrt{t_i - t_{i-1}}}$, where k, m, p, q are four positive constants, derive the mean and standard deviation of the deterioration function $G(t)$ [50].

5.8 In Problem 5.4, what are the mean value and standard deviation of the structural service life?

5.9 In Problem 5.4, if the threshold for the failure probability is 0.01, what is the predicted service life of the structure?

5.10 The load occurrence is modeled by a Poisson process with a constant occurrence rate. The load effect, conditional on occurrence, follows an identical Extreme Type I distribution. For a reference period of $[0, T]$, show that the CDF of the maximum load effect within $[0, T]$ can be reasonably modeled by an Extreme Type I distribution at the upper tail.

5.11 Consider a structure subjected to the combined effect of live load L and dead load D. The dead load is deterministically $1.05 D_n$, where D_n is the nominal dead load. The live load occurrence is a stationary Poisson process with an occurrence rate of 1/year. Conditional on occurrence, the live load effect is an Extreme Type I variable with a mean value of $0.4 L_n$ and a COV of 0.3, where L_n is the nominal live load. The structure was designed according to the criterion $1.0 R_n = 1.25 D_n + 1.75 L_n$, where R_n is the nominal initial resistance. The initial resistance is a lognormal variable with mean $1.1 R_n$ and COV 0.1. The resistance deterioration is modeled by a Gamma process with a linearly-varying mean, and the deterioration function evaluated at the end of 40 years has a mean 0.8 and a standard deviation 0.2. Compute the time-dependent failure probabilities of the structure for reference periods up to 40 years.

5.12 Suppose that the initial resistance of a structure is deterministically 3. The resistance degrades linearly with time from time T_i according to

$$G(t) = \begin{cases} 1, & t \le T_i \\ 1 - K(t - T_i), & t > T_i \end{cases} \tag{5.297}$$

where K is a normal variable with mean 0.006 and COV 0.2. The time of deterioration initiation, T_i, is a normal variable with mean 5 years and COV 0.15. The load occurrence is a Poisson process with a constant occurrence rate of 0.3/year. The load effect, conditional on occurrence, is an Extreme Type I variable with a mean of 0.5 and a COV of 0.6. Compute the time-dependent failure probabilities for reference periods up to 50 years.

5.13 Consider the time-dependent reliability of a 5-out-of-8 system. Suppose that the performance of each component is identical and independent of each other. The annual failure probability of each component increases with time according to $p(t) = 0.01(1 + 0.05t)$, where $t = 1, 2, \ldots$ is in years. What is the system failure probability for a reference period of 10 years?

5.14 Consider a structure subjected to wind hazard. The post-hazard damage state can be classified into four categories: none (D_0, when $V < 20$ m/s), moderate (D_1, when $20 \text{ m/s} \le V < 35$ m/s), severe (D_2, when $35 \text{ m/s} \le V < 50$ m/s) and total (D_3, when $50 \text{ m/s} \le V$), where V denotes the wind speed. Suppose that the occurrence of wind events is a Poisson process with an occurrence rate of $\lambda = 0.5$/year. Conditional on occurrence, the wind speed is a Weibull variable with mean 40 m/s and COV 0.3. The economic costs associated with states D_0, D_1, D_2, D_3 are 0, 1, 5 and 25 respectively. Suppose that the structure is restored to the pre-damage state before the occurrence of the next wind event. What are the mean and standard deviation of the accumulative wind-induced economic costs for a reference period of 20 years?

5.15 In Problem 5.14, if the occurrence of wind events is modeled by a renewal process, where the time interval between two adjacent wind events is a Gamma variable with mean 2 (years) and COV 0.3, reevaluate the mean and standard deviation of the accumulative wind-induced economic costs for a reference period of 20 years.

5.16 In Problem 5.14, let Δ_i be the time interval between the $(i - 1)$th and the ith wind events for $i = 1, 2, \ldots$ (suppose that the 0th event occurs at the initial time with zero wind speed). If Δ_i and Δ_j have a correlation efficient of $0.8^{|i-j|}$ for $i, j \in \{1, 2, \ldots\}$, reevaluate the mean and standard deviation of the accumulative wind-induced economic costs for a reference period of 20 years.

5.17 In Problem 5.14, let t_i (in years) be the occurring time of the ith successful event for $i = 1, 2, \ldots$. If the wind speeds associated with the ith and the jth wind events are correlated with a correlation coefficient of $\exp\left(-\frac{|t_i - t_j|}{10}\right)$ for $i, j \in \{1, 2, \ldots\}$, reevaluate the mean and standard deviation of the accumulative wind-induced economic costs for a reference period of 20 years.

5.18 Consider a structure subjected to wind hazard. The post-hazard damage state can be classified into four categories: none (D_0, when $V < 20$ m/s), moderate (D_1, when 20 m/s $\leq V < 35$ m/s), severe (D_2, when 35 m/s $\leq V < 50$ m/s) and total (D_3, when 50 m/s $\leq V$), where V denotes the wind speed. In terms of the annual extreme winds actioned on the structure, we classify the wind load into two levels (1 and 2). Suppose that the wind level in the $(k + 1)$th year depends on that in the kth year only for $k = 1, 2, \ldots$. Given a level 1 wind in the kth year, the probability of a level 1 wind in the $(k + 1)$th year is 0.7; if the wind load is level 2 in the kth year, then the probability of a level 1 wind in the $(k + 1)$th year is 0.2. The level 1 wind speed is a Weibll variable with mean 30 m/s and COV 0.2, while the level 2 wind speed is a Weibull variable with mean 50 m/s and COV 0.35. In the first year, the wind is classified as level 1. The economic costs associated with states D_0, D_1, D_2, D_3 are 0, 1, 5 and 25 respectively. Suppose that the structure is restored to the pre-damage state before the occurrence of the next wind event. What are the mean and standard deviation of the accumulative wind-induced economic costs for a reference period of 20 years?

5.19 Consider a continuous load process $X(t) = \sum_{j=1}^{n} A_j \sin(\omega t) + \sum_{k=1}^{n} B_k \cos(\omega t)$, where ω is a constant, n is a positive integer, and $\{A_1, A_2, \ldots A_n, B_1, B_2, \ldots B_n\}$ are $2n$ independent standard normal variables. Show that $X(t)$ is a covariance stationary process.

5.20 For a weakly stationary process $X(t)$ with an autocorrelation function of $\mathbb{R}(\tau) = \frac{a}{k\tau^2 + a}$, where $a > 0, k > 0$, find the PSDF of $X(t)$.

5.21 Reconsider Example 5.23. For a zero-mean stationary Gaussian load process $S(t)$ whose PSDF is as in Eq. (5.140) with $k = 1$, use a simulation-based approach to discuss the accuracy of Eq. (5.152). Use $T = 10$ and $T = 50$ respectively.

5.22 For the case in Problem 5.21, what are the mean value and variance of the maximum load effect within $[0, T]$, $S_{\max}$, conditional on $S_{\max} > 2$? Use $T = 10$ and $T = 50$ respectively.

5.23 If the hazard function is

$$h(t) = \begin{cases} \dfrac{a_1 - a_2}{t_1} t + a_2, & 0 \leq t < t_1 \\ a_1, & t_1 \leq t < t_2 \\ \dfrac{a_2 - a_1}{t_3 - t_2}(t - t_2) + a_1, & t_2 \leq t \leq t_3 \end{cases} \tag{5.298}$$

where $0 < a_1 < a_2$ and $0 < t_1 < t_2 < t_3$, derive the time-dependent reliability $\mathbb{L}(t)$ for $t \in [0, t_3]$.

5.24 For a k-out-of-n system with independent component performance, if the hazard function for each component is identically $h(t)$, what is the hazard function for the system?

Table 5.3 Statistics of resistance and load effects of a single component

Item	Mean (kN·m)	COV	Distribution
Initial resistance	3600	0.2	Lognormal
Dead load	700	/	Deterministic
Live load 1	500	0.35	Extreme type I
Live load 2	Equation (5.299)	0.35	Extreme Type I

5.25 If the service life follows a Gamma distribution with a shape parameter of $a > 0$ and a scale parameter of $b > 0$, and the corresponding hazard function represents a DFR, what are the ranges for a and b?

5.26 Consider a series or parallel system with four components having the same physical configuration and load conditions as summarized in Table 5.3. Two live load models are considered, with an occurrence rate of 1.0/year. The first represents a stationary load process with a constant mean value and a constant COV with time, while the second is a nonstationary load process with an increasing trend of load magnitude, whose mean value, $\mu_2(t)$, increases with time as

$$\mu_2(t) = \mu_2(0) \cdot (1 + \epsilon \cdot t) \tag{5.299}$$

in which ϵ is the scale factor, $\mu_2(0)$ is the initial load intensity, which equals 500 kN·m so that the mean initial load intensity of model 2 is the same as that of model 1. It is assumed that each load event will induce identical load effect to each component, with which $c_i = 1$ for $\forall i$ in Eq. (5.195) or (5.199).

The deterioration of resistance for each component is assumed to be deterministic and is given by $g(t) = 1 - 0.004t$, where t is in years. The resistances of the components are mutually correlated, and it is assumed that the resistances are identically distributed and are equally correlated pairwise with a correlation coefficient of $\rho = 0.6$. In the presence of the two live load models in Table 5.3, compute the time-dependent failure probabilities for reference periods up to 50 years, considering a series or parallel system.

5.27 We reconsider the rigid-plastic portal frame as shown in Fig. 2.1a, which is subjected to horizontal load H and vertical load V. The structure may fail due to one of the following three limit states,

$$G_1 = M_1 + 2M_3 + 2M_4 - H - V$$
$$G_2 = M_2 + 2M_3 + M_4 - V$$
$$G_3 = M_1 + M_2 + M_4 - H$$

in which M_1 through M_4 are the plastic moment capacities at the joints. It is assumed that at the initial time, M_1 through M_4 are independent normal variables with mean 1 and COV 0.3. Furthermore, M_1 through M_4 degrade with time independently with a constant rate of 0.003/year. The vertical load V is constantly applied to the frame, which is a normal variable with mean 1.2 and COV 0.3. The occurrence of the horizontal load H is a Poisson process with occurrence rate $\lambda = 0.5$/year. The magnitude of H, conditional on occurrence, follows an Extreme Type I distribution with mean 1 and COV 0.4. Compute the time-dependent failure probabilities of the frame for reference periods up to 50 years.

5.28 Consider a bridge girder that has an initial resistance (moment) of 2500 kN·m and a successful service history of 20 years. The *in-situ* inspection suggests that the current resistance, having served for 20 years, is a lognormal variable with mean 2000 kN·m and COV 0.2. The annual live load causes a moment that follows an Extreme Type I distribution with mean 900 kN·m and COV 0.3. The dead load causes a deterministic moment of 600 kN·m. Find the mean and standard deviation of the updated current resistance taking into account the impact of successful service history.

5.29 We use a compound Poisson process to model the accumulation of structural damage $Y(T)$ with a reference period of $[0, T]$, that is, $Y = \sum_{i=1}^{N(T)} X_i$, where X_i is the magnitude of the ith shock deterioration, and $N(T)$ is the number of shock deteriorations within $[0, T]$, which is a Poisson variable with a mean occurrence rate of λ. Assume that each X_i is independent of $N(T)$. Suppose that each X_i is statistically independent and identically Gamma-distributed, with a shape parameter of $a > 0$ and a scale parameter of $b > 0$. Define a function $W(x, z) = \sum_{n=1}^{\infty} \frac{x^n}{n!\Gamma(zn)}, x \geq 0, z > 0$. Derive the PDF of $Y(T)$ for a reference period of $[0, T]$ [48].

5.30 In Problem 5.29, if the occurrence of shock deterioration is a nonstationary Poisson process with mean occurrence $\lambda(t)$, what is the PDF of $Y(T)$?

5.31 Consider a structure that has successfully served for T years. The PDF of the initial resistance, R_0, is $f_{R_0}(r)$. The resistance deterioration function, $G(t)$, takes a form of $G(t) = 1 - A \cdot t^\alpha$, where $\alpha > 0$ is a constant while $A > 0$ is a random variable that reflects the uncertainty associated with $G(t)$. Let $f_A(x)$ be the PDF of A. For the past service history of T years (and the future years), let $F_{S,i}$ be the CDF of the maximum load effect within the ith year ($i = 1, 2, \ldots T, \ldots$). Suppose that the structural resistance is constant within each year, and that the load process is independent of the resistance deterioration. At the current time (with a service history of T years), if the structure survives from a proof load with a deterministic magnitude of w, derive the time-dependent reliability of the structure for a subsequent reference period of T' years [56].

5.32 In Problem 5.31, we further assume that the annual maximum load effect, S_i, is identically distributed for each year. Each S_i ($i = 1, 2, \ldots T$) is correlated with $G(T)$, and the correlation is modeled by a copula function $C(u, v)$. What is the updated PDF of the current resistance (with a service history of T years) taking into account the successful service history (without considering the impact of the proof load) [49]?

5.33 Reconsider Example 5.35. Taking into account the impact of discount rate, denoted by r, the cumulative damage index is rewritten as $D_c^* = \sum_{j=1}^{N} \frac{Q(t_j)}{(1+r)^{t_j}}$. With this, derive the mean and variance of D_c^*.

5.34 Reconsider Example 5.11. If the resistance deterioration is described by the model in Problem 5.7 with $q = 0.5$, recalculate the time-dependent failure probabilities for reference periods up to 40 years.

5.35 In Problem 5.4, if we additionally consider the impact of Type II failure mode (c.f. Sect. 5.4.5.2) with a permissible level of 1.2 for the degraded resistance, recompute the time-dependent failure probabilities for reference periods up to 50 years.

5.36 Reconsider Example 5.38. If the PSDF of $S(t)$ is as in Eq. (5.140) with $k = 1$, recompute the time-dependent failure probability for T being up to 50.

References

1. Basu AK (2003) Introduction to stochastic processes. Alpha Science International Ltd
2. Choe DE, Gardoni P, Rosowsky D (2010) Fragility increment functions for deteriorating reinforced concrete bridge columns. J Eng Mech 136(8):969–978. https://doi.org/10.1061/(ASCE)EM.1943-7889.0000147
3. Chung L, Najm H, Balaguru P (2008) Flexural behavior of concrete slabs with corroded bars. Cement and Concr Compos 30(3):184–193. https://doi.org/10.1016/j.cemconcomp.2007.08.005
4. Ciampoli M (1998) Time dependent reliability of structural systems subject to deterioration. Comput Struct 67(1–3):29–35. https://doi.org/10.1016/S0045-7949(97)00153-3
5. Comenetz M (2002) Calculus: the elements. World Scientific Publishing Co Inc, Singapore
6. Dey A, Mahadevan S (2000) Reliability estimation with time-variant loads and resistances. J Struct Eng 126(5):612–620. https://doi.org/10.1061/(ASCE)0733-9445(2000)126:5(612)
7. Dieulle L, Bérenguer C, Grall A, Roussignol M (2003) Sequential condition-based maintenance scheduling for a deteriorating system. Eur J Oper Res 150(2):451–461. https://doi.org/10.1016/S0377-2217(02)00593-3
8. Ellingwood BR (1996) Reliability-based condition assessment and lrfd for existing structures. Struct Saf 18(2–3):67–80. https://doi.org/10.1016/0167-4730(96)00006-9
9. Ellingwood BR (2005) Risk-informed condition assessment of civil infrastructure: state of practice and research issues. Struct Infrastruct Eng 1(1):7–18. https://doi.org/10.1080/15732470412331289341
10. Ellingwood BR, Lee JY (2016) Life cycle performance goals for civil infrastructure: intergenerational risk-informed decisions. Struct Infrastruct Eng 12(7):822–829. https://doi.org/10.1080/15732479.2015.1064966
11. Engelund S, Rackwitz R, Lange C (1995) Approximations of first-passage times for differentiable processes based on higher-order threshold crossings. Probab Eng Mech 10(1):53–60. https://doi.org/10.1016/0266-8920(94)00008-9
12. Enright MP, Frangopol DM (1998) Service-life prediction of deteriorating concrete bridges. J Struct Eng 124(3):309–317. https://doi.org/10.1061/(ASCE)0733-9445(1998)124:3(309)
13. Fujino Y, Lind NC (1977) Proof-load factors and reliability. J Struct Div 103(4):853–870
14. Hall WB (1988) Reliability of service-proven structures. J Struct Eng 114(3):608–624. https://doi.org/10.1061/(ASCE)0733-9445(1988)114:3(608)

15. Hong H (2000) Assessment of reliability of aging reinforced concrete structures. J Struct Eng 126(12):1458–1465. https://doi.org/10.1061/(ASCE)0733-9445(2000)126:12(1458)
16. Iervolino I, Giorgio M, Chioccarelli E (2013) Gamma degradation models for earthquake-resistant structures. Struct Saf 45:48–58. https://doi.org/10.1016/j.strusafe.2013.09.001
17. Knutson TR, McBride JL, Chan J, Emanuel K, Holland G, Landsea C, Held I, Kossin JP, Srivastava A, Sugi M (2010) Tropical cyclones and climate change. Nat Geosci 3:157–163. https://doi.org/10.1038/ngeo779
18. Kumar R, Gardoni P (2014) Effect of seismic degradation on the fragility of reinforced concrete bridges. Eng Struct 79:267–275. https://doi.org/10.1016/j.engstruct.2014.08.019
19. Kumar R, Cline DB, Gardoni P (2015) A stochastic framework to model deterioration in engineering systems. Struct Saf 53:36–43. https://doi.org/10.1016/j.strusafe.2014.12.001
20. Li CQ, Lawanwisut W, Zheng JJ (2005) Time-dependent reliability method to assess the serviceability of corrosion-affected concrete structures. J Struct Eng 131(11):1674–1680. https://doi.org/10.1061/(ASCE)0733-9445(2005)131:11(1674)
21. Li H, Wen D, Lu Z, Wang Y, Deng F (2016a) Identifying the probability distribution of fatigue life using the maximum entropy principle. Entropy 18(4):111. https://doi.org/10.3390/e18040111
22. Li K, Li Q, Zhou X, Fan Z (2015a) Durability design of the hong kong-zhuhai-macau sea-link project: principle and procedure. J Bridge Eng 20(11):04015001. https://doi.org/10.1061/(ASCE)BE.1943-5592.0000741
23. Li Q, Wang C (2015) Updating the assessment of resistance and reliability of existing aging bridges with prior service loads. J Struct Eng 141(12):04015072. https://doi.org/10.1061/(ASCE)ST.1943-541X.0001331
24. Li Q, Ye X (2018) Surface deterioration analysis for probabilistic durability design of RC structures in marine environment. Struct Saf 75:13–23. https://doi.org/10.1016/j.strusafe.2018.05.007
25. Li Q, Wang C, Ellingwood BR (2015b) Time-dependent reliability of aging structures in the presence of non-stationary loads and degradation. Struct Saf 52:132–141. https://doi.org/10.1016/j.strusafe.2014.10.003
26. Li Q, Wang C, Zhang H (2016b) A probabilistic framework for hurricane damage assessment considering non-stationarity and correlation in hurricane actions. Struct Saf 59:108–117. https://doi.org/10.1016/j.strusafe.2016.01.001
27. Ma Y, Zhang J, Wang L, Liu Y (2013) Probabilistic prediction with bayesian updating for strength degradation of RC bridge beams. Struct Saf 44:102–109. https://doi.org/10.1016/j.strusafe.2013.07.006
28. Melchers RE (2003) Probabilistic model for marine corrosion of steel for structural reliability assessment. J Struct Eng 129(11):1484–1493. https://doi.org/10.1061/(ASCE)0733-9445(2003)129:11(1484)
29. Melchers RE, Beck AT (2018) Structural reliability analysis and prediction, 3rd edn. Wiley. https://doi.org/10.1002/9781119266105
30. Mohd MH, Paik JK (2013) Investigation of the corrosion progress characteristics of offshore subsea oil well tubes. Corros Sci 67:130–141. https://doi.org/10.1016/j.corsci.2012.10.008
31. Möller B, Beer M, Graf W, Sickert JU (2006) Time-dependent reliability of textile-strengthened RC structures under consideration of fuzzy randomness. Comput Struct 84(8–9):585–603. https://doi.org/10.1016/j.compstruc.2005.10.006
32. Mori Y, Ellingwood BR (1993) Reliability-based service-life assessment of aging concrete structures. J Struct Eng 119(5):1600–1621. https://doi.org/10.1061/(ASCE)0733-9445(1993)119:5(1600)
33. Newland DE (1993) An introduction to random vibrations, spectral & wavelet analysis, 3rd edn. Pearson Education Limited, Edinburgh Gate, Harlow, England
34. Nowak AS, Lutomirska M, Sheikh Ibrahim F (2010) The development of live load for long span bridges. Bridge Struct 6:73–79. https://doi.org/10.3233/BRS-2010-006
35. OBrien EJ, Bordallo-Ruiz A, Enright B (2014) Lifetime maximum load effects on short-span bridges subject to growing traffic volumes. Struct Saf 50:113–122. https://doi.org/10.1016/j.strusafe.2014.05.005

36. Otsuki N, Miyazato S, Diola NB, Suzuki H (2000) Influences of bending crack and water-cement ratio on chloride-induced corrosion of main reinforcing bars and stirrups. Mater J 97(4):454–464 https://doi.org/10.14359/7410
37. Pan P, Li Q, Zhou Y, Li Y, Wang Y (2011) Vehicle survey and local fatigue analysis of a highway bridge (in Chinese). China Civ Eng J 44(5):94–100 https://doi.org/10.15951/j.tmgcxb.2011. 05.006
38. Pang L, Li Q (2016) Service life prediction of RC structures in marine environment using long term chloride ingress data: Comparison between exposure trials and real structure surveys. Constr Build Mater 113:979–987. https://doi.org/10.1016/j.conbuildmat.2016.03.156
39. Saassouh B, Dieulle L, Grall A (2007) Online maintenance policy for a deteriorating system with random change of mode. Reliab Eng Syst Saf 92(12):1677–1685. https://doi.org/10.1016/ j.ress.2006.10.017
40. Sanchez-Silva M, Klutke GA, Rosowsky DV (2011) Life-cycle performance of structures subject to multiple deterioration mechanisms. Struct Saf 33(3):206–217. https://doi.org/10. 1016/j.strusafe.2011.03.003
41. Sha Y, Hao H (2012) Nonlinear finite element analysis of barge collision with a single bridge pier. Eng Struct 41:63–76. https://doi.org/10.1016/j.engstruct.2012.03.026
42. Stewart MG (1997) Time-dependent reliability of existing RC structures. J Struct Eng 123(7):896–902. https://doi.org/10.1061/(ASCE)0733-9445(1997)123:7(896)
43. Stewart MG, Mullard JA (2007) Spatial time-dependent reliability analysis of corrosion damage and the timing of first repair for RC structures. Eng Struct 29(7):1457–1464. https://doi.org/ 10.1016/j.engstruct.2006.09.004
44. Stewart MG, Rosowsky DV (1998) Time-dependent reliability of deteriorating reinforced concrete bridge decks. Struct Saf 20(1):91–109. https://doi.org/10.1016/S0167-4730(97)00021-0
45. Svensson H (2009) Protection of bridge piers against ship collision. Steel Constr: Des Res 2(1):21–32. https://doi.org/10.1002/stco.200910004
46. Val DV, Stewart MG, Melchers RE (1998) Effect of reinforcement corrosion on reliability of highway bridges. Eng Struct 20(11):1010–1019. https://doi.org/10.1016/S0141-0296(97)00197-1
47. Vickery PJ, Masters FJ, Powell MD, Wadhera D (2009) Hurricane hazard modeling: the past, present, and future. J Wind Eng Ind Aerodyn 97(7–8):392–405. https://doi.org/10.1016/j.jweia. 2009.05.005
48. Wang C (2020) An explicit compound poisson process-based shock deterioration model for reliability assessment of aging structures. J Traf Transp Eng (English Edition)
49. Wang C (2020b) Resistance assessment of service-proven aging bridges incorporating deterioration-load dependency. Infrastructures 5(1):10. https://doi.org/10.3390/ infrastructures5010010
50. Wang C (2020c) A stochastic process model for resistance deterioration of aging bridges. Adv Bridge Eng 1:3. https://doi.org/10.1186/s43251-020-00003-w
51. Wang C, Li Q (2016a) Simplified method for time-dependent reliability analysis of aging bridges subjected to nonstationary loads. Int J Reliab Q Saf Eng 23(01):1650003. https://doi. org/10.1142/S0218539316500030
52. Wang C, Li Q (2016b) Time-dependent reliability of existing bridges considering non-stationary load process. Eng Mech 33(3):18–23. https://doi.org/10.6052/j.issn.1000-4750.2014.05.0451
53. Wang C, Zhang H (2017) Probability-based cumulative damage assessment of structures subjected to non-stationary repeated loads. Adv Struct Eng 20(11):1784–1790. https://doi.org/10. 1177/1369433217713927
54. Wang C, Zhang H (2018a) Probability-based estimate of tropical cyclone damage: an explicit approach and application to hong kong, china. Eng Struct 167:471–480. https://doi.org/10. 1016/j.engstruct.2018.04.064
55. Wang C, Zhang H (2018b) Roles of load temporal correlation and deterioration-load dependency in structural time-dependent reliability. Comput Struct 194:48–59. https://doi.org/10. 1016/j.compstruc.2017.09.001

56. Wang C, Zhang H (2020) A probabilistic framework to optimize the target proof load for existing bridges. Innov Infrastruct Sol 5(1):12. https://doi.org/10.1007/s41062-020-0261-9

57. Wang C, Li Q, Zou A, Zhang L (2015) A realistic resistance deterioration model for time-dependent reliability analysis of aging bridges. J Zhejiang Univ Sci A 16(7):513–524. https://doi.org/10.1631/jzus.A1500018

58. Wang C, Li Q, Ellingwood BR (2016a) Time-dependent reliability of ageing structures: an approximate approach. Struct Infrastruct Eng 12(12):1566–1572. https://doi.org/10.1080/15732479.2016.1151447

59. Wang C, Li Qw, Pang L, Zou Am (2016b) Estimating the time-dependent reliability of aging structures in the presence of incomplete deterioration information. J Zhejiang Univ-Sci A 17(9):677–688. https://doi.org/10.1631/jzus.A1500342

60. Wang C, Zou A, Zhang L, Li Q (2016c) A method for updating reliability of existing bridges considering historical vehicular loading information. J Highway Transp Res Dev 33(2):67–72. https://doi.org/10.3969/j.issn.1002-0268.2016.02.011

61. Wang C, Li Q, Zhang H, Ellingwood BR (2017a) Modeling the temporal correlation in hurricane frequency for damage assessment of residential structures subjected to climate change. J Struct Eng 143(5):04016224. https://doi.org/10.1061/(ASCE)ST.1943-541X.0001710

62. Wang C, Zhang H, Li Q (2017b) Reliability assessment of aging structures subjected to gradual and shock deteriorations. Reliab Eng Syst Saf 161:78–86. https://doi.org/10.1016/j.ress.2017.01.014

63. Wang C, Zhang H, Li Q (2017c) Time-dependent reliability assessment of aging series systems subjected to non-stationary loads. Struct Infrastruct Eng 13(12):1513–1522. https://doi.org/10.1080/15732479.2017.1296004

64. Wang C, Zhang H, Beer M (2018) Computing tight bounds of structural reliability under imprecise probabilistic information. Comput Struct 208:92–104. https://doi.org/10.1016/j.compstruc.2018.07.003

65. Wang C, Feng K, Zhang L, Zou A (2019a) Estimating the resistance of aging service-proven bridges with a gamma process-based deterioration model. J Traff Transp Eng (English Edition) 6(1):76–84. https://doi.org/10.1016/j.jtte.2018.11.001

66. Wang C, Zhang H, Beer M (2019b) Structural time-dependent reliability assessment with a new power spectral density function. J Struct Eng 145(12):04019163. https://doi.org/10.1061/(ASCE)ST.1943-541X.0002476

67. Zhong J, Gardoni P, Rosowsky D (2009) Stiffness degradation and time to cracking of cover concrete in reinforced concrete structures subject to corrosion. J Eng Mech 136(2):209–219. https://doi.org/10.1061/(ASCE)EM.1943-7889.0000074

Index

Printed in the United States
by Baker & Taylor Publisher Services